Lecture Notes in Physics

The Editorial Policy for Proceedings

The series Lecture Notes in Physics reports new developments in physical research and teaching – quickly, informally, and at a high level. The proceedings to be considered for publication in this series should be limited to only a few areas of research, and these should be closely related to each other. The contributions should be of a high standard and should avoid lengthy redraftings of papers already published or about to be published elsewhere. As a whole, the proceedings should aim for a balanced presentation of the theme of the conference including a description of the techniques used and enough motivation for a broad readership. It should not be assumed that the published proceedings must reflect the conference in its entirety. (A listing or abstracts of papers presented at the meeting but not included in the proceedings could be added as an appendix.)

When applying for publication in the series Lecture Notes in Physics the volume's editor(s) should submit sufficient material to enable the series editors and their referees to make a fairly accurate evaluation (e.g. a complete list of speakers and titles of papers to be presented and abstracts). If, based on this information, the proceedings are (tentatively) accepted, the volume's editor(s), whose name(s) will appear on the title pages, should select the papers suitable for publication and have them refereed (as for a journal) when appropriate. As a rule discussions will not be accepted. The series editors and Springer-Verlag will normally not interfere with the detailed editing except in fairly obvious cases or on technical matters.

Final acceptance is expressed by the series editor in charge, in consultation with Springer-Verlag only after receiving the complete manuscript. It might help to send a copy of the authors' manuscripts in advance to the editor in charge to discuss possible revisions with him. As a general rule, the series editor will confirm his tentative acceptance if the final manuscript corresponds to the original concept discussed, if the quality of the contribution meets the requirements of the series, and if the final size of the manuscript does not greatly exceed the number of pages originally agreed upon. The manuscript should be forwarded to Springer-Verlag shortly after the meeting. In cases of extreme delay (more than six months after the conference) the series editors will check once more the timeliness of the papers. Therefore, the volume's editor(s) should establish strict deadlines, or collect the articles during the conference and have them revised on the spot. If a delay is unavoidable, one should encourage the authors to update their contributions if appropriate. The editors of proceedings are strongly advised to inform contributors about these points at an early stage.

The final manuscript should contain a table of contents and an informative introduction accessible also to readers not particularly familiar with the topic of the conference. The contributions should be in English. The volume's editor(s) should check the contributions for the correct use of language. At Springer-Verlag only the prefaces will be checked by a copy-editor for language and style. Grave linguistic or technical shortcomings may lead to the rejection of contributions by the series editors. A conference report should not exceed a total of 500 pages. Keeping the size within this bound should be achieved by a stricter selection of articles and not by imposing an upper limit to the length of the individual papers. Editors receive jointly 30 complimentary copies of their book. They are entitled to purchase further copies of their book at a reduced rate. As a rule no reprints of individual contributions can be supplied. No royalty is paid on Lecture Notes in Physics volumes. Commitment to publish is made by letter of interest rather than by signing a formal contract. Springer-Verlag secures the copyright for each volume.

The Production Process

The books are hardbound, and the publisher will select quality paper appropriate to the needs of the author(s). Publication time is about ten weeks. More than twenty years of experience guarantee authors the best possible service. To reach the goal of rapid publication at a low price the technique of photographic reproduction from a camera-ready manuscript was chosen. This process shifts the main responsibility for the technical quality considerably from the publisher to the authors. We therefore urge all authors and editors of proceedings to observe very carefully the essentials for the preparation of camera-ready manuscripts, which we will supply on request. This applies especially to the quality of figures and halftones submitted for publication. In addition, it might be useful to look at some of the volumes already published. As a special service, we offer free of charge LATEX and TEX macro packages to format the text according to Springer-Verlag's quality requirements. We strongly recommend that you make use of this offer, since the result will be a book of considerably improved technical quality. To avoid mistakes and time-consuming correspondence during the production period the conference editors should request special instructions from the publisher well before the beginning of the conference. Manuscripts not meeting the technical standard of the series will have to be returned for improvement.

For further information please contact Springer-Verlag, Physics Editorial Department V, Tiergartenstrasse 17, D-69121 Heidelberg, FRG

F. Ehlotzky (Ed.)

Fundamentals of Quantum Optics III

Proceedings of the Fifth Meeting
on Laser Phenomena
Organized by the Institute for Theoretical Physics
University of Innsbruck, Austria, 7-13 March 1993

Springer-Verlag Berlin Heidelberg GmbH

Editor

Fritz Ehlotzky
Institute for Theoretical Physics, University of Innsbruck
Technikerstraße 25, A-6020 Innsbruck, Austria

ISBN 978-3-662-13932-5   ISBN 978-3-540-47974-1 (eBook)
DOI 10.1007/978-3-540-47974-1

© Springer-Verlag Berlin Heidelberg 1993
Originally published by Springer-Verlag Berlin Heidelberg New York in 1993

58/3140-543210 - Printed on acid-free paper

FOREWORD

The Seminar on Fundamentals of Quantum Optics III was the fifth meeting on Laser Phenomena organized by the Institute for Theoretical Physics of the University of Innsbruck and held this time at the Congress Center in Kühtai, Tyrol. It was attended by 50 physicists from Austria, Canada, Denmark, the Federal Republic of Germany, Finland, France, Italy, New Zealand, Poland, Russia, Spain and the United States, who work actively in the rapidly developing field of quantum optics.

The present Seminar offered the opportunity to discuss at leisure problems of mutual interest to theoreticians and experimentalists who are working on various aspects of the field of quantum optics. The intention was to bring together people who are doing research on atomic interferometry, physics of cooled and trapped particles, cavity quantum electrodynamics, quantum statistics of light and other fundamentals.

At the seminar 25 Invited Lectures were given by:

E. Arimondo (Pisa)

R. Blatt (Hamburg)

J.I. Cirac (Ciudad Real)

M. Collett (Auckland)

J. Dalibard (Paris)

W. Ertmer (Bonn)

C.W. Gardiner (Hamilton)

E. Giacobino (Paris)

R.J. Glauber (Cambridge)

Yu.M. Golubev (St. Petersburg)

P. Grangier (Orsay)

S. Haroche (Paris)

A. Hemmerich (München)

J. Javanainen (Storrs)

W. Ketterle (Cambridge)

P. Meystre (Tucson)

G. Rempe (Konstanz)

C. Savage (Canberra)

J. Schmiedmayer (Cambridge)

M. Scully (Houston)

M. Sigel (Konstanz)

S. Stenholm (Helsinki)

C. Westbrook (Gaithersburg)

K. Wodkiewicz (Warzawa)

P. Zoller (Boulder)

In addition, there were 18 contributed papers (posters) presented at the meeting.

The following pages present the full text of the invited lectures. The invited lecture of J. Peřina (Olomouc) was not presented at the Seminar but has been accepted for publication in the Proceedings. For the invited lecture of R.J. Glauber (Cambridge) no typescript has been received in time for publication. The editor is grateful to the contributors for their collaboration in preparing their typescripts for rapid publication.

The active yet relaxed atmosphere of the Sport Hotel and Congress Center at Kühtai, surrounded by the snow—capped peaks of the Stubai Alps, provided a congenial setting for a very stimulating and rewarding meeting. It is a pleasure to thank all participants for their interest and enthusiasm. The most valuable secretarial assistance of Miss I. Wiedermann is gratefully acknowledged.

Innsbruck, June 1993   F. Ehlotzky

For the convenience of the reader we quote below the proceedings of the previous meetings on laser phenomena held at Obergurgl:

Fundamentals of Quantum Optics
(February 26 — March 3, 1984)
Acta Physica, Austriaca, Vol. 56, No. 1–2, 1984.

Fundamentals of Laser Interactions
(February 24 — March 2, 1985)
Lecture Notes in Physics, Vol. 229, 1985.

Fundamentals of Quantum Optics II
(February 22 — 28, 1987)
Lecture Notes in Physics, Vol. 282, 1987.

Fundamentals of Laser Interactions II
(February 26 — March 4, 1989)
Lecture Notes in Physics, Vol. 339, 1989.

ACKNOWLEDGEMENTS

THE SEMINAR ON FUNDAMENTALS OF QUANTUM OPTICS III
HAS BEEN SUPPORTED BY:

Bundesministerium für Wissenschaft und Forschung
Amt der Tiroler Landesregierung
Rektor der Universität Innsbruck
Bundeskammer der Gewerblichen Wirtschaft
Österreichische Forschungsgemeinschaft
Landes–Hypothekenbank Tirol
Raiffeisenkasse Silz–Haiming
Tourismusverband Kühtai

LIST OF PARTICIPANTS

1. G. Alber, Univ. Freiburg, FRG
2. B. Appasamy, Univ. Hamburg, FRG
3. E. Arimondo, Univ. di Pisa, Italy
4. N. Bjerre, Univ. Aarhus, Denmark
5. R. Blatt, Univ. Hamburg, FRG
6. J. Bolle, Univ. Hamburg, FRG
7. M. Brune, ENS, Paris, France
8. A.Z. Capri, Univ. of Alberta, Edmonton, Canada
9. J.I. Cirac, Univ. de Castilla—La Mancha, Ciudad Real, Spain
10. M. Collett, Univ. of Auckland, New Zealand
11. J. Dalibard, ENS, Paris, France
12. F. Ehlotzky, Univ. Innsbruck, Austria
13. K. Ellinger, Univ. Innsbruck, Austria
14. W. Ertmer, Univ. Bonn, FRG
15. R. Gaggl, TU Graz, Austria
16. C.W. Gardiner, Univ. of Waikato, Hamilton, New Zealand
17. Elisabeth Giacobino, ENS, Paris, France
18. R.J. Glauber, Harvard Univ., Cambridge, USA
19. Yu.M. Golubev, St. Petersburg State Univ., Russia
20. P. Grangier, Orsay, France
21. S. Haroche, ENS, Paris, France
22. H. Helm, SRI Internat., Menlo Park, USA
23. A. Hemmerich, Univ. München, FRG
24. J. Javanainen, Univ. of Connecticut, Storrs, USA
25. W. Ketterle, MIT, Cambridge, USA
26. G. Krenn, Atominstitut d. Öst. Univ., Wien, Austria
27. W. Lange, MPQ, Garching, FRG
28. P. Meystre, Univ. of Arizona, Tucson, USA
29. R. Paschotta, Univ. Konstanz, FRG
30. T. Pellizzari, Univ. Innsbruck, Austria
31. G. Rempe, Univ. Konstanz, FRG
32. C. Richy, ENS, Paris, France
33. H. Ritsch, Univ. Innsbruck, Austria
34. C.M. Savage, Australian National Univ., Canberra, Australia
35. K.J. Schernthanner, Univ. Innsbruck, Austria
36. J. Schmiedmayer, MIT, Cambridge, USA
37. M.O. Scully, Texas A & M Univ., USA
38. W. Seifert, Univ. Konstanz, FRG
39. M. Sigel, Univ. Konstanz, FRG
40. I. Siemers, Univ. Hamburg, FRG
41. S. Stenholm, Univ. of Helsinki, Finland
42. K.-A. Suominen, Univ. of Helsinki, Finland
43. C. Wagner, MPQ, Garching, FRG
44. H. Wallis, Univ. Bonn, FRG
45. R. Walser, Univ. Innsbruck, Austria
46. C. Westbrook, Orsay, France; NIST, Gaithersburg, USA
47. J. Werner, Univ. Bonn, FRG
48. L. Windholz, TU Graz, Austria
49. K. Wodkiewicz, Warzawa Univ., Poland
50. P. Zoller, JILA, Boulder, USA

CONTENTS

XI

Part V: Other Fundamentals

PART I: Atomic Interferometry

Optical elements for atoms: A beamsplitter and a mirror

M. Sigel, T. Pfau, C. S. Adams, C. Kurtsiefer,
W. Seifert, C. Heine, and J. Mlynek
Fakultät für Physik, Universität Konstanz, D-78434 Konstanz, Deutschland
R. Kaiser and A. Aspect
Institut d' Optique, F-91403 Orsay, France

Introduction

Atom optics, in analogy to electron or neutron optics, is concerned with the manipulation of atomic matter waves [1-3]. Experiments on atomic beams have been performed since the nineteen-twenties, but recently the improvement of tools, like free-standing microstructures and tunable laser light, has lead to a rapid development of the field. Beams of atoms have been split coherently, focussed and reflected from mirrors. "Optical instruments" for atoms have a number of potential applications: Atom interferometers could be well suited as gravitational and inertial sensors and may allow precision measurements of atomic properties. A number of such measurements have already been carried out. An "atom-microscope" - investigating the interaction of atoms with surfaces with high spatial resolution - may provide an interesting new tool for surface physicists. "Atom lithography" may allow the deposition of atoms on surfaces with high precision. Cavities for atoms could be interesting to store cold atoms and for the study of quantum-statistical effects.

In the first part of this article we attempt to provide a very brief introduction to atom optics. In the second and third part we report on recent experiments in our group on two specific atom-optical elements that may be useful in atom interferometers and atom cavities: the demonstration of a new beam splitter based on the diffraction of atomic matter waves from a "magneto-optical grating" and the investigation of a mirror based on the reflection of atoms from an evanescent light field.

1 Atom optics

Particle beams exhibit many phenomena well known from classical optics like diffraction and refraction. Particle optics based on electrons and neutrons [4] are well established fields. Experiments with these particles have played an important role in exploring the puzzling wave-particle duality. Both neutron and electron interferometers have been used in many beautiful experiments, which have improved our understanding of quantum mechanics. The interactions of neutron and electron beams with matter have become indispensable tools in the analysis of structures and surfaces, consider e.g. neutron scattering and the electron microscope.

As atoms are different from electrons and neutrons, particle optics with atoms promises a wealth of new effects and applications. Like neutrons, atoms are neutral and therefore cannot be manipulated as easily as electrons or ions using static electromagnetic fields. In return beams of atoms and neutrons are less susceptible to electromagnetic stray fields. Atoms are heavier, which makes them more suitable as gravitational and inertial sensors and means that very short de Broglie wavelengths can be obtained for lower particle energies. Atoms come in many species, which can be either bosons or fermions, and may have a total angular momentum or magnetic moment much larger than neutrons or electrons. Beams of atoms are easier and cheaper to produce than neutron beams as there is no need for a nuclear reactor. And maybe most importantly, atoms have a complex internal structure that can be probed and manipulated using resonant laser light or static electromagnetic fields.

As in classical optics, the building blocks of any experiment are optical elements e.g. beamsplitters, mirrors and lenses. Light can be manipulated by transmission through interfaces between materials with different refractive indices. Electrons, due to their charge, couple strongly to static electric or magnetic fields. Neutrons penetrate through solids and can be diffracted from crystal lattices. As atoms experience much smaller forces in static fields and do not penetrate through matter, none of these approaches are suitable for atom optics.

So far optical elements for atoms have been demonstrated using diffraction, refraction or the recoil due to the stimulated emission or absorption of a single photon. "Diffractive optical elements" achieve the desired distortion of an incoming wave front by interaction with small structures, thus exploiting the wave nature of the atomic centre of mass motion. "Refractive optics" is based on introducing phase shifts by a spatial modulation of the index of refraction experienced by the atoms. In analogy to standard optics the index of refraction is defined as the ratio of a particle's group velocity and the vacuum group velocity. The index of refraction can be adjusted by shifting the potential energy of the atom. Intuitively this means: If the internal energy of an atom is e.g. decreased, energy conservation demands an acceleration of the centre of mass motion (i.e. an increase in the group velocity of the atomic matter wave). The phase velocity on the other hand is decreased in this case. The phase fronts are consequently retarded with respect to the unperturbed propagation. The principle of refractive optical elements is to choose the spatial dependence of these phase shifts to obtain the desired distortion of the incoming wave fronts. In static electric or magnetic fields the atomic internal energy can be changed by the Zeeman or Stark shift, in near-resonant light fields by the light shift.

The diffraction of atoms from the surface of single crystals was studied by Stern et al. as early as 1929 [5]. In 1969 Leavitt and Bills [6] observed a single-slit diffraction pattern using a thermal potassium beam. The progress in micro-fabrication technology now permits the production of structures sufficiently fine to diffract thermal atoms, whose de Broglie wavelengths are typically below 1 Å. The diffraction of thermal atoms from free-standing micro-fabricated gratings has been studied by several groups [7,8]. The first deflection of an atomic beam by the light pressure force was already demonstrated by Frisch in 1933 [9]. In recent years the availability of intense tunable laser radiation has again focussed interest on the mechanical effects of atom-light interactions. The feasibility of such experiments has stimulated considerable theoretical interest. In addition to the conservative interaction using the dipole force, the spontaneous decay of excited electronic states can be utilized for the preparation of atomic beams by laser cooling techniques to obtain well-collimated, dense atomic beams with a well defined velocity.

2 The magneto-optical beamsplitter

2.1 Introduction

A coherent beam splitter couples an incoming momentum state to a coherent superposition of outgoing momentum states. Beam splitters for atoms are not only important for atom interferometry, they also could be used for correlation measurements or as in/outcouplers for atomic cavities. Beam splitters have been produced using the magnetic [10] or optical Stern-Gerlach effect [11], reflection from crystalline surfaces [12], the recoil due to a single absorption or stimulated emission [13,14], diffraction from micro-fabricated transmission gratings (amplitude gratings) [7,8] and the diffraction from standing wave light fields (phase gratings) [15,16].

We present a new type of beam splitter which is a modification of the diffraction of atoms from an optical standing wave. This diffraction can be looked upon as a beam splitter both in the regime of long and short interaction times. For short interaction times the change in "transverse" kinetic energy can be neglected. In the following we are concerned with this regime, which is called the Raman-Nath regime.

As for the diffraction of two-level atoms from a standing wave the magneto-optical beamsplitter is based on introducing position dependent phase shifts. The advantages of the diffraction from a phase grating over diffraction from an amplitude grating are that there is no absorption and that by changing the phase modulation amplitude it is possible to change the envelope of the diffraction pattern. For a two-level atom the eigenenergies of the two energy eigenstates in a near-resonant light field with Rabi frequency $\omega_R = -\langle e | \mathbf{d} \cdot \mathbf{E} | g \rangle E_0 / \hbar$ and resonance detuning Δ are

$$E_{\pm}(x) = \pm \frac{\hbar}{2} \sqrt{\omega_R(x)^2 + \Delta^2} \ .$$

In a standing wave light field with large detuning these eigenergies display a roughly sinusoidal modulation. In the Raman-Nath regime and if neither spontaneous emission nor non-adiabatic following induce transitions between the eigenstates the phase modulation $\Delta\Phi(x) = \int E(x,t)/\hbar \, dt$ displays the same sinusoidal modulation. The momentum distribution and therefore the far-field spatial distribution (for a monochromatic beam) is given by the Fourier transform of $e^{i\Delta\Phi(x)}$. For a sinusoidal phase modulation the population of the nth diffraction peak is given by the square of the nth order Bessel function $J_n(\Delta\Phi_{max})$ wherein the argument is the amplitude of the phase modulation [17]. Diffraction from a standing wave can therefore be looked upon as an effective beam splitter only for momentum transfers of a few $\hbar k$. In order to obtain a phase grating which diffracts all intensity into a few high order momentum states a triangular phase modulation is necessary. Such a triangular phase modulation can be obtained if a combination of light shifts and Zeeman shifts are used to control the spatial modulation of the eigenenergies [18,19]. We consider the experimental situation shown in figure 1: a V-type three-level system interacts with two counter-propagating crossed linearly polarized light beams and a static magnetic field applied parallel to the light beams. The level scheme contains the Zeeman shifts but not the light shifts.

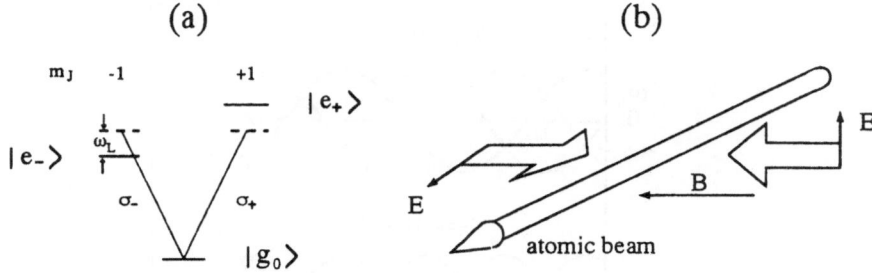

Figure 1: (a) Level scheme for a $J=0$ to $J'=1$ transition with the quantization axis chosen parallel to the magnetic field. (b) Configuration of the counterpropagating light beams which orthogonally intersect the atomic beam. The light fields are linearly polarized and the polarization vectors enclose an angle of 90°. A magnetic field is aligned parallel to the light beams.

In figure 2 the energy eigenvalues of this three-level atom are displayed as a function of the transverse position in the light field for three ratios of the Larmor frequency $\omega_L = g_J\mu_B B/\hbar$ and the Rabi frequency $\omega_R = -\langle e_\pm|d\cdot E|g_0\rangle E_0/\hbar$.

For small Rabi frequencies the levels are split by the Zeeman effect, for large Rabi frequencies by the light shift – for $\omega_R = 2\omega_L$ the Zeeman shift and the light shift at positions of circularly polarized light are equal and the eigenstate $|2\rangle$ displays the desired triangular modulation. For the V-system the ground state $|g\rangle$ evolves adiabatically into this eigenstate $|2\rangle$. If nonadiabatic processes and spontaneous emission do not populate the two other eigenstates the diffraction pattern obtained from this phase grating should display two distinct peaks.

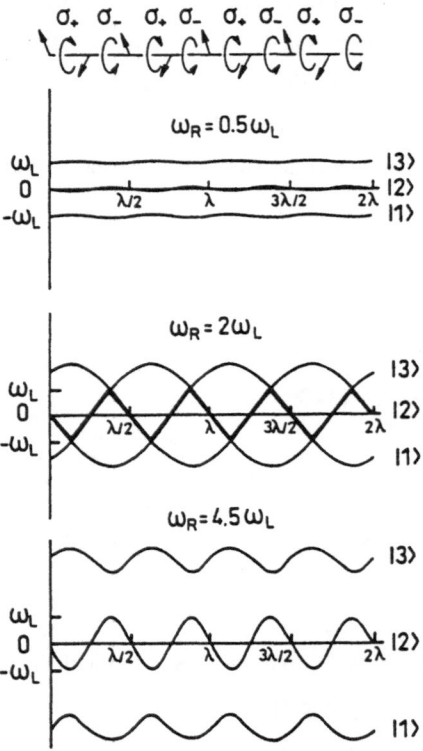

Figure 2:   Plots of the spatial dependence of the eigenenergies of a three-level atom in a light field created by two counterpropagating orthogonally linear polarized light beams and a static magnetic field for three different ratios of Rabi frequency ω_R and Larmor frequency ω_L.

9

2.2 Experiment

We investigated the magneto-optical beamsplitter in an experiment using helium atoms in the triplet state. The light field was resonant with the transition at 1083 nm from the metastable 2^3S_1 state to the 2^3P_1 state. If the incoming metastables are optically pumped into the $m_J=0$ substate a V-type three-level system as depicted in figure 1a is obtained. A schematic of our experimental setup is displayed in figure 3. The laser beam, which was derived from a Ti:sapphire ring laser, was focussed to a waist of 21 µm in the beam direction. The polarization of the back-reflected beam was rotated by double pass through a λ/4-plate. As a consequence of the large mean velocity in the supersonic beam of 1900 m/s and the long lifetime of the excited state of 100 ns it was possible to largely avoid spontaneous emission. The superimposed magnetic field was produced by a pair of Helmhotz coils. The far-field diffraction pattern of the supersonic helium beam was observed with a scanning slit detector.

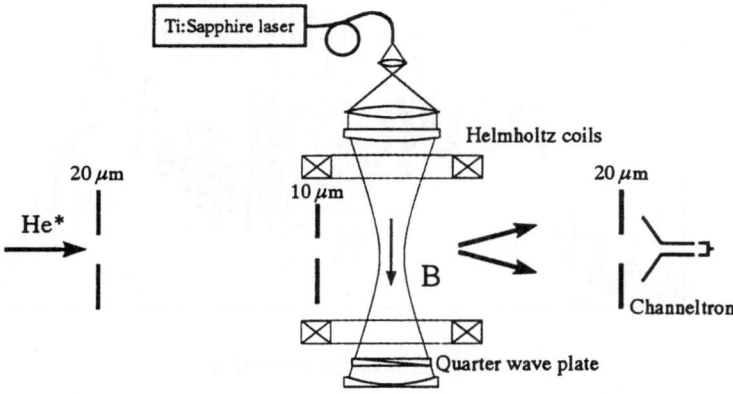

Figure 3: Experimental setup of the magneto-optical beamsplitter. A beam of triplet helium atoms interacted with two counterpropagating orthogonally linear polarized laser beams and a static magnetic field. The far-field diffraction pattern was detected with a scanning slit detector and a channeltron detector.

In Figure 4 the number of detected atoms is plotted as a function of transverse position in the detection plane 0.85 m downstream from the interaction zone. The choice of parameters was such that the Rabi frequency ω_R in the centre of the

10

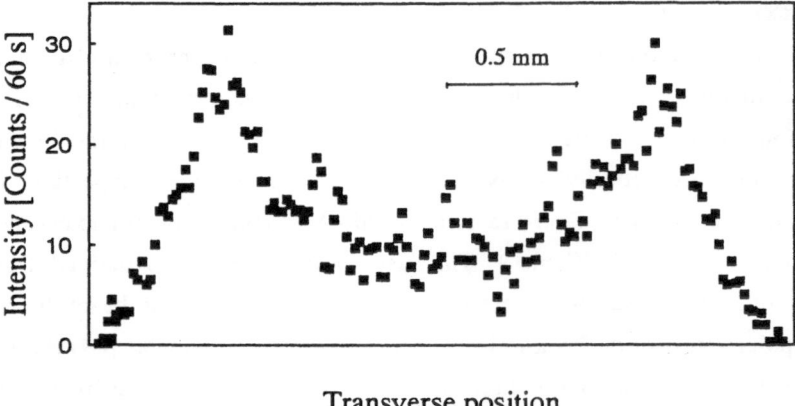

Figure 4: Plot of the far-field intensity distribution for the diffraction of triplet helium atoms from a magneto-optical grating for $\omega_R = 2\,\omega_L$.

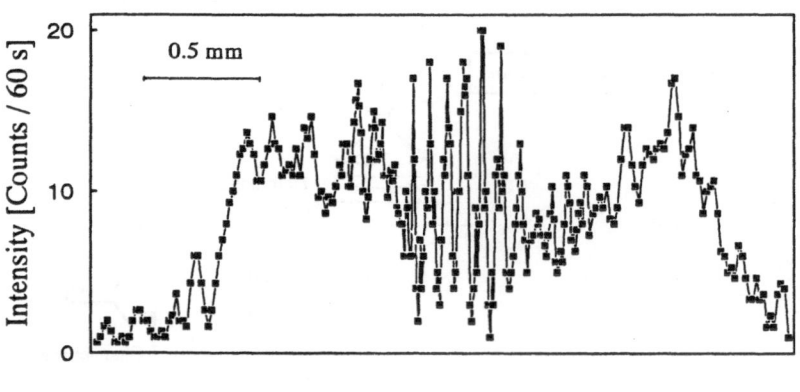

Figure 5: Plot of the far-field intensity distribution for the diffraction of triplet helium atoms from an optical standing wave.

Gaussian beam was twice the Larmor frequency ω_L. It can be seen that the incoming atomic matter wave is largely diffracted into two groups of diffraction orders as expected from theory. The finite velocity distribution leads to a broadening of the split beams in the real space distribution. The splitting of

1.5 mm between the two peaks observed in the detection plane corresponds to a momentum splitting of 42 $\hbar$k. This is a large splitting compared to the splitting typically achieved by diffraction from an amplitude grating or compared to the 1 $\hbar$k achieved with a single $\pi/2$-pulse.

For comparison we also examined the diffraction of atoms from an "ordinary" standing wave. Figure 5 shows a far-field diffraction pattern we obtained for a laser intensity comparable to the one used in figure 4. It displays the expected broad, Bessel function shaped envelope. In the real space distribution the 2 $\hbar$k structure is only resolved for low diffraction order. For larger diffraction order the finite width of the velocity distribution smears out distinct diffraction peaks.

3 A mirror for atoms

3.1 Introduction

Another crucial element to do optics with atoms is a mirror. As in classical optics, mirrors could be used in interferometers or in cavities for atoms. Such cavities may allow the storage of very cold atoms, which e.g. is interesting to examine quantum statistical effects. Cavities could be realized to sustain standing waves between two mirrors or, due to the gravitational acceleration, on top of one mirror.

Atomic matter waves can be reflected directly from surfaces if the projection of the surface roughness onto the direction of the incoming beam is smaller than the de Broglie wavelength and if the atoms do not tend to stick to the surface. As early as 1929 Knauer and Stern reported a 5 % reflection of thermal beams of H_2 and He from a metal surface for 1 mrad incidence angle Θ [20]. In 1930 Estermann and Stern reflected helium atoms from the surface of cleaved ionic crystals for much larger angles [12]. Berkhout et al. used a concave mirror coated with superfluid helium to focus a divergent hydrogen beam [21]. In 1986 Anderson et al. reported the reflection of thermal cesium atoms from a polished glass surface [22].

As for the magneto-optical beamsplitters, the dipole force may be used to provide coherent interaction. Spontaneous emission in the interaction process tends to decrease the coherence of the incoming beam due to the photon recoil as well as due to fluctuations in the dipole force. In order to minimize spontaneous emission by minimizing the interaction time, large gradients of the light intensity are desirable. Such large gradients can be obtained in evanescent light fields as suggested in 1982 by Cook and Hill [23]. Evanescent waves are created by total internal reflection of a light beam inside a dielectric medium; the electric field amplitude E(z) in an evanescent wave produced by a plane wave of angular frequency ω for incidence angle φ (measured against the surface plane) is given by

$$E(z) = E(0) \cdot e^{-\alpha z} \ , \quad \text{where} \quad \alpha = \frac{\omega}{c}\sqrt{n^2 \cos^2 \varphi - 1} \ .$$

If there is no spontaneous emission two-level atoms will be reflected in an evanescent light field of Rabi frequency ω_R and positive detuning Δ if the potential barrier $V(0) - V(\infty)$ $(V_\pm(z) = \pm \frac{\hbar}{2}\sqrt{\omega_R(z)^2 + \Delta^2}\,)$ is larger than the transverse kinetic energy of the atoms $\frac{1}{2}mv\sin\Theta$. Spontaneous emission leads to statistical jumps between the two eigenstates of the atom and therefore to fluctuations of the force as well as a modified mean value (along with random photon recoil); in the steady state limit the mean force is given by the gradient of the eigenenergies weighted by the steady state population of the eigenstates [24].

In 1987 Balykin et al. reported such a reflection experiment using sodium atoms [25]. Kasevich et al. have demonstrated an "atomic trampoline" by bouncing slow sodium atoms off an evanescent wave under normal incidence [26]. Aminoff et al. recently have observed four bounces in this configuration using cesium atoms [27]. In our group we have investigated the reflection properties of an evanescent wave using metastable argon atoms.

3.2 Reflection of metastable argon atoms from an evanescent wave

In the first experiment we produced an evanescent wave on the surface of a glass prism in the setup shown in figure 6. The glass prism could be rotated to examine the reflection properties for various angles of incidence. A beam of metastable argon atoms in the metastable triplet state was directed on the glass prism such

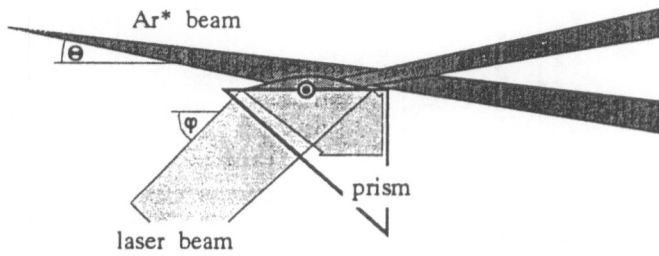

Figure 6:   Schematic of the experimental setup for the reflection of atoms. The glass
            prism with the evanescent wave clipped the incoming atomic beam such that
            part of the beam passed the prism and part of the beam was reflected. The
            angle of incidence of the atomic beam Θ could be adjusted by rotating the
            prism around the displayed axis.

that part of the beam passed the prism while part of the beam could interact with
the evanescent wave. We used a channeltron detector, which is only sensitive to
metastable states. The width of collimation slits and the width of the scanning slit
were chosen such that the angular resolution was around 0.1 mrad. A Ti:sapphire
ring laser was used to drive the two-level transition $1s_5 - 2p_9$ (Paschen notation)
at 812 nm. The waists of the elliptical evanescent wave were 5.25 mm along the
beam direction and 0.32 mm perpendicular to the beam direction. A set of
experimental results for the atomic intensity as a function of the detector position
for angles of incidence Θ between 0 and 3 mrad are shown in figure 7. The peak
at zero deflection angle is due to unreflected atoms which pass the prism. It can
be seen that the reflection law is obeyed nicely and that there are no atoms
scattered towards angles between the trespassing and the reflected beam. The
decrease of the zero peak with increasing angle of incidence was caused by the
increase of the "shadow" of the prism. The increasing area of the evanescent
wave "seen" by the incoming beam led to the initial increase in the number of
reflected atoms.   For larger angles the increasing normal velocity led to a
decrease in the area of the "Gaussian" evanescent wave where the intensity was
large enough to deflect atoms. This effect caused the number of reflected atoms to
ultimately drop to zero.

14

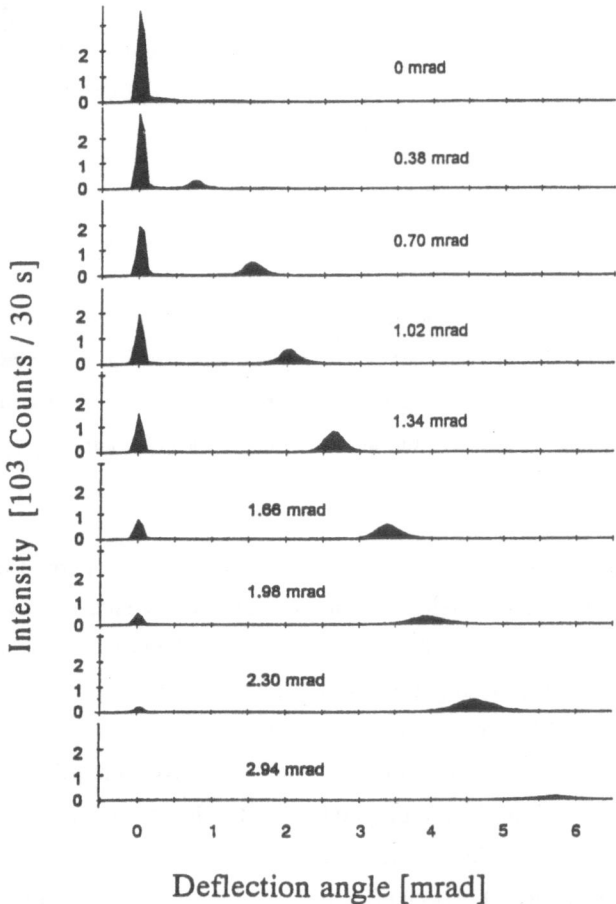

Figure 7: Experimental results for the number of detected metastable argon atoms over the detector position for various angles of incidence Θ on the evanescent wave. The peaks at zero deflection angle are due to unreflected atoms which pass the glass prism while the right peaks are due to reflected atoms.

3.3 Enhancements of evanescent waves using planar waveguides

In order to reflect larger transverse velocities or in order to decrease the number of spontaneous emission by increasing the detuning very intense evanescent light waves are important. This is true despite the fact that the maximum transverse

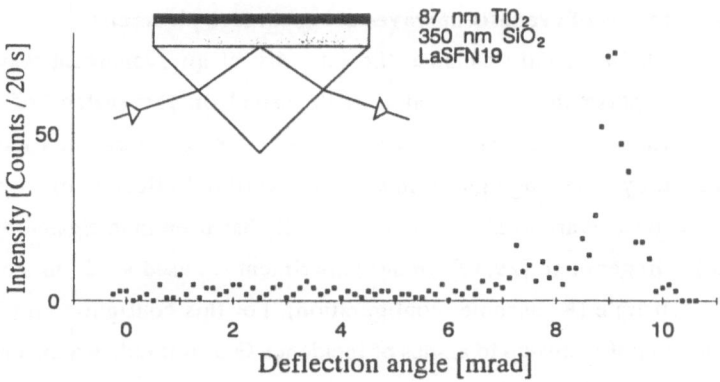

Figure 8: Experimental result for the reflection from an evanescent light field outside of a planar waveguide structure. The experimental configuration of the planar waveguide is displayed in the inset.

velocity and therefore the maximum angle of incidence scales only with the fourth root of the light intensity. The intensity of the evanescent field can be enhanced by resonantly increasing the light intensity at the surface. One possibility would be to use the surface as a mirror in a resonator. A specific type of such a resonator is a planar waveguide as shown in the inset of figure 8. In our experiment a TiO_2 layer was used as the planar waveguide. The light was coupled in from a high index of refraction glass substrate ($LaSFN_{18}$) through a SiO_2 spacer layer using frustrated total reflection. The evanescent field above the waveguide was enhanced by a factor of about 130 compared to the total internal reflection in the same prism for the same choice of polarization and angle of incidence (this is not the optimum choice of polarization for the reflection from a "bare" glass surface) [28].

We observed reflection of triplet argon atoms from this enhanced evanescent light field for angles of incidence Θ up to 6 mrad. This corresponds to transverse velocities of 3.4 m/s. An experimental result for the atomic intensity as a function of deflection angle is displayed in figure 8.

3.3 Enhancements of evanescent waves using surface plasmons

Another way to resonantly enhance the intensity of an evanescent wave is to excite surface plasmons in a metal film deposited on the surface of a glass substrate. Surface plasmons are surface waves of the electron gas in a metal that can be excited by reflecting light from a thin metal film. Reflection of atoms from plasmon enhanced evanescent waves very recently has been investigated by Feron et al. and Esslinger et al. [29,30]. In our experiment we used a 52 nm silver film on a glass substrate (Kretschmer configuration). For this configuration we again found reflection of atoms up to angles of incidence Θ of 6 mrad. An experimental result is displayed in figure 9.

We have performed numerical simulations on the reflection from the three types of evanescent waves. According to these simulations the transverse broadening of the reflected beams is largely due to fluctuations of the dipole force. The effect of the recoil in the spontaneous emission process is less important. The reflected beam is broader for the plasmon enhanced field than for the waveguide enhanced field because the "decay length" α of the evanescent wave was larger such that the interaction time was longer. In addition, variations in the intensity of the evanescent wave caused by variations in the thickness of the silver film may contribute to the broadening.

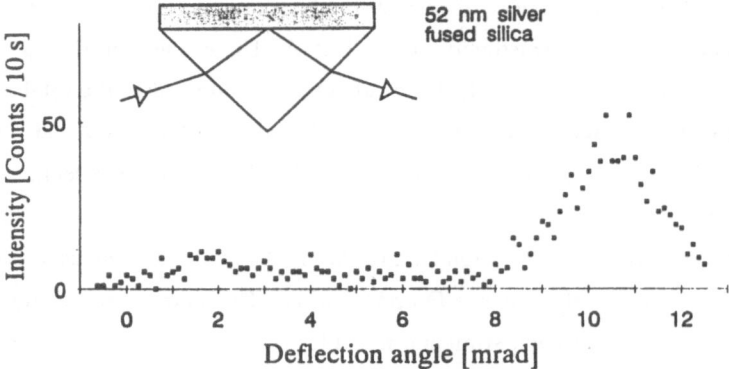

Figure 9:   Experimental result for the reflection from an evanescent light field enhanced by surface plasmons on a silver film. A schematic of the glass prism with the silver film on top is displayed in the inset.

3.4 Estimate on the importance of spontaneous emission

Spontaneous emission in the reflection process contributes to a decrease in the coherence of the incoming atomic matter wave. For applications like interferometers or cavities for atoms there must be no spontaneous emission. We investigated the importance of spontaneous emission by using the "open" transition $1s_5$ - $2p_8$ at 802 nm in triplet argon. 71% of the atoms excited in the $2p_8$ state decay into the argon ground state, which is not detected by our channeltron detector. The number of detected reflected atoms compared to the result for a closed transition therefore contains information about the number of spontaneous emission processes. In figure 10 the maximum atomic intensity reflected from the plasmon enhanced evanescent wave is plotted as a function of the resonance detuning for both the "closed" and the "open" transition for the same angle of incidence of 1.1 mrad and equal laser intensity. The detuning is given relative to the resonance frequency of the mean atomic velocity class. It can be seen that for the closed transition the reflected intensity rises from zero to a constant value for a positive detuning, the width of the rise being due to the finite velocity distribution in the atomic beam. For the open transition this rise is

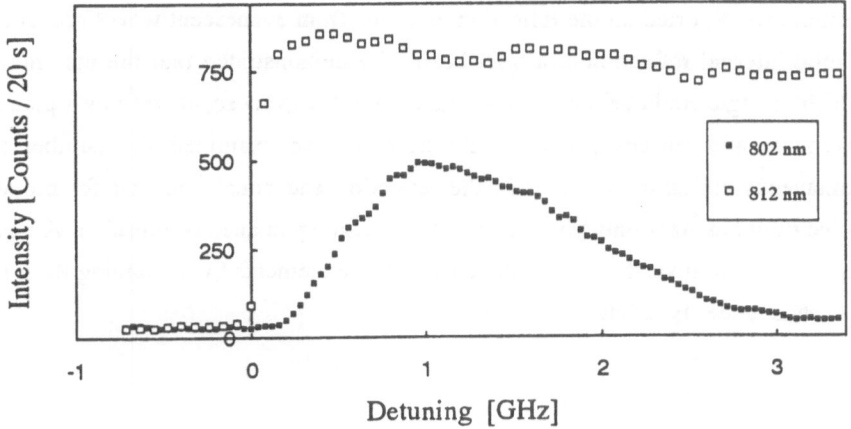

Figure 10: Plot of the reflected intensity as a function of the resonance detuning for the "closed" two level transition at 812 nm and the "open" transition at 802 nm in triplet argon. The "open" transition allows part of the population of the excited state to decay to the argon ground state which is invisible to our detector.

hampered by the decay into the undetected argon ground state such that the maximum is only reached for a detuning large enough to suppresses spontaneous emission. The decrease for large detunings sets in earlier than in case of the closed transition because the 812 nm transition is weaker. For the detuning at the maximum of this reflection curve the number of atoms reflected with no spontaneous emission can be estimated to larger than 90 %. This is a promising result in that it demonstrates that for a proper choice of experimental parameters the coherent reflection of atoms from an evanescent light field may be useful in atom interferometers and atom cavities.

4 Conclusion

In this contribution we presented experimental results for a new type of beamsplitter and an investigation of the reflecting properties of an evanescent light field. The beamsplitter relies on the diffraction of atomic matter waves from a phase grating with triangular phase modulation produced by a combination of a light field and a magnetic field. We observed a momentum splitting of 42 $\hbar k$. In addition, we reported on the reflection of atoms from evanescent waves produced by total internal reflection of a laser beam. We demonstrated that the maximum reflection angle could be increased by enhancing the evanescent wave by a planar waveguide and surface plasmons. Furthermore, we examined the number of spontaneous emission processes in the reflection and concluded that for proper choice of parameters only few atoms will undergo spontaneous emission. A next step could be to investigate the coherence of these elements by combining them in a Mach-Zehnder type interferometer.

Acknowledgement
This work was supported by the Deutsche Forschungsgemeinschaft and the European Community through the SCIENCE programme under contract no. SC1-CT92-0778.

References

[1]   Special Issue of Appl. Phys. B (April 1992), *Optics and Interferometry with Atoms*, J. Mlynek, V. Balykin and P. Meystre (eds.).

[2]   M. Sigel, C. S. Adams and J. Mlynek, to be published in *Frontiers in LaserSpectroscopy*, T. W. Hänsch, M. Inguscio (eds.), Proceedings of the International School of Physics "Enrico Fermi", Course CXX, Varenna, 1992.

[3]   M. Sigel and J. Mlynek, *Atom Optics*, Physics World, February 1993.

[4]   V. Sears, *Neutron Optics* (Oxford University Press, New York, Oxford, 1989).

[5]   O. Stern, Naturwissensch. **17**, 391 (1929).

[6]   J. A. Leavitt and F. A. Bills, Am. J. Phys. **37**, 905 (1969).

[7]   D. W. Keith, M. L. Schattenburg, H. I. Smith and D. E. Pritchard, Phys. Rev. Lett. **61**, 1580 (1988).

[8]   O. Carnal, A. Faulstich and J. Mlynek, Appl. Phys. B **53**, 88 (1991).

[9]   R. Frisch, Zeits. f. Phys. **86**, 42 (1933).

[10]   W. Gerlach und O. Stern, Zeits. f. Phys. **8**, 110 (1922).

[11]   T. Sleator, T. Pfau, V. Balykin, O. Carnal and J. Mlynek, Phys. Rev. Lett. **68**, 1996 (1992).

[12]   I. Estermann und O. Stern, Zeits. f. Phys. **61**, 95 (1930).

[13]   F. Riehle, Th. Kister, A. Witte, J. Helmcke and Ch. J. Bordé, Phys. Rev. Lett. **67**, 177 (1991).

[14]   M. Kasevich and S. Chu, Phys. Rev. Lett. **67**, 181 (1991).

[15]   P. E. Moskowitz, P. L. Gould, S. R. Atlas and D. E. Pritchard, Phys. Rev. Lett. **51**, 370 (1983).

[16]   P. L. Gould, G. A. Ruff and D. E. Pritchard, Phys. Rev. Lett. **56**, 827 (1986).

[17]   R. J. Cook and A. F. Bernhardt, Phys. Rev. A **18**, 2533 (1978).

[18]   T. Pfau, C. S. Adams and J. Mlynek, Europhys. Lett. **21**, 439 (1993).

[19]   C. S. Adams, T. Pfau, Ch. Kurtsiefer and J. Mlynek, "Interaction of atoms with a magneto-optical potential" to be published in Phys. Rev. A.

[20]   F. Knauer und O. Stern, Zeits. f. Physik **53**, 779 (1929).

[21]   J. J. Berkhout, O. J. Luiten, I. D. Setija, T. W. Hijmans, T. Mizusaki and J. T. M. Walraven, Phys. Rev. Lett. **63**, 1689 (1989).

[22]   A. Anderson, S. Haroche, E. A. Hinds, W. Jhe, D. Meschede and L. Moi, Phys. Rev. A **34**, 3513 (1986).

[23]   R. J. Cook and R. K. Hill, Opt. Commun. **43**, 258 (1982).

[24]   J. Dalibard and C. Cohen-Tannoudji, J. Opt. Soc. Am. B **2**, 1707 (1985).

[25]   V. I. Balykin, V. S. Letokhov, Yu. B. Ovchinnikov and A. I. Sidorov, Phys. Rev. Lett. **60**, 2137 (1988).

[26]   M. A. Kasevich, D. S. Weiss and S. Chu, Opt. Lett. **15**, 607 (1990).

[27]   C. G. Aminoff, P. Bouyer and P. Desbiolles, C. R. Acad. Sc., Paris, 15 Mars 1993.

[28]   R. Kaiser et al., *Resonant enhancement of evanescent waves with a thin dielectric waveguide*, submitted to Opt. Commun.

[29]   S. Feron et al., *Reflection of metastable neon atoms by a surface plasmon wave*, submitted to Europhys. Lett.

[30]   T. Esslinger. M. Weidenmüller, A. Hemmerich and T. W. Hänsch, Opt. Lett. **18**, 450 (1993).

Atom Interferometry

Jörg Schmiedmayer, Christopher R. Ekstrom, Michael S. Chapman,
Troy D. Hammond and David E. Pritchard

Department of Physics and Research Laboratory of Electronics
Massachusetts Institute of Technology, Cambridge, Massachusetts 02139, USA.

Abstract. We have demonstrated an atom interferometer with completely separated beams using three 200 nm period transmission gratings. A stretched 10 cm long and 10 μm thick metal foil, was inserted in the interferometer so that separated portions of the atom wave went on opposite sides of the foil. A fringe amplitude of up to 900 cps was observed, which allows us to determine the phase to 15 milliradians in 1 minute. We have performed several experiments by applying different interactions to one or both arms of the interferometer. We have determined the electric polarizability of the ground state of sodium, the coherence length of our beam, and the differential phase shifts and coherences of the different Zeeman ground states in a magnetic field. Some future prospects of separated beam experiments are discussed.

1 Introduction:

Interference has a long tradition that takes us from Young's demonstration of two slit diffraction of light [1] in the 19th century to the application of optical interferometers to a number of important scientific problems by Fizeau [2], Michelson [3] and Rayleigh [4] among others. The rise of quantum mechanics, with its central tenet of wave-matter duality led directly to the idea of matter wave interference devices such as the electron interferometer [5,6] and the perfect crystal neutron interferometer [7]; both being well known for their contribution to precision experiments and fundamental physics [8].

Given the many demonstrations of matter wave interference devices and interferometers with electrons and neutrons, the concept of atom interferometers requires only a small conceptual step. Indeed, an atom interferometer was patented in 1973 [8]. The billiard ball model [9] of coherent transients emphasised the spatial separation of the two coherent, internal states, and several papers discussed the close similarity between multiple pulse laser spectroscopy and atom interferometers [10, 11]. About half a dozen experiments that demonstrate atom interferometers have been performed during the last five years [12-18] A good overview can be found in [19, 20] and [21]. It is our judgement that we have now entered a period in which the important advances involving atom interferometers will be new applications. Therefore it is appropriate in this paper about the new subfield of atom interferometers to stress these areas of application, rather than to dwell at length on the various recent demonstrations. We will, as an example for new measurements, focus on recent experiments performed at MIT and begin with a brief discussion of our own atom interferometer.

2 The MIT Atom Interferometer:

Our interferometer is of a Mach-Zender type built with three nanofabricated transmission gratings [22] mounted on separate translation stages inside the vacuum envelope (Fig.1). This configuration produces a robust white fringe [23]. Our setup has been considerably improved since the first demonstration experiment [12]. We now use 200 nm period gratings which completely separate the centers of the interfering beams (a trapezoid with 20 μm flat top, 30 μm FWHM and 40 μm width on the bottom) by 55 μm at the position of the second grating. Our new finer and taller gratings do, however, require much better alignment and vibration control. We have a supersonic sodium beam with argon as carrier gas (v = 1000 m/s, $\lambda_{dB} \sim 0.16$ Å). By varying the gas pressure in our new source we can now change

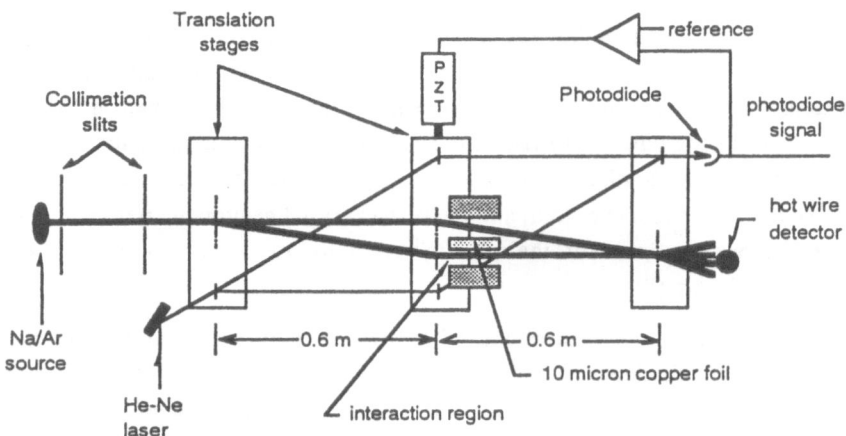

Figure 1: A schematic, not to scale, of our atom interferometer (thick lines are atom beams). The 0th and 1st order beams from the first grating strike the middle grating where they are diffracted in the 1st and -1st orders so that they form an interference pattern in the plane of the third grating, which acts as a mask to sample this pattern. The detector, located beyond the third grating, records the flux transmitted by the third grating. The 10 cm long interaction region with the 10 μm thick copper foil between the two arms of the interferometer is positioned behind the 2nd grating. An optical interferometer (thin lines are laser beams) measures the relative position of the 200 nm period atom gratings (which are indicated by vertical dashed lines).

the FWHM of our velocity distribution from $\Delta v/v = 70\%$ (no seed gas) to $\Delta v/v < 8\%$ at 3 bar Ar pressure. The mean velocity (deBroglie wavelength) of the beam can be changed by using different seed gases. A new Re hot wire detector has higher detection efficiency, faster time response (~1 msec response time) and background as low as <10 cps.

We built an interaction region consisting of a stretched metal foil positioned symmetrically between two side electrodes, each spaced 2 mm from this septum. It was inserted behind the 2nd grating so that portions of the atom wave in the two arms of the interferometer pass on opposite sides of the foil. The septum is up to 10 cm long and 10 µm thick, but the shadow it cast on the detector was typically 20 - 30 µm wide due to slight deviations of the stretched foil from perfect flatness and imperfections at the ends. Since we have a conducting, physical barrier between the separated beams, we can apply different, nominally uniform electric and magnetic fields to the portions of the atom wave on each side of the interferometer. The fringe amplitude with the septum between the two arms of the interferometer is up to 900 cps, which allows us to determine the phase to 15 milliradians in 1 minute (Fig.2). It is interesting to note that the deBroglie wavelength $\lambda_{dB} \sim 0.16$ Å and the coherence length $l_{coh} \sim 0.65$ Å (1.6 Å FWHM) were both smaller than the size of the atom (3 Å) whose motion they characterise. The use of a compound particle with such large dimensions in an interferometer was unprecedented before the advent of atom interferometers.

With the two beams now completely separated, let us consider the effect of an interaction in only one arm of the interferometer. Applying a potential $U(x)$ in one arm changes the k-vector $k(x) = \frac{1}{\hbar}\sqrt{2m(E_{kin} - U(x))}$. The phase difference, $\Delta\varphi$, between the two arms will be:

$$\Delta\varphi(k_0) = \int k(x)ds - \int k_0 ds = \int \Delta k(x)ds$$

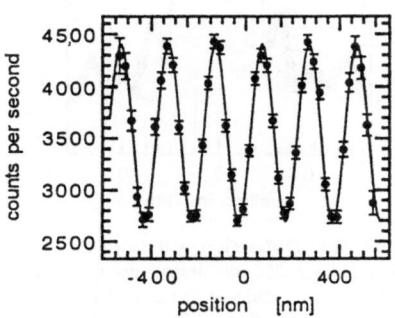

where $k_0 = \frac{1}{\hbar}\sqrt{2mE_{kin}}$ is the unperturbed k-vector. If the potential U is much smaller than the energy of the atom E_{kin} (as is in all the cases discussed here), the phase shift can be evaluated using an Eikonal approximation, yielding

$$\Delta\varphi(k_0) = \frac{m}{\hbar^2 k_0}\int U(x)dx = \frac{1}{\hbar v}\int U(x)dx.$$

Figure 2: Interference pattern from 40 seconds of data (1 second per point). The contrast is 25% and the phase uncertainty is 17 milliradians.

The action of the potential $U(x)$ can be also viewed as an refractive index $n = \frac{k_0}{k}$.

Using the Eikonal approximation ($U << E_{kin}$) one finds $n = \frac{k_0}{k} = 1 - \frac{U}{2E}$ and the phase shift $\Delta\varphi(k) = k\int(1 - n(k))dx$. In our interferometer a phase shift of 1 rad corresponds to $|1 - n| = 2.7 \times 10^{-11}$.

The discussion above shows that the phase shift for a velocity and time independent potential depends on k_0. Since our beam is not monochromatic, the phases shift averages over the initial k-vector distribution $f(k)$:

$$\Delta\varphi = \arctan\left(\frac{\int f(k)\sin(\Delta\varphi(k))dk}{\int f(k)\cos(\Delta\varphi(k))dk}\right)$$

This average over the final velocity width of the beam leads to a reduction of the interference contrast for dispersive phase shifts like the Stark phase shift or the phase shift from the magnetic interaction. In the most general case, the interference contrast $C = \frac{I_{max} - I_{min}}{I_{max} + I_{min}}$ is given by:

$$C = C_0\sqrt{\left[\int f(k)\sin(\Delta\varphi(k))dk\right]^2 + \left[\int f(k)\cos(\Delta\varphi(k))dk\right]^2}$$

where C_0 is the contrast at zero phase shift. This reduction of the interference contrast can be viewed as being caused by the coherence length l_{coh} of the beam (see Fig.3). The above equation reduces to $C = C_0 e^{-\frac{1}{2}(\Delta\varphi)^2 \sigma_k^2/k^2}$ in the case of linear dispersion, and $f(k)$ is a gaussian distribution with width σ_k. The coherence length is given by $l_{coh} = \frac{1}{\sigma_k}$.

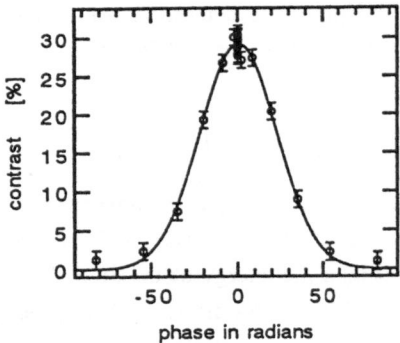

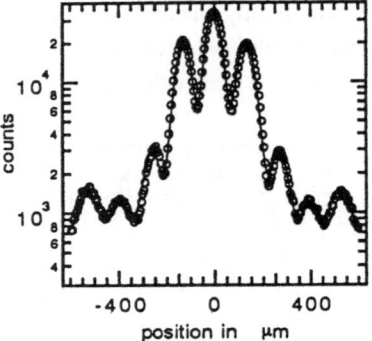

Figure 3a: Reduction of the interference contrast with applied Stark phase. From the with of the contrast curve we evaluated a coherence length of 0.65(3) Å in good accordance with an evaluation of the velocity from a measurement of the diffraction pattern.

Figure 3b: Diffraction pattern of sodium atoms from a 200 nm period grating. The solid line is calculated for a grating with average open fraction of 40 % with 9% rms relative variation. The sodium atoms have a mean $\lambda_{dB} = 0.165$ Å and a rms velocity distribution of 3.7(3)%.

It is important to note that, in contrast to photons in a coherent state in vacuum, the coherence length and wave packet size for matter waves are never the same, except perhaps at specific points in time. This is because matter waves, even in a vacuum, are dispersive. In our beam the coherence length is about 0.65 Å (1.6 Å FWHM), but the "size" of our wave packet at the 3rd grating (where the interference "occurs") is on the order of 10 cm FWHM.

To conclude this cursory description of our interferometer, let us compare its sensitivity with spectroscopic methods. A good comparison is with separated oscillatory field resonance experiments, where the resonance pattern may be viewed as interference between two internal states of the atom. In both these instruments the fringe separation is set by the inverse of the interaction time. Our 0.1 m long interaction region and the 1000 ms^{-1} velocity result in an intrinsic line width of 10 kHz. Or in other words, a potential of 4×10^{-11} eV gives a phase shift of 2π radian. In both instruments, neglecting systematic errors, the fringe can be split by a factor proportional to $\sqrt{N}C$ where C is the contrast and N is the total number of atoms counted in the measurement. For our interferometer a typical one minute measurement gives a sensitivity better than 15 mrad (2 millifringes) corresponding to an energy sensitivity of roughly 10^{-13} eV or about 20 Hz.

In contrast to an oscillatory field measurement, with the separated beam interferometer one can do experiments using only *one* state and does not have to rely on difference measurements between the *two different* states. This will be very important in our first applications of the atom interferometers to measurements of ground state atomic and molecular properties. In addition the interferometer gives us the opportunity to subject part of the atom wave to an interaction which causes a phase shift without a classical force and then to measure this phase shift by interference with the unshifted part of the atom wave in the same state. The latter is important for clear demonstrations of topological quantum phases.

3 Talbot Experiment:

Before going into the applications of our interferometer let us discuss first an atom optics experiment we performed to insure the quality of our diffraction gratings. In a Mach-Zender type grating interferometer the first and second grating form an interference pattern in the plane of the third grating, which acts as a mask to sample this pattern. Therefore the gratings must be phase coherent over their entire area. One way to test them is to look for Moiré patterns between two gratings. From simple geometric optics, this requires the two gratings to be separated by much less than $\frac{d}{\theta_{coll}}$ (~10 mm in our case) where d is the grating period and θ_{coll} is the collimation angle of the beam. From the viewpoint of optics with matter waves the distance between the gratings has to be much less than $\frac{d}{\theta_{diff}}$ where θ_{diff} is the diffraction angle for a single slit in the grating, otherwise the single slit diffraction pattern overlap. This distance is about 1 mm in our case, which is too small for a practical setup with our grating holders.

On the other hand, the interference between the single slit diffraction patterns of the grating openings results in periodic self imaging of the grating in the near field. The images are separated by the Talbot length, $l_{Talbot} = \frac{d^2}{\lambda}$. This effect was discovered in light optics experimentally by Talbot [24] and later explained by Rayleigh [25] as interference of the single slit diffraction patterns of the grating openings. A 3-grating atom interferometer using this effect was proposed by Clauser [26]. We used this effect, demonstrating it for the first time with matter waves, to test our new 200 nm and 300 nm gratings. By choosing a distance of ~5 mm (corresponding to about 2 l_{Talbot} for the 200 nm and l_{Talbot} for the 300 nm gratings), we could test both sets at once. A typical test pattern is shown in Fig.4. We achieved nearly the theoretical contrast for all the tested gratings over their whole area. This gives us great confidence in our nanofabrication process and the possibility to make even larger gratings, or ones with significantly smaller period.

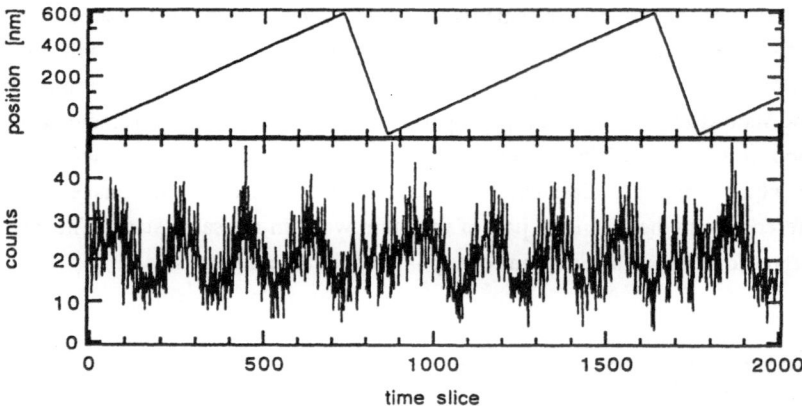

Figure 4: Talbot experiment with a pair of 200 nm gratings (50 μm wide and 1.5 mm high). The upper graph shows the relative position of the 2 gratings. The lower graph shows intensity transmitted through both gratings (counts per time slice of 10 ms) as measured by a detector behind the second grating. The pattern has a contrast of 30% close to the calculated maximum of 36% achievable with our geometry.

4 Atomic Properties:

Separated beam atom interferometers are ideally suited for absolute measurements of atomic and molecular properties for *single* states. The key point is that the separated beams in the interferometer present the opportunity to subject part of the atom wave to an interaction which causes a phase shift, and then to measure this phase shift by interference with the unshifted freely evolving part of the atom wave in the *same* state.

4.1 The Electric Polarizability of the Sodium Ground State

An obvious application of our interferometer is a precision measurement of the polarizability of the ground state of the sodium atom, by subjecting one part of the atom wave to a uniform electric field. This is intrinsically a higher precision approach than measuring the deflection in a field gradient, whether this is measured by conventional [27] or interference techniques [28]. We also note that the polarizability *differences* between *two* states of an atom have long been measured using optical resonance techniques, and can also be measured in Chebotayev type interferometers, even though the two legs of the interferometer are not spatially resolved [29, 18].

A homogeneous electric field on one side of the interaction region induces a potential due to the DC Stark effect $U_{Stark} = -\frac{1}{2}\alpha E^2$, where α is the electric polarizability, resulting in an differential phase accumulation relative to the wave that passes on the other side of the septum where there is no field. Since U_{Stark} is about eight orders of magnitude smaller than the kinetic energy E_{kin}, we can use the Eikonal approximation, and the phase shift is $\Delta\varphi_{Stark} = -\frac{1}{\hbar}\frac{l}{v}\frac{1}{2}\alpha E^2 = \frac{t}{\hbar}U_{Stark}$ where l is the length of the interaction region, v is the velocity of the atoms, giving a transit time t. The measured phase shift is quadratic in the applied electric field (Fig.5), allowing us to determine the polarizability of the ground state with a statistical error of less than 0.1%. We are currently investigating several systematic errors which limit our determination of an absolute value of the polarizability.

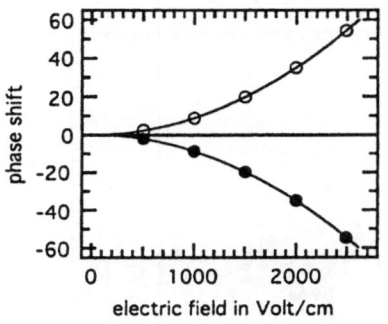

Figure 5: Stark phase shifts for voltages applied to the left (o) and the right (•) side of the interaction region. The phase shift per applied electric field square in (volt/cm)^2 is 1.220(7)×10^{-5} for the left and 1.224(7)×10^{-5} for the right side.

28

4.2 Magnetic Rephasing

We have also investigated the magnetic interaction of the sodium ground state with a magnetic field. By running a current down the metal septum, perpendicular to the plane of the interferometer, we generate a magnetic field in the interaction region along the beam direction with the same magnitude, but opposite sign for the two arms of the interferometer. When an additional uniform magnetic field is applied along the beam axis, it adds to the field from the current in the septum giving different field magnitudes on both sides of the septum. In this case one gets a differential Zeeman energy, $\vec{\mu} \cdot \vec{B}$, and therefore different accumulated phases, for the two paths around the septum. The phase difference is proportional to the current passing through the septum and the projection of the magnetic moment $g_F \mu_o$ along the quantization axis.

$$\Delta\varphi = \frac{1}{\hbar}\frac{l}{v}g_F\mu_o m_F B$$

Because our beam is unpolarized, the observed interference pattern is a sum of interference patterns for each of the eight sodium ground states. Since the g-factors of the F = 1 and F = 2 hyperfine levels have equal magnitude (but opposite sign), there are only five different projected magnetic moments. At low fields, these are proportional to 0, $\pm\frac{1}{2}$, and ±1 times a Bohr magneton. Consequently, the independent interference patterns periodically rephase constructively to produce a high contrast interference pattern with the same phase as the pattern at zero field.

We have observed this periodic rephasing of the 8 independent magnetic substates in an unpolarized sodium beam (see Fig.6) as an increasing differential magnetic field is applied to opposite sides of the septum. The first revival of contrast is the point where the phase shifts are 4π for the $|m_F| = 2$ states, 2π for the $|m_F| = 1$ states, and 0 for the $m_F = 0$ states. Therefore in this experiment the important

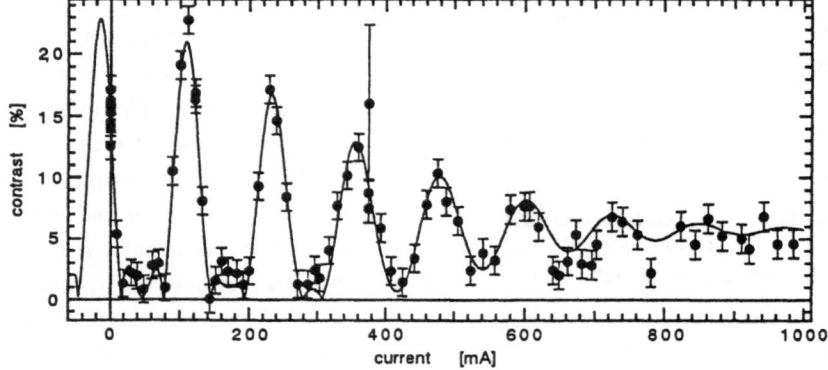

Figure 6: Contrast revivals from constructive rephasing of the independent interference patterns of the 8 different magnetic sub-states of sodium. These patterns are dephased by a current flowing down the septum which alters the magnitude of the uniform magnetic field on the two sides.

observable is the contrast of the interference pattern vs. magnetic field. The fit to the data is from models which account for the 8 magnetic states and the effects of our finite velocity distribution and the misalignment of the uniform magnetic field that determines the quantization axis. Not only are the relative positions of the contrast maxima, but also their width and the degradation of contrast of subsequent rephasings due to the finite coherence length are well accounted for by the model. An application of this rephasing experiment is that (since the value of the Bohr magneton is accurately known), one of the fit parameters is the average velocity of the atoms that contribute to the interference pattern. This can be exploited to eliminate systematic effects arising from processes, like the clipping of part of the beam by the septum, which cause this velocity to differ from the average velocity of the atoms in the beam upstream of the interferometer as determined by a measurement of the diffraction pattern (Fig.3b).

For large currents through the foil, the average over the velocity distribution of the atom beam reduces the contrast in the interference pattern of all atoms except those in the two $m_F = 0$ states, which experience no Zeeman phase shift. This results in a contrast one-fourth of that observed for no current. At this point, any small phase shifts observed from additional interactions are those of only the $m_F = 0$ states. By applying a large Stark phase shift to all of the substates, the contrast of these $m_F = 0$ states could be reduced to nearly zero while another polarisation state was shifted back into coherence with itself (Fig.7). This allows experiments to be performed with polarised atoms in a unpolarized beam without the difficulty of optical pumping, (but without the gain in intensity which such optical pumping could bring).

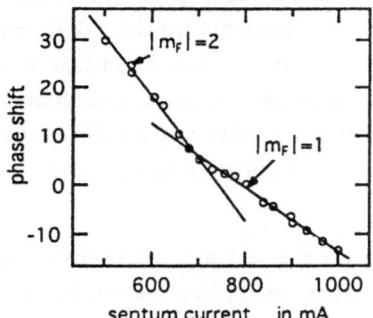

Figure 7: Magnetic phase shift for $|m_F|=2$ and $|m_F|=1$ states as observed with an unpolarized beam. in a *"magnetic rephasing"* experiment. An additional DC stark phase shift of 65 rad is applied in one arm of the interferometer. For 400 - 700 mA septum current one of the $|m_F|=2$ states, and between 700 - 1050 mA two of the $|m_F|=1$ states are shifted back in coherence with itself.

4.3 Refractive Index of an Atom in a Gas:

An other experiment currently pursued with our interferometer is investigating the effect of a medium like a gas on the atomic wave function. From the perspective of wave optics, the passage of a wave through a medium is described in terms of an index of refraction $n = 1 + \frac{2\pi}{k^2} N \cdot f(k,0)$ [30] where N is the density of the medium and $f(k, \vartheta)$ is the scattering amplitude. The real part of n is proportional to the phase shift of the wave and the imaginary part is proportional to the attenuation of the wave by scattering into undetected channels. The atom wave function evolving in a gas with density N is given by (x is the distance from entering the gas)

$$\Psi_{Atom}(x) = \Psi_{Atom}(0) e^{ikx} e^{-i\frac{2\pi}{k}Nx\,\mathrm{Re}(f(k,0))} e^{-\frac{2\pi}{k}Nx\,\mathrm{Im}(f(k,0))}$$

We see that for one the amplitude of Ψ_{Atom} is attenuated as one would expect from the total scattering cross section $\sigma_{tot} = \frac{4\pi}{k}\mathrm{Im}(f(k,0))$. In addition, there is a phase shift $\Delta\varphi(k,x) = -\frac{2\pi}{k}Nx\,\mathrm{Re}(f(k,0))$ proportional to the real part of the forward scattering amplitude. This is in analogy to optics with photons or neutrons. In contrast to the neutron optics [31] where scattering is dominantly s-wave, up to a few hundred partial waves participate atom-atom scattering at thermal energies.

For our experiment we fill one side of our interaction region with gas (Ne, Ar, Kr, Xe). We have three observables: The averaged transmission through our sample, the averaged interference amplitude, and the phase shift of the interference pattern. From the measured attenuation and the phase shift, the phase and amplitude of $f(k,0)$ can be determined. Furthermore, the scattering cross section is not a constant with respect to the center of mass energy, and therefore, different velocities have different attenuation. Therefore the average velocity of the transmitted wave differs for different gas densities. This may enable us to gain additional information on the energy dependence of $f(k,0)$.

Interferometry is the only direct way to measure the real part of the scattering amplitude. Therefore the separated beam atom interferometer offers the possibility to obtain additional information about the atom-atom scattering process. This is especially interesting for low energy collisions, because knowing $f(k,0)$ means that both the magnitude and sign of the scattering length can be determined. A knowledge of the sign of the scattering length, hitherto not measurable, is critical to predicting low temperature collective behaviour [32].

5 Fundamental Measurements

The inherent precision available with phase sensitive measurements makes atom interferometers ideal instruments for a variety of fundamental measurements like inertial [33, 14] and gravitational effects [33, 34], a precision measurement of the photon recoil [35] or "null" tests like a search for a residual charge of neutral atoms. The sensitivity to phase will allow atom interferometers to probe physical processes that generate phase shifts such as Berry's (and other) topological phases (a recent related proposal is [36]), the passage of atoms through a wave guide, or the phase shift which accompanies surface bounces. In general, it has not previously been possible to observe these phase-generating effects.

5.1 The Aharonov Casher effect:

A recent proposal by Anandan [37] and Aharonov and Casher [38], an analogy to the Aharonov-Bohm effect, but for the magnetic moment, is a good example: it is a topological phase which tests a fundamental tenet of quantum mechanics, that a phase shift can occur in the absence of any classical force. A study of this effect using neutron interferometers has been conducted [39]. The two advantages of using atoms are the greater magnetic moment and the greater intensity. Care has to be taken to compensate for the large DC Stark phase shift which may limit the practical size of the fields which can be applied. The total phase measured in the neutron experiment was only 3 mrad [39]. In contrast we will be able to achieve up to 1 radian Aharonov-Casher phase for sodium atoms in our interferometer. This should greatly reduce the statistical error and will also allow us to study, the predicted dependence on the dipole orientation. Varying the velocity and the velocity distribution of our Na beam will allow us to explicitly investigate the nondispersive character of the Aharonov Casher effect and therefore give an direct evidence of its topological nature [40].

5.2 Measurement of the Geometric Phase:

As another example of the possibilities of our separated beam interferometer we will describe here a new method to explicitly measure the Berry phase of the wave function.

When a quantum system evolves (even adiabatically) around a cyclic path in phase space it gains an additional phase, as first described by Berry [41-43]. Various demonstrations of this effect have been performed with photons [44], spin rotation experiments with neutrons [45], and atomic hydrogen [46]. With our separated beam interferometer, we can demonstrate the geometric phase of the wave function directly by transporting the atom on the two arms of the interferometer on different paths in phase space. Such an explicit determination of the geometric phase acquired by the wave function of a massive particle was prohibited in previous experiments, because of an unknown and much larger dynamic phase shift between the two paths. We plan the following experiment based on our magnetic rephasing setup. This time the quantization axis is defined by a guide field in y- direction, normal to the atomic beam which travels in z- direction. Adding a screw coil around the interaction region, we get the following magnetic field:

$$\mathbf{B}_{total} = \mathbf{B}_{septum} + \mathbf{B}_{screw} + \mathbf{B}_{bias} = \begin{pmatrix} 0 \\ 0 \\ \pm \frac{b}{1+a^2x^2} \end{pmatrix} + \begin{pmatrix} B_{screw}\sin(x) \\ B_{screw}\cos(x) \\ 0 \end{pmatrix} + \begin{pmatrix} 0 \\ B_{bias} \\ 0 \end{pmatrix}$$

where a,b are constants that describe the septum magnetic field coming from the current in the metal sheet connected at two points and the +/- stands for the left/right arm of the interferometer. One can show that when the septum field is normal to the other fields, $|\mathbf{B}_{total}|$ is the same for both arms of the interferometer and therefore the dynamic phase $\varphi_{dynamic} = \frac{1}{\hbar}\frac{l}{v}\int \vec{\mu}\cdot\vec{B}_{total}dx$ acquired in either arm of the interferometer will be the same. Hence there will be no dynamic phase shift between the two arms of the interferometer, as we have shown experimentally using a combination of bias field in the y-direction and our septum field.

The spinor wave function itself will evolve on different trajectories in phase space on different sides of the septum and acquire a phase shift $\Delta\varphi_{geometric} = m_F\Delta\Omega_{evolution}$ that is purely geometric and is given by the solid angle which the state vector subtends ($\Delta\Omega_{evolution} = \Omega_{left} - \Omega_{right}$). If the motion is adiabatic $\Omega_{evolution}$ is given

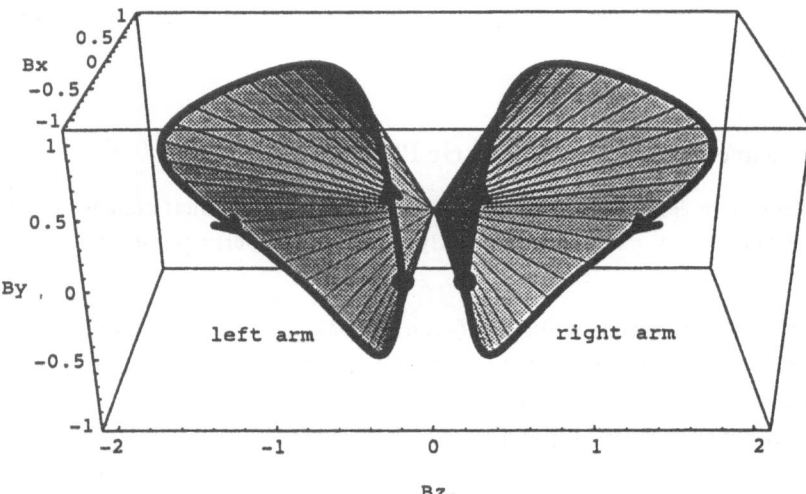

Figure 7: Movement of the magnetic field vectors on both sides of the septum with distance through the interaction region.. Entering the interaction region the atom finds the magnetic field vector at the marked location (●). While the atom moves through the interaction region the magnetic field vector moves along the two trajectories (____) as given by the arrows, and returns to the starting point when the atom leaves the interaction region. The septum field is twice the size of the B_{screw} and $B_{bias}=0$. The solid angle *between* the two curves is a measure of the geometric phase obtained by the adiabatic evolution of the state.

by the solid angle subtended by the magnetic field vector B_{total} as shown in Fig. 8. We can vary $\Omega_{evolution}$ by changing the relative field strengths for the various components of our field. We can also change the adiabaticity in our experiments by varying the total field strengths. By optical pumping to a $|m_F| = 1$ or $|m_F| = 2$ state we can demonstrate the spin dependence of the geometric phase.

The geometric phase is only seen in a combination of at least $B_{septum} + B_{screw}$. Any small additional dynamic phase shift of each component alone can be measured independently. Therefore we will be able to separate the dynamic phase from the geometric phase, and can make a direct interferometric measurement of the geometric phase of the wave function for massive particles. It is interesting to point out that an experiment like this can also be seen as a demonstration of geometric forces [36]. The phase shift has the same relation to the geometric force as the Aharonov Bohm phase shift has to the Lorentz force.

6 Conclusion

The future of atoms interferometers looks bright: atom beams are relative inexpensive and intense. Several techniques have now been demonstrated to make interferometers, and the atoms which may be used in them come with a wide range of parameters such as polarizability, mass, and magnetic moment. One can even imagine molecular interferometers. We think that we will see a lot of applications of atom and molecular interferometers to a wide range of measurements of both fundamental and practical interest.

Our recent work on atom interferometers and atom optics is supported by the Army Research Office contracts DAAL03-89-K-0082, and ASSERT 29970-PH-AAS, the Office of Naval Research contract N00014-89-J-1207, and the Joint Services Electronics Program contract DAAL03-89-C-0001. J.S. was supported in part by an *Erwin Schrödinger Fellowship* of the *Fond zur Förderung der Wissenschaftilchen Forschung* in Austria.

REFERENCES

[1] Th. Young, Phil. Trans. Roy. Soc., London XCII **12** 387 (1802).
[2] H. Fizeau Ann. Phys. Chem. 3, 457 (1853); Ann. Chem. Phys. **B66**, 429 (1862).
[3] A.A. Michelson, Am. J. Sci., (3) **22**, 20 (1881); and E.W. Morley, Am. J. Sci. **31**, 377 (1886).
[4] Lord Rayleigh, Proc. Roy. Soc., **59**, 148 (1896).
[5] L. Marton, J. Arol Simpson, and J.A. Suddeth, Phys. Rev. **90**, 490 (1954), Rev. Sci. Inst. **25**, 1099 (1954).

[6]   G. Mollenstedt, and H. Duker Naturwissenschaften **42**, and Z. Phys. **145**, 377 (1956).

[7]   H. Rauch, W. Treimer, and U. Bonse, Phys. Lett. **A47**, 369 (1974).

[8]   Matter Wave Interferometry, a special review issue of Physica B&C **151**, (1988).

[8]   S. Altschuler and L.M. Franz, U.S. Patent #3,761,721.

[9]   R. Beach, S.R. Hartmann, and R. Friedberg, Phys. Rev. **A25**, 2658 (1982).

[10] V.P. Chebotayev, B. Ya. Dubertsky, A.P. Kasantsev, and V.P.Yakovlev, J. Opt. Soc. Am. **2** (11), 1791 (1985).

[11] Ch. J. Borde, Phys. Lett. **A140**, 10 (1989).

[12] D.W. Keith, C.R. Ekstrom, Q.A. Turchette, and D.E. Pritchard, Phys. Rev. Lett. **66**, 2693 (1991).

[13] O. Carnal, J. Mlynek, Phys.Rev.Lett. **66**, 2689 (1991).

[14] F. Riehle, Th. Kisters, A. Witte, J. Helmcke, J. Borde, Phys. Rev. Lett. **67**, 177 (1991).

[15] M. Kasevich, S. Chu, Phys. Rev. Lett. **67**, 181 (1991).

[16] J.Robert et.al. Eur. Phys. Lett. **16** 29 (1991).

[17] F. Shimizu, K. Shimizu and H. Takuma, Phys. Rev. **A46**, 46 (1992).

[18] U. Sterr, K. Sengstock, J.H. Muller, D. Bettermann and W. Ertmer, Appl. Phys. **B54**, 341 (1992).

[19] D.E. Pritchard, *"Atom Interferometry"*, Proceedings of ICAP 16 (Munich 1992)

[20] B. Levy, Physics Today, **17** (1991).

[21] For an overview of recent demonstration experiments see: *"Optics and Interferometry with Atoms"* Special issue Applied Physics **B54** (1992), Plenum New York (1986) editors: J. Mlynek, V. Balykin, P. Meystre

[22] D.W. Keith, R.J. Soave, and M.J. Rooks, J. Vac. Sci. Technol. **9**, 2846 (1991); C.R. Ekstrom, D.W. Keith, and D.E. Pritchard, Appl. Phys. **B54** , 369 (1992).

[23] B.J. Chang, R. Alverness, and E.N. Leith, Appl. Optics **14**, 1597 (1975).

[24] H. Talbot, Phil.Mag. **9**, 401 (1836).

[25] Lord Rayleigh, Phil.Mag. **11**, 196 (1881).

[26] J.F. Clauser, M.W. Reinisch, Appl. Phys. **B54**, 380 (1992).

[27] R. Molof, H. Schwartz, T. Miller and B. Bederson, Phys. Rev. **A10**, 1131 (1974).

[28] F. Shimizu, K. Shimizu, and H. Takuma, J. Appl. Phys. **B31**, L436 (1992).

[29] F. Riehle, A. Witte, Th. Kisters, and J. Helmcke, Appl. Phys. **B54**, 383 (1992).

[30] J.D. Jackson *"Classical Electrodynamics"* John Wiley & Sons (1975)

[31] V.F.Sears, *"Neutron Optics"*, (Oxford Uinversity Press) (1990).

[32] V. Bagnato, D. Pritchard and D. Kleppner, Phys. Rev. **A35**, 4357 (1987); E. Tiesinga, A.J. Moerdijk, B.J. Verhaar, and H.T.C. Stoof, Phys. Rev. **A46**, R1167, (1992)

[33] J.F. Clauser, Physica B&C **151**, 262 (1988).

[34] M. Kasevich, S. Chu, App. Phys. **B54**, 321 (1992).

[35] D. Weiss and S. Chu, Phys.Rev.Lett (in press) (1993)

[36] A Stern, Phys. Rev. Lett. **68**, 1022 (1992); Y. Aharonov and A. Stern, Phys. Rev. Lett. **69**, 3593, (1992).

[37] J. Anandan, Phys. Rev. Lett. **48**, 1660 (1982).

[38] Y. Aharonov and A. Casher, Phys. Rev. Lett. **53**, 319 (1984).

[39] A. Cimmino, G. I. Opat, A.G. Klein, H. Kaiser, S.A. Werner, M. Arif, and R. Clothier, Phys. Rev. Lett. **63**, 308 (1989).

[40] A. Zeilinger in: *"Fundamental Aspects of Quantum Theory"* NATO ASI Series B, Vol.144, p. 311, editors: Vittorio Gorini and Alberto Frigerio, (Plenum, New York 1986)

[41] M.V. Berry, Proc.Roy.Soc. (London) A392, 45 (1984).

[42] M.V. Berry, Physics Today, Dec 1990 p.34.

[43] *"Geometric Phases in Physics"*, Advanced Series in Mathematical Physics Vol.5, editors: F.Wilczek and A.Shapere (World Scientific, Singapore, 1989).

[44] R.Y. Chiao, Y.S. Wu Phys.Rev.Lett. 57, 933 (1986); P.G. Kwiat, R.Y. Chiao .Phys. Rev. Lett. 66, 588 (1991).

[45] T. Bitters, D. Dubbers, Phys.Rev.Lett. 59, 251 (1987); H. Weinfurter, G. Badurek, Phys.Rev.Lett. 64, 1318 (1990).

[46] Ch. Miniatura et.al., Phys.Rev.Lett 69, 261 (1992)

Optical Ramsey Interferometry with Magnesium Atoms

K. Sengstock, U.Sterr, D. Bettermann, J.H. Müller, V. Rieger and W. Ertmer

Institut für Angewandte Physik der Universität Bonn, Wegelerstr. 8, W-53115 Bonn 1, Germany

1. Introduction

With the recent developments of atom interferometers [1-7], matter wave interferometry has been extended to "particles" with many internal degrees of freedom. This opens up new experimental fields not or hardly accessible to neutron or electron interferometers taking advantage of additional interactions, e.g., with light fields.

In atom interferometers the atomic wavefunction is split either by mechanical, microfabricated beamsplitters (wavefront splitting in terms of optical interferometry) [1,2] or by the interaction with light waves (amplitude splitting) [3-5]. The splitting due to the redistribution of photons in a standing wave divides the wavefunction in partial beams of ground state atoms, differing in momentum by even multiples of the photon momentum $\hbar k$ (k: wavevector of the light field).

On the other hand, the splitting due to the absorption of a single photon from a travelling wave, exciting atoms to a reasonably long living state, will result in two partial beams, differing in momentum as well as in the internal states. This simplifies extraordinarily the selective access to one interferometer arm by state selective interactions, without the need of a spatial separation between the two arms (typically a few μm in atom interferometers based on mechanical beamsplitters) or a spatial limitation of the interaction to one arm.

This development is strongly aided by the recent developments in cooling and manipulation of atomic samples [8]. These preparation methods offer widely tuneable sources of atomic waves, ranging from typical pm-wavelength (thermal) up into the thousands of Å, and allow - concerning interferometry - totally new arrangements by trapping, as will be explained below.

One of the earliest motivations for the advance of laser manipulation methods were possible applications of cold atoms for ultra high resolution spectroscopy and frequency standards [9,10]. Optical high resolution spectroscopy of thermal atomic samples is mainly limited by the second order Doppler shift which broadens and shifts spectral lines by an amount in the order of $\Delta\nu/\nu = 10^{-12}$; in addition transit time broadening reduces the line Q-value.

Recently several groups demonstrated microwave Ramsey spectroscopy on cold atoms in atomic fountains [12-15]. On their ballistic flight due to earth gravitation the slow atoms travel twice through a microwave region, which reduces systematic field phase errors, usually present in Ramsey spectroscopy [16]. In the optical domain the compensation of the first order Doppler effect demands for more elaborate techniques, as, e.g., the four zone optical Ramsey-spectroscopy [17,18]. But for beam experiments, laser cooled or thermal, the problem of residual phase errors in the Ramsey zones remains unresolved respectively is one of the main uncertainties [19].

In one of the experiments described here magneto-optically trapped atoms are directly utilized for a new spectroscopic scheme, based on a pulsed Ramsey-excitation cycle, which is inherently free of phase errors. The method uses the advantages of a three level V-system with a fast transition for trapping and cooling and a slow transition for spectroscopy, like the $^1S_0$- $^1P_1$,$^3P_1$ system in ^{24}Mg.

The atoms are captured, stored and cooled on the fast $^1S_0$-$^1P_1$ transition ($\lambda=$ 285,2nm, $\tau(^1P_1)=$2,02ns). To avoid light shifts and incoherent scattering due to the trap laser light the trapping beams are switched off for a short time to allow the undisturbed interaction with the clock transition, the intercombination line $^1S_0$-$^3P_1$ ($\lambda=$457nm, $\tau(^3P_1)=$ 5,1ms). In analogy to an atomic beam crossing four Ramsey beams this interaction is achieved by two pairs of counterpropagating laser pulses.

This scheme of periodic sequences of trapping and probing is different to schemes which separate the spectroscopic process from the trap in space, as e.g., in microwave spectroscopy on atomic fountains [14,15]. Atoms *shelved* in the metastable $^3P_1$-state leave the trap volume, not influenced by the trap forces, when switching on the trap laser beams again. The subsequent decrease in trap fluorescence is a signature for the Ramsey excitation easy detectable on the strong fluorescence. In addition this detection is strongly amplified by the dynamics of the trapping process itself, as will be explained below. Besides the high resolution this detection scheme opens up new possibilities for investigations of cold collisions, analysis of atom traps , and in particular *pulsed interferometer* experiments.

2. The experimental setup

An overview of the experimental setup is shown in Fig.1. It shows two Ramsey interferometer, one formed by a thermal atomic beam and a second one formed by a laser cooled and trapped ensemble of cold magnesium atoms. Both sections are supported by the same laser beam tuned to the 457nm $^1S_0$-$^3P_1$ clock transition. The laser light is generated by a frequency stabilized stilbene 3 dye laser with a linewidth of 5 Hz relative to a stable reference cavity.

This scheme of simultaneous recording of Ramsey fringes on a thermal atomic beam using a four-zone excitation and a laser cooled ensemble with pulsed excitation allows to measure the second order Doppler shift of the thermal atomic

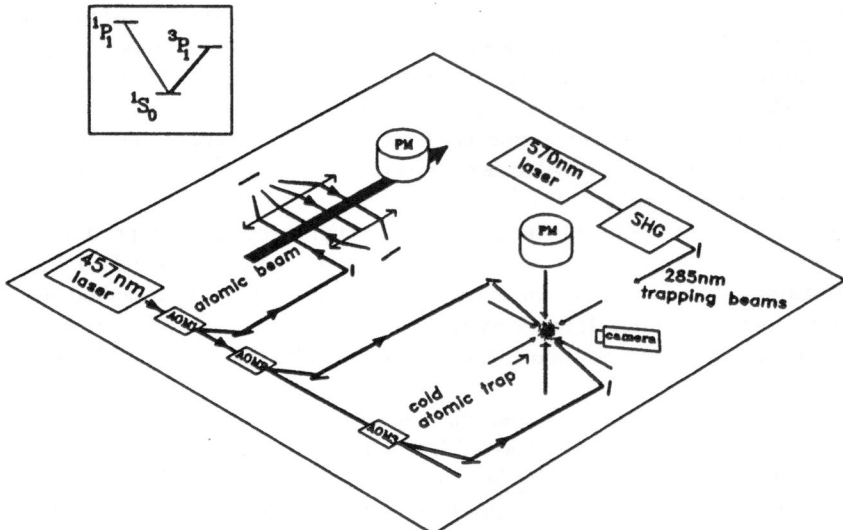

Fig.1: Experimental setup

beam. In addition by stabilizing the laser to the beam signal it is possible to correct for the frequency drift of about 10Hz/s of the reference cavity, thus allowing longer integration times for the trap signal. An acousto optical modulator (AOM1) is switching the available laser power between the thermal beam apparatus and the trap. Two further AOM's (AOM2/3) form the pulses for the trap excitation scheme, discussed in detail in a later paragraph.

We capture and cool an ensemble of Magnesium atoms in a magnetooptic trap similar in concept to that first demonstrated by Raab et al.[20]. The trap consists of a spherical magnetic quadrupole field and three perpendicular pairs of σ^+-σ^- polarized laser beams. The trap can be filled either from an atomic beam, decelerated to an adjustable velocity of about 50m/s by the Zeeman tuning technique [21], or, more simply but about a factor of 100 less efficiently, from the low velocity tail of the Maxwell-Boltzmann distribution of a thermal atomic beam. The trapping beams with diameters of 2,6mm and intensities of 2mW/mm^2 as well as the stopping laser beam are obtained by frequency doubling the 570nm light of two rhodamine 6G dye lasers in ADA crystals in external cavities. The frequency of the trap laser is tuned about one linewidth below the $^1S_1$-$^1P_0$ transition ($\lambda = 285$nm). The magnetic field is provided by two coils in anti-Helmholtz configuration (diameter 5cm) which generate an adjustable gradient of up to 80 G/cm along their axis.

For magnetic field gradients of 50 G/cm 10^5 atoms (with atomic beam deceleration) or 10^3 atoms (without beam deceleration) are stored within a diameter of 300μm. In most of the experiments presented here the trap was filled directly from a thermal beam . The temperature of the trapped atoms reached nearly the Doppler limit as expected for the closed two level system ($^1S_0$-$^1P_1$) of ^{24}Mg.

3. The *Beam* Ramsey Interferometer

The interferometer setup consists of a thermal atomic beam of Magnesium atoms (T=700K) crossed perpendicularly by four travelling laser waves resonant with the intercombination transition $^1S_0$-$^3P_1$ of ^{24}Mg. This setup constitutes an atom interferometer as was first shown by Bordé [22].

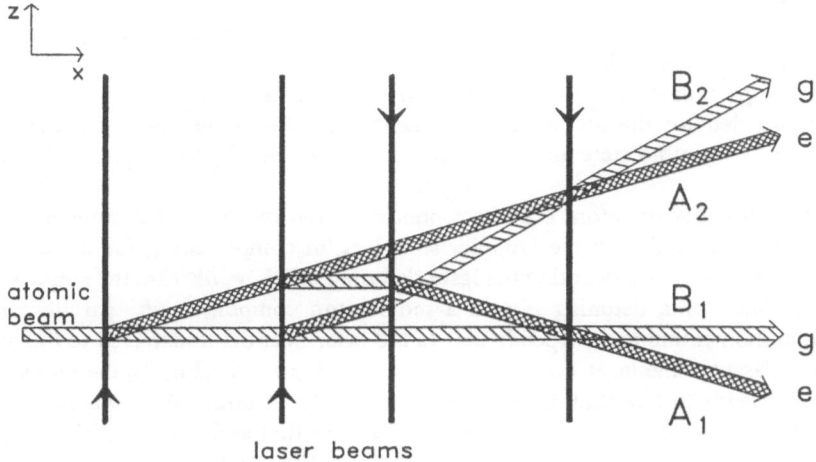

Fig.2: Scheme of the Ramsey interferometer

The interactions with the four laser fields split and subsequently recombine the atomic wavefunction by the exchange of photon momenta and photon energy between the laser fields and the atom. Fig.2 shows those atomic trajectories which contribute to the interference. Each closed loop represents an interferometer of the Mach-Zehnder type corresponding to the two recoil components. The two exit ports of both interferometers differ in momentum and in the internal state (e: excited state, g: ground state). Therefore the interferometer can simply be read out by monitoring the excited state population, e.g., by detecting the fluorescence.

3.1 Mechanical Interpretation of the Ramsey Interferometer

Compared to atom interferometers working with material beamsplitters Ramsey interferometers offer the advantage that the beamsplitting can be affected by the laser frequency. In the following we will explain the frequency dependence of the interference pattern as effect of a momentum transfer perpendicular to the direction of the laser beams. This momentum transfer, which is widely neglected when treating the scattering of atoms by a laser field, plays a crucial role in the Ramsey interferometer. Due to the high interferometric sensitivity, small longitudinal displacements caused by different momenta in both arms strongly affect the interference. The momentum transfer with a component perpendicular to the laser beam is possible according to the Heisenberg minimum uncertainty between the rms waist radius Δw and the angular spread of photon momenta $\hbar \Delta k_x$, $\Delta w \, \hbar \Delta k_x = \hbar/2$ of a Gaussian beam. For simplicity we will consider only one interferometer (the blue recoil component, corresponding to interferometer 1 in Fig.2).

Energy and momentum conservation on the absorption of a single photon with frequency ω_L demands

$$p_0^2/2m + \hbar\omega_L = (p_0+\hbar k)^2/2m + \hbar\omega_0 \, , \quad |k| = \omega_L/c. \tag{1}$$

Here ω_0 denotes the atomic eigenfrequency, p_0 the momentum of the incident atom in the ground state and m the mass of the atom. Eq.1 can be simplified to

$$k \cdot p_0/m = \Delta - \delta \tag{2}$$

This means that the atom absorbs a photon of the momentum decomposition of the localized field with the Doppler shift $k \cdot p_0/m$ compensating for the detuning $\Delta - \delta$, $\Delta = \omega_L - \omega_0$ denoting the laser detuning and $\delta = \hbar k^2/2m$ the single recoil shift. Thus for a detuning $\Delta \neq \delta$ a momentum component $\hbar k_x$ parallel to the atomic momentum and perpendicular to the laser beam is transferred to the atom.

If we assume the laser beams to be parallel to the z-axis (Fig.2), the momentum transfer leads to a spatial displacement Δx and Δz of both partial waves $|g,p_0\rangle$ and $|e,p_0+k\rangle$ during the dark zone between the first and the second laser beam:

$$\Delta z = T \, \hbar k/m, \quad \Delta x = T \, \hbar(\Delta - \delta)/p_x. \tag{3}$$

Here $T = D/v_x$ denotes the time of flight between first and second resp. 3rd and 4th laser beam, p_x and p_z the atomic momentum along x and z and k_z the mean laser wavevector component along z. During the central dark zone between 2nd and 3rd laser beam atoms in both partial waves are in the same internal and external state $|g,p_0\rangle$ ($|e,p_0+\hbar k\rangle$ for the red recoil component), leading to no additional displacement. In the third dark zone a momentum difference between both partial atomic waves leads to a final displacement at the exit ports of $\Delta z = 0$, $\Delta x = 2T \, \hbar(\Delta-\delta)/p_x$ ($\Delta x = 2T \, \hbar(\Delta+\delta)/p_x$ for the red recoil component). With our experimental parameters a laser detuning of one kHz corresponds to a longitudinal separation of the atomic wavepackets by $\Delta x = 300$ fm. In each interferometer this displacement leads to an interference term in the probability of finding the atom in one exit port which harmonically varies with the phase difference $\phi = 2\pi \, \Delta x/\lambda_{DB}$ (λ_{DB} being the de Broglie wavelength $\lambda_{DB} = h/p_x$) between both interfering arms. As the two exit ports correspond to different internal states, the probability P_e of finding an atom behind the interferometer in the excited state is given by

$$P_e \propto \cos(\phi + \phi_L) = \cos(2T \, (\Delta \pm \delta) + \phi_L) \tag{4}$$

neglecting a constant background. The $+\delta$ corresponds to the red recoil component, the -δ to the blue one. Here $\phi_L = \phi_2 - \phi_1 + \phi_4 - \phi_3$ is a phase depending on the relative phases $\phi_1, \phi_2, \phi_3, \phi_4$ of the four laser beams. This oscillation in the excitation probability describes the well known Ramsey fringe pattern.

The resulting amplitude of the interference fringes is given by the product of the individual excitation amplitudes in the subsequent interaction with all four laser beams [18].

3.2 Application as Frequency Discriminator

The sensitive dependence of the interference signal on the laser frequency has been the basis of the optical Ramsey spectroscopy since several years. Typical widths of the central fringe range for thermal atomic beams and moderate laser beam separations D of some centimeters from several kHz up to several ten kHz depending on experimental conditions and according to (2) on the separation D between the copropagating pairs of laser beams. Fig.3 shows a typical Ramsey pattern for Mg atoms (the incoherent background has been subtracted). The figure demonstrates the increasing resolution with increasing distance D, ranging from 4.5mm up to 21mm. Only the lowest order fringes are visible due to the low coherence length of the thermal atomic beam of about two de Broglie wavelengths (λ_{DB} = 23pm). With a width (FWHM) of the lowest curve of Fig.3 of 4kHz and a signal to noise ratio of 10 a stability of $3.4 \cdot 10^{-13} 1/\sqrt{\tau}$ (sec) is possible.

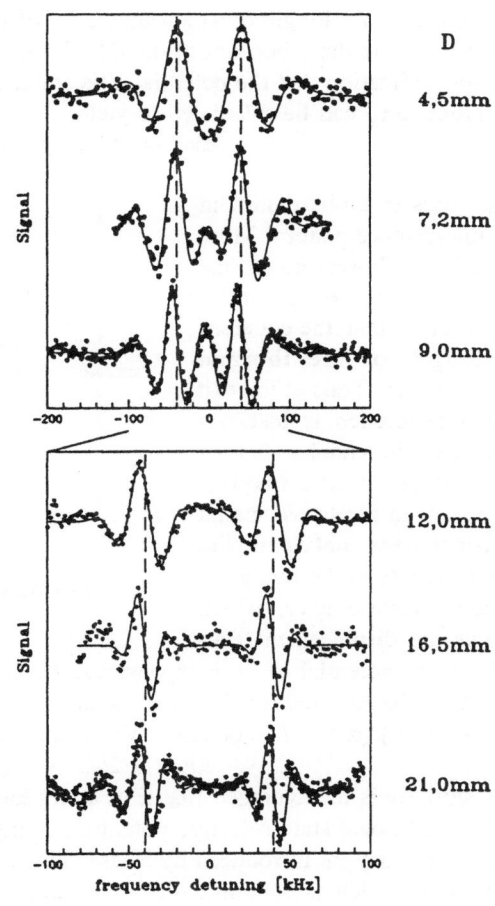

Fig 3: Ramsey signals for different values D

3.3 DC- and AC- Phase Shifter for Atomic Waves

In subsection 3.1 it has been shown that the interferometer may be explained by a longitudinal splitting of the atomic wave by laser induced longitudinal momentum transfer to "parts of the atom", which are labeled by the internal state. Because of this labeling it is rather obvious that by state selective interaction one partial wave can be selectively addressed, e.g. by an additional laser beam. If this beam is e.g. an off-resonant Gaussian beam, it will be possible to decelerate or accelerate one partial wave by dipole forces in the inhomogeneous intensity profile.

This leads to a longitudinal advance Δx_V of this part of the wave packet compared to that in the other arm of the interferometer. Using Ehrenfest's theorem in the approximation that the potential V is small compared to the total energy E of the atom, Δx_V can be calculated to yield:

$$\Delta x_V = -1/2E \int V(x) \, dx \qquad (5)$$

This leads to a phase shift in the interference pattern. More significant, however, is the fact that Δx_V is generally much larger than the coherence length of the thermal atomic beam. Thus at the exit port interference is lost. As explained in chapter 3.1, a detuning of the laser frequency results in a relative spatial displacement between the wave packets in the two arms of the interferometer, which allows the displacement cau-

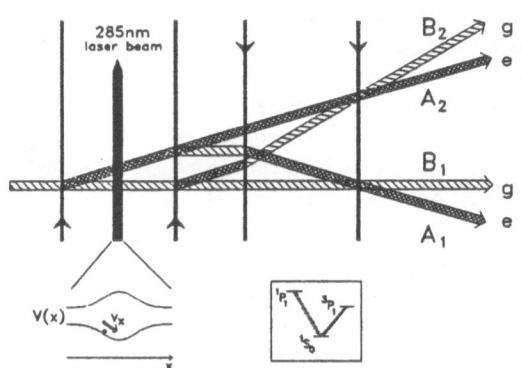

Fig.4: The AC-Stark phase shifter

sed by the potential to be compensated. Maximum contrast of the interference signal can be restored by changing the laser frequency by an amount $\Delta\nu$ given by $\Delta\nu = -(\Delta x_V E)/(hD)$. The potential V and the frequency shift $\Delta\nu$ are related by

$$\Delta\nu = 1/2hD \int V(x) \, dx \qquad (6)$$

The shift $\Delta\nu$ is the averaged line shift along the path of the atom.

This additional state selective potential, acting only on one arm of each interferometer, may be introduced by a laser beam, nearly resonant with the $^1S_0$-$^1P_1$ transition, which is sent into the dark zone between the first and the second Ramsey laser beam (Fig.4). The corresponding potential V is the local energy of the ground state (the "dressed state" [23]) shifted by the interaction with the additional light field (AC-Stark effect)

$$V = \hbar \Delta_{UV} / 2 \left(\left(1 + \Omega^2 / \Delta_{UV}^2 \right)^{1/2} - 1 \right). \qquad (7)$$

$\Omega = \Gamma (\text{I}/2\text{I}_{SAT})^{1/2}$ denotes the Rabi frequency of the $^1S_0\text{-}^1P_1$ transition, $\text{I}_{SAT} = \pi \, h \, c \, / \, (3\tau \, \lambda^3)$ its saturation intensity and. For a sufficiently large detuning Δ_{UV} the spontaneous emission from the excited level $^1P_1$ can be neglected. Inserting this potential in eq.6 gives the resulting compensating frequency shift $\Delta\nu$; it is proportional to P_{UV}/Δ_{UV} (P_{UV} denotes the UV laser power)

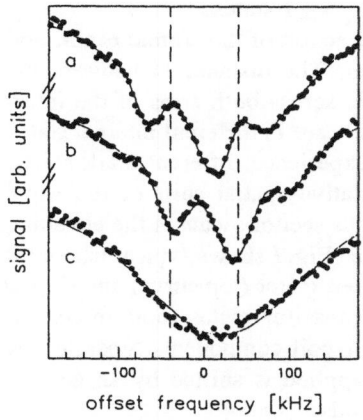

Fig.5: Shift of the undisturbed Ramsey signal (a) due to an additional UV-laser beam with detuning 2GHz(b) resp.0Hz(c)

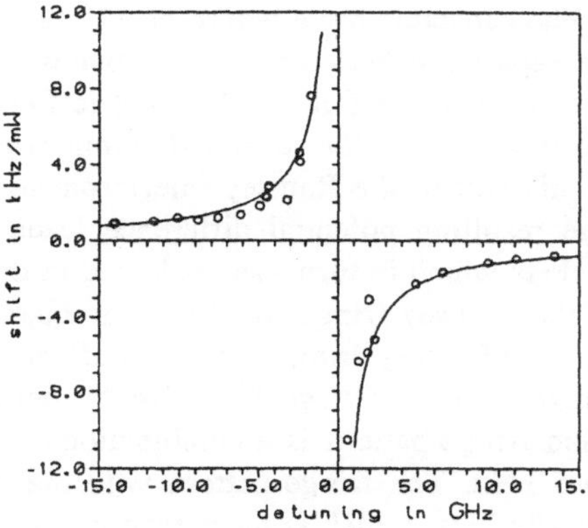

Fig.6: Shift of the Ramsey signal for different values of the Uv-laser detuning and power

A typical Ramsey spectrum demonstrating the light induced shift is shown in Fig.5 and the shift for various detunings and intensities of the UV laser is presented in Fig.6.

The probabiliy of spontaneous emission increases with smaller detuning and higher power of the UV laser. This results in a decrease of the fringe contrast, as each spontaneous emission destroys the coherence; it would in principle allow to detect the atom path in the interferometer.

With a fringe amplitude of 3,000 counts per second and total signal strength of 20,000 counts per second we can measure a shift of 0.15 rad within one second.

A further extension of this experimental method is to use the static polarizability of the atoms. This also opens the way to Aharonov-Casher type experiments as will be discussed below.

Applying an electric field $\mathscr{E}$ to the arms of the Ramsey interferometer influences the internal energy of the atoms through the dipole interaction er$\mathscr{E}$. Since the atom enters the electric field slowly compared to its internal time scale this leads - analogous to the Born-Oppenheimer approximation - to a potential V for the external motion. For the Stark effect, V is given through the atomic polarizability α:

$$V(x) = -1/2 \, \alpha \, \mathscr{E}(x)^2 \qquad (8)$$

We have placed a capacitor between two co-propagating laser beams. In addition to the electric field there is a magnetic field parallel to the atomic beam direction across the interferometer region. It separates the magnetic sublevels of the excited state. With the polarization of the laser beams parallel to the magnetic field, the interferometer is only utilizing the $\Delta m_x = 0$ transition. Note that with regard to the perpendicular electric field in the y-direction, the atoms in the excited state are in a coherent superposition of $m_y = \pm 1$ states.

The capacitor has a length of 5.0 mm in the direction of the atomic beam, and the separation between the two plates is 1.9 mm. The distance D between the laser beams is 8.6 mm. The electric field extends across both arms of the interferometer. Since the separated atomic wave packets are in different internal states in this part of the Ramsey interferometer, they experience different Stark shifts. The resulting potential difference leads to a relative spatial shift of the wave packets which in turn - as explained in the previous section - causes the envelope of the Ramsey fringes to shift in frequency space. Fig.7 shows typical examples for both Ramsey fringes of the uninfluenced interferometer (top curve) and shifted fringes with a voltage U of 1.4 kV across the capacitor plates (bottom curve). Each fringe pattern is a combination of the two recoil components separated by 79.5 kHz. The fringe pattern with the voltage applied is shifted by an amount $\Delta\nu = 82$ kHz, which is more than two fringe periods.

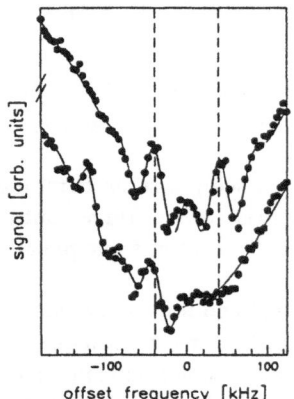

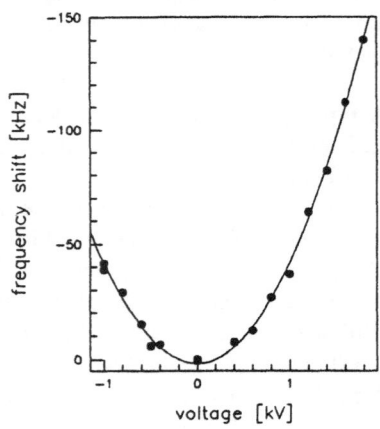

Fig.7: DC-Stark shift of the Ramsey signal (see text)

Fig.8: DC-Stark shift for different values of the electric field strength

The results are presented in Fig.8.

We have numerically calculated the electric field distribution of the capacitor including fringe fields and performed the integration of the potential in the region between the two laser beams. We thus deduce a difference in the polarizabilities of the $3s^2(^1S_0)$ and the $3s3p(^3P_1,\ |m|=1)$ states of

$$\alpha(^3P_1) - \alpha(^1S_0) = 8.0\pm1.0\text{kHz/(kV/cm)}^2 \qquad (9)$$

The main contributions to the error are statistical uncertainties in the fitting of our data as in Fig. 7 (10%), plus our incomplete knowledge of the field distribution (5%).

There are several theoretical calculations of the polarizability of the ground state, the most accurate yielding $\alpha(^1S_0)=(17.7\pm0.4)$ kHz/(kV/cm)^2. If one accepts the theoretical result for the ground state polarizability as fairly accurate, this measurement can be used to derive a value of the polarizability of the $^3P_1(m=\pm1)$ excited state or the corresponding element of the polarizability tensor. The value is $\alpha(^3P_1,m=\pm1) = 25.7\pm1.1$ kHz/(kV/cm)^2, with the error resulting from the uncertainties in the theoretical and the experimental value.

Several authors have pointed out the intriguing possibility of measuring the Aharonov-Casher effect (ACE) with an atom interferometer. Such an experiment requires placing an electric field across the arms of the interferometer, too. For the Ramsey interferometer one has to consider the large fringe shifts associated with the Stark effect in addition to ACE phases. From our result, one can estimate that for thermal velocities and moderate to high field strength the ACE results in a phase shift that is at least one million times smaller than the Stark phase shift. One possible solution is to extend the electric field into the interaction regions with the laser beams. In this case the ACE would be detectable as a fringe shift relative to the dc-Stark shifted incoherent background.

Furthermore, with the setup of Fig. 1 the Stark effect can be used advantageously to perform a scalar Aharonov-Bohm experiment similar to the one performed by Allman et al. [24] with neutrons. Localized and velocity selected wave packets can be created using pulsed excitation. Applying a voltage to the capacitor only while such an atomic wave packet is present between the electrodes is expected to result in a dispersion free phase shift of the interference signal. Corresponding experiments are in progress.

4. The *Trap* Ramsey Interferometer

The coherence length L_C of an atomic beam, e.g., the width of the atomic wavepackets, is given by $L_C = h\ /\ (m\ \Delta v_x)$. For a thermal atomic beam of Mg atoms (T = 700 K) the coherence length is 33 pm. In terms of optical interferometry this beam corresponds to white light and thus only low order fringes are visible. This situation can be substantially improved by laser cooling. For Mg atoms cooled to the Doppler limit the corresponding velocity spread Δv reads $\Delta v =$ 81 cm/s. This is equivalent to a coherence length of 20 nm which is a factor of 600 larger than for a thermal beam.

Furthermore, also the mean atomic velocity can be reduced and tuned over a wide range by means of laser cooling techniques. This offers also a large tuning range of the corresponding de Broglie wavelength. In a matter wave interferometer slow atoms interact longer with a potential, hence they will experience a larger phase shift compared to fast atoms; this leads to a higher sensitivity for small potentials.

For trapping of neutral magnesium atoms a magnetooptical trap [20] is well suited in which ensembles of atoms with temperatures near the Doppler limit are attainable. If such a cold and dense ensemble of atoms is irradiated by successive, coherent pulses of two counterpropagating laser beams tuned to the intercombination transition, we have a situation similar to the spatial sequence of four Ramsey laser beams in the atomic beam experiment (Fig.9). Thus the atomic waves are split and recombined "in place", neglecting a slight motion due to the photon recoil.

Such a cycle could be started with the filling of the trap during a short time. Then the trapping lasers and the magnetic field would be switched off to allow the application of a 457 nm $\pi/2$ pulse to the atoms. After a time

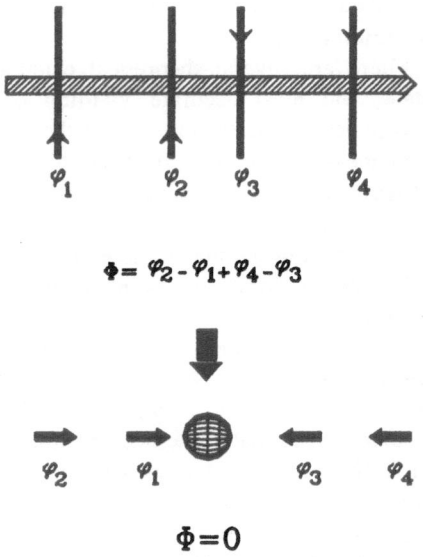

$$\Phi = \varphi_2 - \varphi_1 + \varphi_4 - \varphi_3$$

$$\Phi = 0$$

Fig.9: Scheme of the Ramsey experiments (see text)

T a second $\pi/2$ pulse would be applied from the same direction, followed by the same sequence from the opposite direction.

4.1 The Pulsed Scheme

The pulse sequence is applied at a repetition rate of typically 70 Hz. Each cycle (Fig.10) is started by shutting off the trapping beams within 50 ns by an UV-acousto-optic modulator (see Fig.1) and an additional mechanical shutter. Simultaneously with the trapping lasers the magnetic field configuration is switched from a quadrupole to a homogeneous field of 15G perpendicular to the direction of the Ramsey pulses in order to shift the $\Delta m = \pm 1$ transitions out of resonance. About $50\mu s$ later the cold atoms were irradiated by two pairs of counterpropagating laser pulses. The pulses with a width of $2\mu s$ each are separated by a time T between 5 and $50\mu s$. Due to this Ramsey excitation a part of the atoms is shelved in the metastable $^3P_1$ state ($\tau = 5,1ms$) thus leaving the trap region undisturbed by the trap laser beams, which are switched on 50 μs after the last excitation pulse. The decrease in the trap fluorescence is monitored by a photomultiplier tube and completes the cycle.

4.2 The Ramsey Signal

The frequency dependence of the excitation probability for a four-beam excitation had been calculated by Bordé et al. [18], but the calculation performed in the atomic rest frame is actually that of a four-pulse experiment; it is hence directly applicable to our trap arangement.

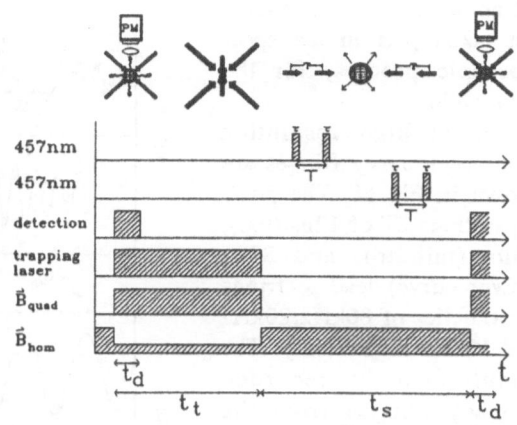

Fig.10: Time scheme of the pulsed Ramsey experiment

One of the most important features of the pulsed scheme is its inherently complete phase compensation. As in our experiment the pulses are simply cut from a cw beam, the phase difference Φ_L (eq.4) exactly cancels to zero.

4.3 Scheme for Signal Amplification

Excited atoms, shelved in the metastable $^3P_1$-state moving at the typical velocity in the trap fly a distance of about 5mm before they return to the ground-state by spontaneous emission. Those atoms are not recaptured and leave the trap region. Therefore the Ramsey excitation constitutes an additional loss mechanism for the trap. As in our trap the number of stored atoms is not limited by the trap density but mainly by collisions with the background-gas, this additional loss will lead to a lower equilibrium number of trapped atoms and hence after reaching a new steady state to a decrease in the trap fluorescence.

The resulting number of trapped atoms N is given by

$$N = N_0 \, 1/(1+P\tau_T/T_C). \qquad (10)$$

Here N_0 denotes the number of atoms without Ramsey excitation, T_C the duration of one cycle and τ_T the undisturbed lifetime of the trap. P is the excitation probability for one pulse sequence. Typical values for the experiment described here are: $T_C=20$ms, $\tau_T=2$s, P=0.01.

With these parameters we obtain an amplification factor of 50 which allowed us to detect a 1% excitation amplitude as a 50% decrease in trap fluorescence. Because of this strong amplification and the efficient detection of the electron shelving via the fast trap transition, we were able to detect Ramsey fringes with integration times of only two seconds per frequency point with only about 1000 trapped atoms.

Compared to schemes which separate trapping and spectroscopy in space this method offers the advantage of accumulating the loss of many cycles and of retrapping atoms, which have not taken part in the spectroscopic process, for the next cycle.

Typical high resolution scans of Ramsey fringes are shown in Fig.11. The pulse separations 2T of 12μs (top), 48μs (middle), and 88μs (lower curve) lead to fringe periodicties of 80kHz,20kHz and 12kHz respectively. The simultaneously recorded Ramsey fringes from the thermal atomic beam (lowest curve) clearly show the relativistic red shift due to the second order Doppler effect measured in a series of scans to (-1.5 ± 0.3)kHz [25]. The error of 300Hz is mainly determined by the strongly non-linear influence

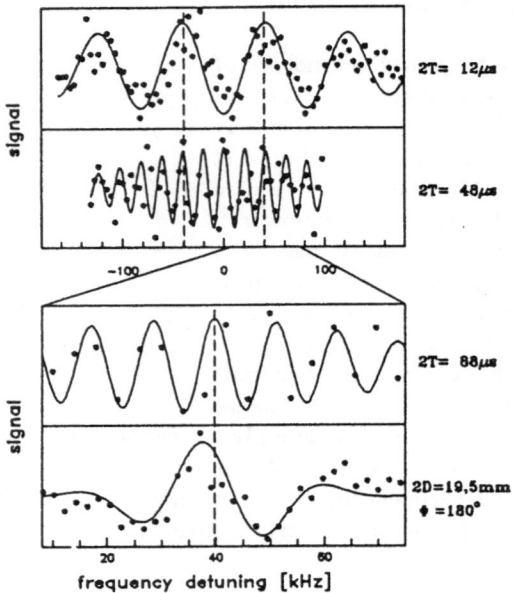

Fig.11: Ramsey fringes on a magnetooptical trap for different resolutions compared to a signal on a thermal beam (bottom)

of the uncertainties in the thermal velocity distribution. The second order Doppler effect of the residual velocity spread of 2m/s in the trap will contribute to a broadening of only $2 \cdot 10^{-17}$. Similarly the phase error in the thermal beam experiment amounts to about 400Hz whereas the residual phase error for the trap experiment contributes less than 1Hz. Residual magnetic and electric fields contribute in both cases less than 1Hz. Thus the main problems of the thermal beam experiment producing an uncertainty in the determination of the line center of 500Hz ($\Delta\nu/\nu \leq 7 \cdot 10^{-13}$) add up in the trap experiment to the Hz-level (10^{-15}) [25].

The experimental resolution of 4 kHz is at the momentary state of the apparatus mainly limited by residual vibrations and fluctuations of the trap apparatus. With a more elaborated setup, a resolution in the order of the natural linewidth of 31Hz seems to be feasable.

As the trap setup forms an atom interferometer too, splitting the atomic wave function in space, it is best suited for atom interferometry like measurements of the scalar Aharonov-Bohm effect [26,27] or of a Berry phase [28,29]. The physical mechanisms inducing these small phase shifts (additional electric, magnetic and/or laser fields) can easily be induced in time in the dark periods between the Ramsey pulses [5]. Furthermore, the small extension of the trapped ensemble will drastically reduce the necessary efforts for shielding and/or compensating external influences.

References

[1]   O. Carnal, J. Mlynek. Phys.Rev.Lett. 66, 2689 (1991)
[2]   D.W. Keith, C.R. Ekstrom, Q.A. Turchette, D.E. Pritchard. Phys.Rev.
      Lett. 66, 2693 (1991)
[3]   F. Riehle, Th. Kisters, A. Witte, J. Helmcke, C.J. Bordé. Phys.Rev.Lett.
      67, 177 (1991)
[4]   M. Kasevich, S. Chu. Phys.Rev.Lett.67, 181 (1991)
[5]   U. Sterr, K. Sengstock, J.H. Müller, D. Bettermann, W. Ertmer. Appl.
      Phys.B 54, 341 (1992)
[6]   C. Miniatura, J. Robert, O. Gorceix, V. Lorent, S. Le Boiteux,
      J. Reinhardt, J. Baudon. Phys.Rev.Lett. 69, 261, (1992)
[7]   F. Shimizu, K. Shimizu, H. Takuma. Phys.Rev.A 46, 17 (1992)
[8]   J.O.S.A.B 2 (special issue on "Mechanical Effects of Light")
      eds. S. Stenholm (1985)
      J.O.S.A.B 6 (special issue on "Laser Cooling and Trapping of Atoms") eds.
      S. Chu, C. Wieman (1989)
[9]   W. Ertmer, R. Blatt, J.L. Hall. in "Laser-Cooled and Trapped Atoms"
      (Proceedings of the Workshop on Spectroscopic Applications of Slow
      Atomic Beams) ed. W.D. Phillips, S.154, Natl.Bur.Stand.(U.S.)
      Spec.Publ. 653 (1983)
[10] W. Ertmer, S. Penselin. Metrologia 22, 195 (1986)
[11] A. Martin, K. Helmerson, V. Bagnato, G. Lafyatis, D.E. Pritchard. Phys.
      Rev.Lett. 61, 2431 (1988)
[12] D.W. Sesko, C.E. Wieman. Opt.Lett. 14, 269 (1989)
[13] M. Kasevich, S. Chu. Phys.Rev.Lett. 63, 612 (1989)
[14] C. Monroe, H. Robinson, C. Wieman. Opt.Lett. 16, 50 (1991)
[15] A. Clairon, C. Salomon, S. Guellati, W.D. Phillips. Europhys.Lett. 16, 165
      (1991)
[16] N.F. Ramsey. Phys.Rev. 78, 695 (1950)
[17] Y.V. Baklanov, B.Y. Dubetski, V.P. Chebotayev. Appl.Phys. 9, 171 (1976)
[18] C.J. Bordé, C. Salomon, S. Avrillier, A. van Lerberghe, C. Bréant,
      D. Bassi, G. Scoles. Phy.Rev.A 30, 1836 (1984)
[19] A. Morinaga, F. Riehle, J. Ishikawa, J. Helmcke. Appl.Phys.B 48, 165
      (1989)
[20] E.L. Raab, M. Prentiss, A. Cable, S. Chu, D.E. Pritchard. Phys.Rev.Lett.
      59, 2631 (1987)
[21] J.V. Prodan, A. Migdall, W.D. Phillips, H.J. Metcalf, J. Dalibard. Phys.
      Rev.Lett. 54, 992 (1985)
[22] C.J. Bordé. Phys.Lett.A. 140, 10 (1989)
[23] J. Dalibard, C Cohen-Tannoudji. J.O.S.A.B 2, 1707 (1985)
[24] B.E. Allman, A. Cimmino, A.G. Klein, G.I. Opat, H. Kaiser,
      S.A. Werner. Phys.Rev.Lett. 68, 2409 (1992)
[25] K. Sengstock, U. Sterr, D. Bettermann, W. Ertmer. to be published

[26] Y. Aharonov, D. Bohm. Phys.Rev. 115, 485 (1959)
[27] Y. Aharonov, A. Casher. Phys.Rev.Lett. 53, 319 (1984)
[28] M.V. Berry. Proc.R.Soc. London A 392, 45 (1984)
[29] M. Reich, U. Sterr, W. Ertmer. Phys.Rev.A, in press (1993)

Momentum Transfer by Adiabatic Passage in a Light Field

C. I. Westbrook*, Lori S. Goldner, C. Gerz, R.J.C. Spreeuw, S. L. Rolston, W. D. Phillips

National Institute of Standards and Technology, Gaithersburg MD, 20899

Abstract. We have demonstrated the transfer of both population and momentum (8ℏk) in a multilevel system by adiabatic following of a slowly evolving light field. Spontaneous emission is suppressed because the atoms remain in a 'dark' state during their evolution. This technique will be useful for the realizaton of coherent, large angle atomic beam splitters. Our results are in good agreement with calculations using the optical Bloch equations.

Keywords. atom optics, atomic beam splitters, coherent population trapping, adiabatic fast passage, laser cooling

1. Introduction and Theoretical Concepts

Atomic physicists have shown great interest in the newly developing field of atom optics[1]. Of especial interest has been the recent development of atomic interferometers, which promise to provide a new sensitive probe of a large variety of phenomena[2]. Certain types of interferometric measurements, such as the detection of rotation, or of long-range forces, benefit greatly from having a large enclosed area in the interferometer. For this purpose a beam splitter capapable of coherent, wide angle separation of an atomic beam is, of course, a necessity. Several groups have made significant progress during the past year in this direction using magneto-optical forces[3] as well as successive applications of π and $\pi/2$ pulses in a two-photon Raman configuration[4]. In this report we discuss our progress toward a different type of wide angle beam splitter using a proposal made by Marte, Zoller and Hall [5].

The technique proposed in Ref. [5] is closely related to one used to transfer population among vibrational sublevels in Na_2 molecules [6]. It is based on the idea that it is possible to prepare a 'dark' state in which the absorption amplitudes from two different laser beams destructively interfere. This dark state can adiabatically follow changes in the two laser intensities thus transferring atomic population and. momentum with out emitting any spontaneous photons. The process is quite robust in the sense that, provided that the conditions for adiabaticity are fulfilled, the transfer is insensitive to the pulse durations or intensities. This may offer some technical advantages when compared to similar processes using Raman π pulses to achieve the momentum transfer.

The simplest example of such a situation is a three level, Λ system, shown in Fig. 1, in which two continuous laser beams separately couple the ground states with a single excited state[6].

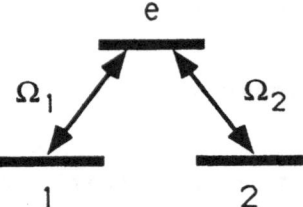

Fig. 1. Three level system with 2 laser induced transitions.

We define Ω_1 and Ω_2 as the Rabi frequencies associated with the transitions $|1\rangle \leftrightarrow |e\rangle$ and $|2\rangle \leftrightarrow |e\rangle$. For any values of Ω_1 and Ω_2 the dark, or non-coupled state:

$$|\psi_{NC}\rangle = \frac{\Omega_2|1\rangle - \Omega_1|2\rangle}{\sqrt{\Omega_1^2 + \Omega_2^2}} \qquad (1)$$

is not coupled to the excited state by the laser fields. The minus sign ensures that the transition amplitudes destructively interfere, assuming that both amplitudes have the same phase. Imagine that the system is prepared in state $|1\rangle$ and $\Omega_1 = 0$ (laser 1 off) and $\Omega_2 \neq 0$ (laser 2 on). If now laser 1 is slowly turned on while laser 2 is slowly turned off, the system will adiabatically evolve into state $|2\rangle$, without ever developing any excited state population. In the high intensity limit, $\Omega_{1,2} \gg \Gamma$, the condition for adiabaticity [6] is that the transfer must take place on a time scale T such that $T \gg (\Omega_1^2 + \Omega_2^2)^{-1/2}$. When the intensity is not in this limit the condition is more complex and we defer discussion of this to a later paper.

For purposes of atom optics, a crucial generalization of the above system is to include an exteral degree of freedom, the momentum state $|p\rangle$ of the atom, in the above considerations[5]. If the two laser beams are counterpropagating, the state $|1,p\rangle$ is coupled by the total laser field to $|2,p+2\hbar k\rangle$ through a Raman transition involving absorption of one photon of laser field 1 and stimulated emission of another photon into laser field 2. Thus, in a counterpropagating laser beam geometry the adiabatic transfer process can be used to coherently transfer momentum to the atom.

Including an external degree of freedom in the problem gives an important technical constraint on the interaction time T. We wish the non-coupled state defined in Eq. 1 to remain uncoupled during the entire interaction time. Because we include momentum of the atom and therefore its kinetic energy, the total energies of the two states $|1,p\rangle$ and $|2,p+2\hbar k\rangle$ can have different values. This means that the non-coupled state can evolve into the coupled state during the interaction time T. The rate at which this happens is related to both the velocity of the atom, as well as the coupling to the excited state. In the low intensity limit this problem has been treated analytically[7]. A detailed discussion of this problem for higher intensities is beyond the scope this article, but it is clear that the if Doppler width of the distribution undergoing the adiabatic transfer is too large, the atoms will not all remain in the dark state during the transfer. Therefore laser cooling of our sample is important in our experiment. Of course higher intensities and shorter pulses permit the use of hotter velocity distributions.

The above idea can be generalised to the case where more than 3 levels are involved in the transfer. For example, any atomic transition between two levels with equal, integer total angular momentum exhibits an analogous dark state which can adiabatically follow the changes in intensity of the two laser fields [5]. When more levels are involved the atoms can undergo more than just one absorption and stimulated emission thus increasing the amount of momentum transferred. For a transition between two levels with angular momentum 4, and with counterpropagating σ^+ and σ^- laser beams, $8\hbar k$ is transferred. Transfer of the atomic population from the F=4, m_F=–4 to the F=4, m_F=+4 state in the Cs D2 ($S_{1/2} - P_{3/2}$) line was recently demonstrated in an experiment using a thermal atomic beam and a copropagating laser beam geometry[8]. Because of the geometry no transfer of momentum occurs. The use of counterpropagating beams in that experiment revealed that, over a small range of velocities, the atoms were indeed in a dark state, but the large velocity width of the atoms in Ref. [8] created a background which rendered the population transfer undetectable.

2. Experiment

Our apparatus is similar to that described in Ref. [9]; we briefly recall it here. Cs atoms from a chirp-cooled atomic beam are loaded into a magneto-optical trap (MOT). The MOT provides a sample of laser cooled atoms confined to less than 1 mm diameter. After loading the MOT for 100 - 500 ms the magnetic field coils are turned off and the atoms are further cooled for less than 100 ms in 3 dimensional optical molasses. The MOT/molasses laser beams are derived from a Ti:Sapph laser tuned slightly below the F=4 to F'=5 transition of the Cs D2 line; these beams are mixed with a repumping laser diode beam tuned to the F=3 to F'=4 transition to prevent optical pumping out of the cycling transition. Using time of flight (TOF) [9] measurements we verify that the the temperature of the atoms is 3 - 6 µK.

After the final cooling phase, the 3D molasses light is turned off and the atoms are subjected to a series of laser pulses which prepare the atoms in a dark state and then effect the transfer. These pulses are resonant with the D2 F=4 to F'=4 transition and no repumping laser beam is included. The pulses derive from the same Ti:S laser, shifted and temporally shaped by appropriately chosen acousto-optic modulators. The first pulse propagates downward and is polarized σ^-. This pulse optically prepares the atoms in the F=4, m_F=–4 sublevel. Atoms (about half the total) which absorb too many photons from this beam are optically pumped into the F=3 hyperfine level, from which they do not escape. Next, the atoms are subject to the adiabatic transfer pulse sequence. A σ^- pulse propagating downwards is followed with a delay τ by a σ^+ polarized pulse propagating upwards. The temporal intensity profile of the two pulses as measured by a fast photodiode is shown in Fig. 2.

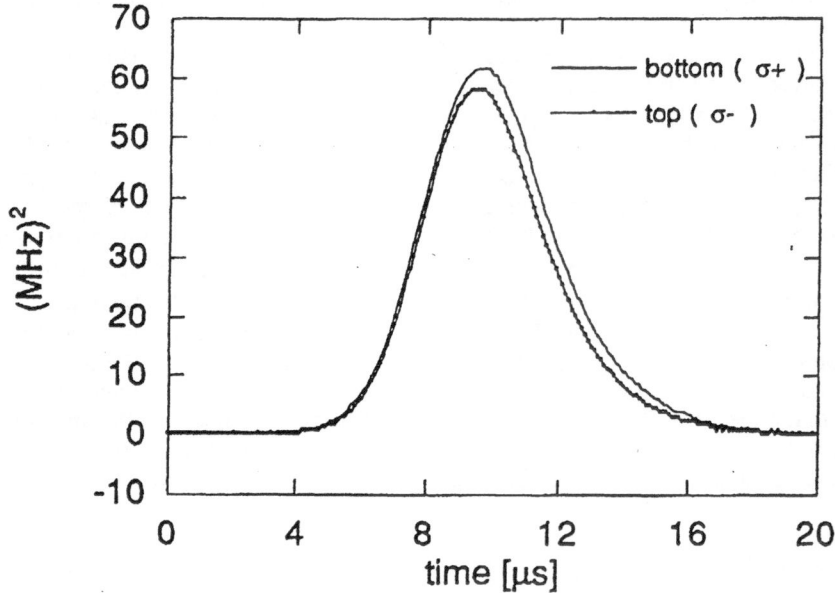

Fig. 2. Intensity profile of the two pulses used for adiabatic transfer. Units on the vertical axis correspond to the Rabi frequency squared for a transition with a Clebsch-Gordan coefficient of $1/\sqrt{2}$.

We observe the transfer by measuring the change in momentum of the atoms after the application of the transfer pulses. This momentum change is measured using the time of flight of the atoms to a probe beam 5 cm below the center of the MOT. The probe beam contains no repumping laser, so that we detect almost exclusively atoms in the F=4 hyperfine level. Figure 3 shows a comparison of the the TOF spectra without the transfer pulses (but including the preparation pulse) and with them. The center of the TOF spectrum is shifted by approximately 3 ms in good agreement with the expected momentum transfer of 2.8 cm/s = 8$\hbar$k for a F=4 to F'=4 transition. Note that the area of the TOF peak after the transfer is about half of that before. This is due to off resonant transitions between the F=4 and F'=3 or 5 states. The F'=3 and 5 levels are separated by only 200 MHz and 250 MHz respectively from the F'=4 level. Thus, with our large pulse areas, some atoms will make transitions to F'=3 or 5 and then decay into the F=3 ground state and escape detection.

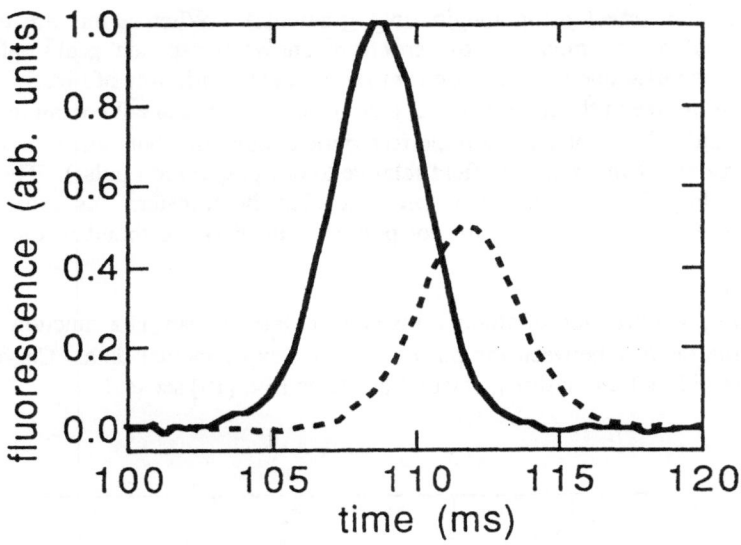

Fig. 3. Demonstration of momentum transfer. Solid line: TOF spectrum without transfer pulses. Dotted line: TOF spectrum with transfer pulses.

We have examined the fraction of atoms which survive the transfer to the $m_F=4$ state as a function of the delay τ between the two pulses. The results are shown in Fig. 4. The optimum transfer fraction occurs for a pulse delay of about 2 μs. For times much different than this the adiabaticity condition is violated. Negative delays correspond to applying the σ^+ pulse first: since the atoms were prepared in the $m_F=-4$ state, they absorb very strongly for negative delays and are pumped to F=3. Note that the large survival fraction when the transfer pulses are in the 'right' order ($\tau>0$) relative to the preparation pulse, is good evidence that a significant fraction of the atoms does remain in the non-coupled state during the transfer. We are able to observe the transfer of momentum in the other direction by applying the transfer pulses without using the preparation pulse. In this case the leading edge of the first transfer pulse optically pumps the atoms into the dark state (or into F=3), and then the transfer proceeds as before. In this case we cannot accurately measure the fraction transferred, but we observe a transfer of momentum in both directions, i.e. to arrival times either 3 ms earlier or later than without the transfer pulses, depending on their order.

To verify our interpretation, we have confirmed that the atomic population is transferred among the internal, Zeeman states as well. We do this by applying a magnetic field gradient of 2 mT/cm along the vertical axis for the first 25 ms after the atoms' release. Different magnetic sublevels of the F=4 manifold are subject to different vertical magnetic forces, resulting in a TOF spectrum which depends on the inital internal state of the atoms (note that this is nothing more than the

Stern-Gerlach effect using longitudinal gradients). When atoms are dropped directly out of the molasses, the spectrum shows 9 separate peaks. after the preparation pulse one sees the spectrum shown in the solid line of Fig. 5; Almost all the atoms are in the $m_F=-4$ level. (We believe that the small contamination of $m_F=-3$ and -2 is not due to imperfect optical pumping but rather to a slight misalignment of the magnetic field relative to our preparation pulse). The dotted curve in Fig. 5 shows the TOF spectrum after the transfer. As expected the majority of the atoms which are not pumped into F=3 are found in the $m_F=+4$ state.

Fig. 4. Fraction of atoms transfered by laser pulses as a function of the delay τ between the pulses. Points: experimental data. Curves (solid, dotted, dashed): calculations from Ref. [10] see text.

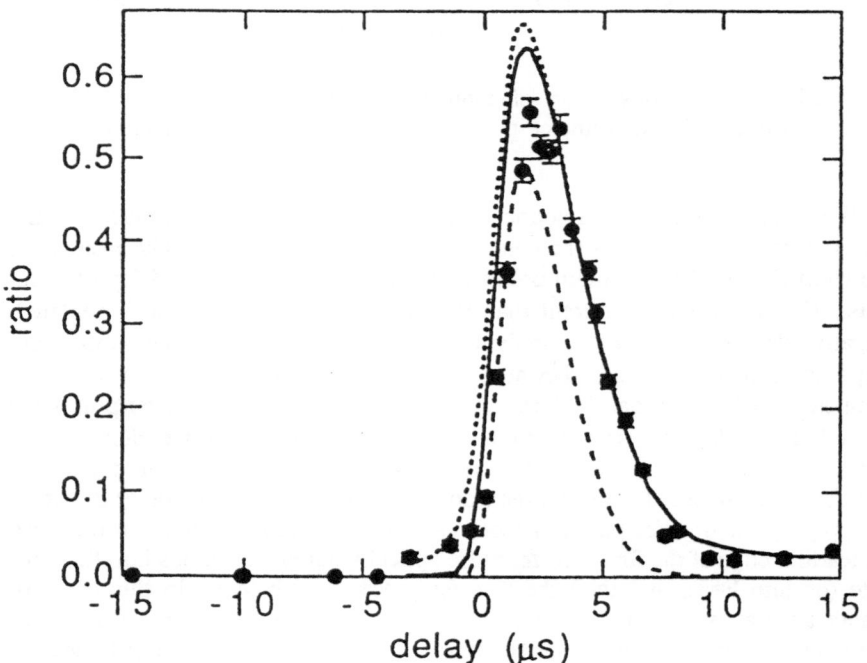

3. Discussion

To further study our data we compare it to a numerical integration of the optical
Bloch equations for our experimental situation [10]. The calculation includes the
the effects of off resonant transitions to the other excited hyperfine states, as well
as the true pulse shapes as measured in our experiment. The results of this
calculation are presented in Fig. 4. The dotted curve shows the fraction of atoms
which the model predicts should remain in the F=4 level during the transfer as a
function of the delay τ between the two pulses. This is the curve which
corresponds to our measurement. We feel the agreement is quite good and it
gives us great confidence that for these experimental parameters, the experiment
is well understood. The solid curve shows the fraction transferred to F=4, m_F=+4
state. The difference between the two curves is because some atoms make non-
adiabatic transitions or absorb photons via the F'=3 or 5 states, and can end up in
other m_F states. Finally, the dashed curve shows the fraction of atoms which
undergo neither spontaneous emission nor nonadiabatic transitions, i.e. are
coherently transferred to the m_F=+4 level. These atoms are those which preserve
their phase during the transfer and could be used in and atomic interferometer.

Fig. 5. Analysis of the internal states of the atoms. Solid line:
without transfer pulses. Dotted line: with transfer pulses.

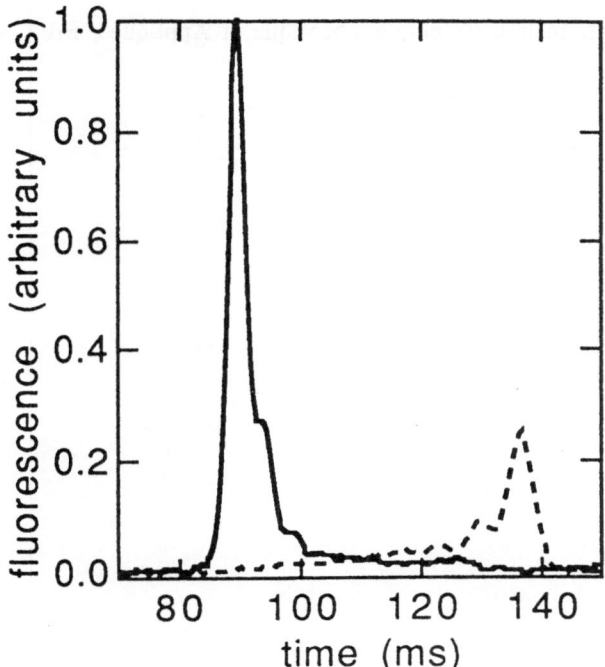

The chief reason for the 49% coherent transfer efficiency is the nearness of the F'=3 and 5 states in the excited state manifold. Calculations analogous to the one shown in Fig. 4 show that for the Cs D1 line ($S_{1/2} - P_{1/2}$), where the separation of F'=4 and F'=3 in the excited state is much larger (1 GHz), the coherent transfer efficiency should be well over 90%. In this case one can easily envision many sets of transfer pulses in order to transfer a much larger momentum. Very recently one group has observed several successive adiabatic transfers in a beam of metastable He [11]. Experiments using several transfer pulses on the Cs or Rb D1 line are planned in our lab.

Finally we note as did the author's of Ref. [5] that this transfer effect can be used as an atomic beamsplitter. In order to do this, one must create a superposition of states, one of which does not undergo the momentum transfer. For example a microwave pulse could be used to produce a superposition of the F=4, m_F=4 and F=3, m_F=3 states. The part of the wavefunction in F=4 could then be subjected to a large number of transfers and then later be recombined with the F=3 part by another microwave pulse. We have high hopes that schemes like this will lead to new generations of interferometers and exciting new measurements.

We wish to thank P. Marte and P. Zoller for permission to show their result in Fig. 4, and for many illuminating discussions. This work was partially supported by the US Office of Naval Research.

*Present address, Institut d'Optique Théorique et Appliquée, B.P. 147, F91403 Orsay, France.

References

1. See Appl. Phys., **B54** (1992), special issue on atom optics, J. Mlynek, V. Balykin and P. Meystre, eds.

2. See e.g. contribution of J. Schmiedmayer in this volume.

3. T. Pfau, C. Adams and J. Mlynek, Europhys. Lett., **21**, 439 (1993), see also contribution of M. Siegel in this volume.

4. D. Weiss, B. Young and S. Chu, submitted to Phys. Rev. Lett.

5. P. Marte, P. Zoller and J. Hall, Phys. Rev., **A44**, R4118 (1991).

6. U. Gaubatz, P. Rudecki, M. Becker, S. Schieman, M. Külz and K. Bergmann, Chem. Phys. Lett, **149**, 463 (1988), J. Kuklinski, U. Gaubatz, F. Hioe and K. Bergman, Phys. Rev. **A40**, 6741 (1989) and references therein.

7. A. Aspect, E. Arimondo, R. Kaiser, N. Vansteenkiste, C. Cohen-Tannoudji, J. Opt. Soc. Am. B, **6**, 2112 (1989), see also C. Cohen-Tannoudji, in *Fundamental Systems in Quantum Optics*, J. Dalibard, J.-M. Raimond and J. Zinn-Justin, eds. (North Holland, 1992), p. 153.

8. P. Pillet, C. Valentin, R. Yuan and J. Yu, submitted to Europhys. Lett.

9. C. Gerz, T. Hodapp, P. Jessen, K. Jones, W. Phillips, C. Westbrook and K. Mølmer, Europhys. Lett., **21**, 661 (1993).

10. P. Marte and P. Zoller, private communication.

11. J. Lavall and M. Prentiss, private communication.

Atomic Waveguides and Cavities from Hollow Optical Fibers

C.M. Savage[1,2], S. Marksteiner[1], and P. Zoller[1]

[1]JILA, University of Colorado, Boulder, CO 80309, USA
[2]Dept. of Physics & Theoretical Physics, Australian National University, Canberra, ACT 0200, Australia.

Keywords. Atomic optics, Atomic waveguides, Atomic cavities, Optical fibers, Evanescent waves

1 Introduction

We investigate the theory of atomic waveguides made from hollow optical fibers and find no fundamental obstacles to their practical realization. The mechanism for atomic guidance is reflection from an evanescent electric field at the interface between the glass and the interior vacuum, figure 1. We suggest the possibility of constructing atomic cavities, analogous to optical cavities, from nonuniform waveguides. The sections following this introduction concern: the optical properties of hollow fibers, the atomic dynamics, "proof of principle" experiments, and speculations on atomic cavities.

A number of experiments have demonstrated reflection of atoms from evanescent fields [1, 2]. For sodium atoms with a velocity perpendicular to the interface of a few ms^{-1}, reflection requires 0.4 W focussed into less than 1mm^2. Such intensities are easily achieved in optical fibers.

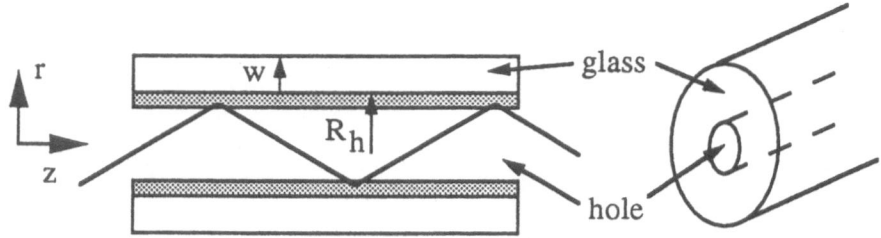

Figure 1: Schematic diagrams of atomic waveguides. Left: longitudinal cross section, showing evanescent field (shaded), and atomic trajectory. Right: transverse cross section showing hole.

Many types of hollow optical fibers could be used. A conventional fiber with a hole allows control over the modal properties by changing the width of the high index doped region. A simple uniform hollow glass tube could also be used. The former could be made single mode while the latter would be multimode since mechanical strength would demand a glass thickness of many tens to hundreds of wavelengths.

We assume that the light is strongly detuned from the atomic resonance. Although this may overestimate the powers required for guidance it makes the theoretical analysis simpler, and spontaneous emission negligible, because the excited state is not significantly populated. In this approximation the evanescent field appears in the Schrödinger equation for the unexcited atom as a potential. All the usual quantum mechanics of potential barriers applies. In particular atoms with transverse kinetic energies less than the barrier's height are reflected. Hence the strength of the evanescent field determines the maximum transverse velocity guided by the fiber.

2   The optical waveguide

We will consider two examples of hollow optical waveguides although many variations are possible: a uniform glass tube with dimensions in the tens of microns, and a doped fiber with dimensions of a few microns. These represent extremes of both fabrication difficulty and atomic guidance properties. The larger waveguide is relatively easy to make but will be multimode. Hence the evanescent field guiding the atoms will fluctuate both spatially and temporally. In contrast the smaller fiber may be hard to make but will have fewer modes. In addition the smaller size increases the spacing between the quantized atomic energy levels (section 3.2).

We do not speculate about how light might be coupled into these fibers. This is a significant experimental problem. However for our calculations we assume only 10 mW of guided power. Given that sources of up to 1 W are available we are allowing for small coupling efficiencies.

The suitability of a waveguide will be determined by the evanescent electric field strength as a function of the guided intensity and by the evanescent field decay length. We denote the width of the light guiding region by w and the radius of the atom guiding hole by R_h, figure 1.

2.1   A large waveguide

A large multimode waveguide can be treated using ray optics [3]. As a further simplification we ignore curvature and approximate the waveguide as a slab. In order to estimate the magnitude of the evanescent electric field we model the field in the waveguide as incident on the glass/vacuum interface with angle of incidence θ, and polarized perpendicular to the plane of incidence. We assume that the intensity of the field on the glass side of the interface I is given by the guided power divided by the cross sectional area of the optical waveguide. The evanescent electric field $e(x)$, as a function of distance x into the vacuum, is then given in terms of the incident electric field e by [4]

$$e(x) = e\alpha \exp(-\kappa x) \tag{1}$$

where the factor α and the decay constant κ are given in terms of the glass refractive index, n, and free space light wavevector, k, by [5]

$$\alpha = 2\sqrt{\frac{n^2}{n^2-1}}\cos(\theta), \quad \kappa = k(n^2\sin^2(\theta)-1)^{1/2} \tag{2}$$

With a glass refractive index of $n = 1.5$, and arbitrarily assuming an angle of incidence of $\theta = 3\pi/8$ radians, the factor $\alpha \approx 1.0$ and the decay constant $\kappa \approx k$.

The incident electric field e is related to the intensity I by

$$e = \sqrt{\frac{2I}{n\epsilon_0 c}} \tag{3}$$

With 10mW of light being guided by $w=50\mu$m thickness of glass with a hole of radius $R_h=50\mu$m the incident intensity $I \approx 0.4$ W mm^{-2}. This is comparable to the intensities already demonstrated to produce atomic reflection [1, 2]. The corresponding electric field is about 1.4×10^4 Vm^{-1}. Similar results follow from a modal analysis of an asymmetric slab waveguide.

2.2 A small waveguide

In this section we report a full modal analysis of a cylindrical hollow fiber in which Na D line light is guided by a high index region of doped glass, as in a conventional optical fiber. This analysis does not require the simplifying assumptions used to analyze the large waveguide. Throughout this section the following parameters are fixed: The refractive indicies of the core and cladding are chosen to be constant and to have the values $n_{co}=1.5$ and $n_{cl}=1.497$ respectively, and the width of the core is chosen to be $w = 1.65\mu$m. With this choice of parameters the fiber is single mode for hole radii $R_h < 1.7\mu$m.

The guided electric field is a superposition of modes of the form [3]

$$\mathbf{E}_\nu(r, \phi, z) = \mathbf{e}_\nu(r, \phi)e^{i(\beta_\nu z - \omega_f t)} + c.c. \tag{4}$$

Here β_ν denotes the propagation constant, ν denotes the angular dependence of the field components to be $\sin(\nu\phi)$ or $\cos(\nu\phi)$, and $\omega_f = kc$ is the angular frequency of the Na D transition. Following [3] we have numerically calculated the normalized propagation constants $\bar{\beta}_\nu$ for the hollow fiber modes,

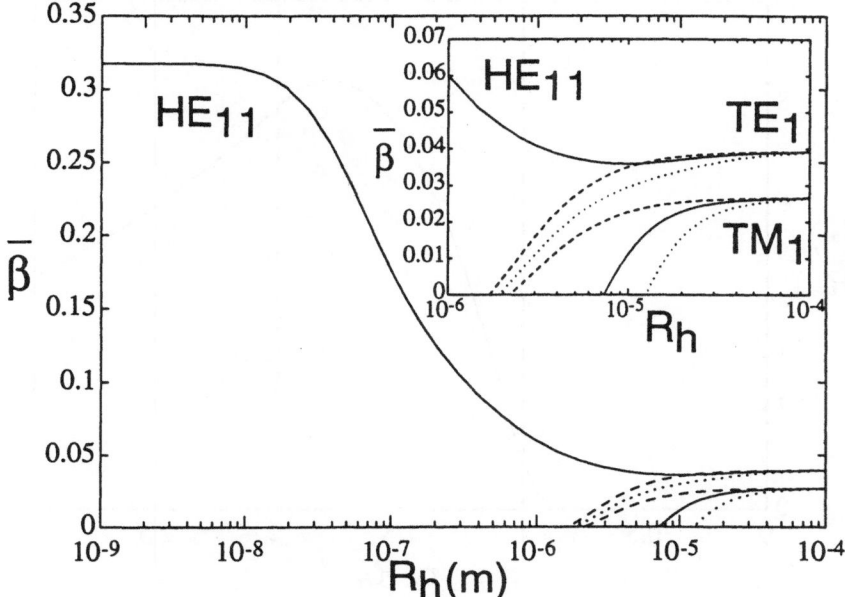

Figure 2: The normalized propagation constants $\bar{\beta}_\nu$, eq.(5), $\nu = 0$ (dashed lines), 1 (solid lines), 2 (dotted lines), versus the radius of the fiber's hole R_h, for $w = 1.65\mu$m. The inset magnifies the region where the fiber becomes multimode.

$\nu = 0,1,2$, as a function of the hole radius, figure 2. The normalized propagation constant is defined to be

$$\bar{\beta}_\nu = \frac{\beta_\nu - n_{cl}k}{(n_{co} - n_{cl})k} \qquad (5)$$

The limits of no hole $R_h = 0$ and an infinite hole $R_h \to \infty$ correspond to a normal fiber and an asymmetric slab waveguide respectively. The $\nu = 1$ mode existing below $R_h = 1.7\mu$m is a hybrid mode HE_{11}. In the limit $R_h \to \infty$ it becomes a TE_1 mode. We may similarly identify a limiting TM_1 mode, figure 2 [3].

We have calculated the electric field in the fiber for the single mode case $R_h = 1.65\mu$m. It has the form

$$\mathbf{E}_1(r, \phi, z) = \mathcal{E}(r)\mathbf{u}(r, \phi)e^{i(\beta_\nu z - \omega_f t)} + c.c \qquad (6)$$

where $\mathbf{u}$ is a complex unit vector. The function $\mathcal{E}$ is plotted in figure 3. It determines the magnitude of the electric field as a function of radius. The electric field for fixed z and t has angular dependence. However for a

64

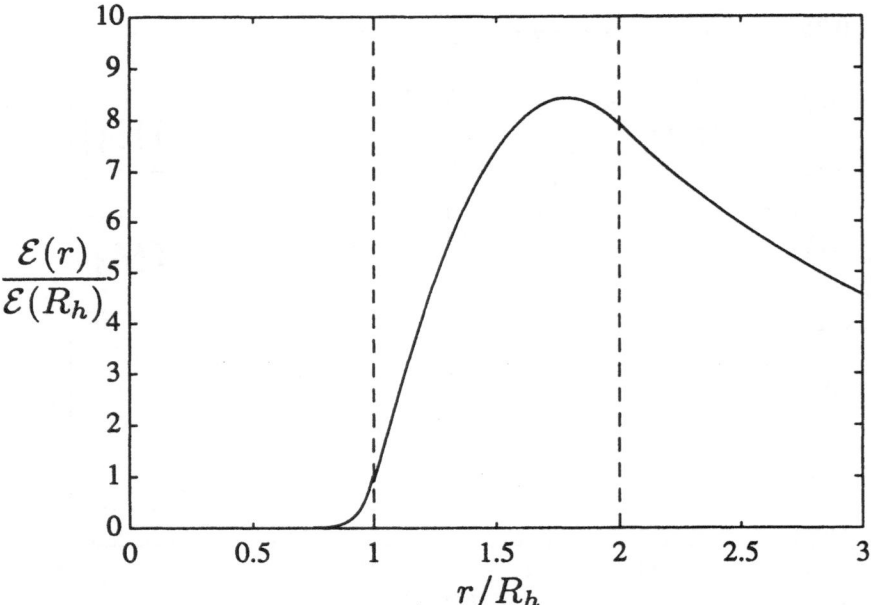

Figure 3: The function $\mathcal{E}$, determining the electric field strength via eq.(6), versus radius for the case $R_h = w = 1.65\mu$m. The first vertical dashed line marks the hole radius and the second the core/cladding boundary.

circularly polarized field the time averaged field is circularly symmetric on the slow time scale of the atomic motion.

The function $\mathcal{E}$ at the hole/core interface depends on the guided power P according to $\mathcal{E}(R_h) = 2.6 \times 10^5\sqrt{P}$. The time averaged root mean square electric field $\bar{E}$ is then $\sqrt{2}$ times bigger

$$\bar{E}(R_h) = 3.7 \times 10^5 \sqrt{P} \qquad (7)$$

The effective evanescent field decay constant $\kappa \approx k$. For a guided power of 10mW the electric field strength is 3.7×10^4 Vm^{-1}. This is nearly three times that found for the large waveguide discussed in section 2.1.

This may seem low since the ratio of linear dimensions for the two waveguides is $50/1.65 \approx 30$. However 70% of the guided power is in the cladding. Furthermore in a ray optics approximation the angle of incidence would be very close to $\pi/2$ so that the factor α, eq.(2), would be small.

Table 1: Physical parameters for Sodium.

parameter	symbol	value
mass of Na atom	m	3.8×10^{-26} kg
Na D line wavevector	k	1.1×10^{7} m^{-1}
Na D line dipole moment	d	2×10^{-29} Asm

3   The atomic waveguide

Let the atom have mass m and momentum operator $\hat{\mathbf{p}}$. Its internal structure is modelled as a two-level electric dipole transition of frequency, ω_a, dipole moment, d, and with raising and lowering operators $\hat{\sigma}_+$ and $\hat{\sigma}_-$. The values of m and d are given in Table 1. The Hamiltonian for the atom is then

$$\hat{H}_S = \frac{\hat{\mathbf{p}}^2}{2m} + \hbar\omega_a\hat{\sigma}_+\hat{\sigma}_- + dE(\hat{\mathbf{r}})(\hat{\sigma}_+ + \hat{\sigma}_-) \tag{8}$$

The electric field $E(\hat{\mathbf{r}})$ is classical but depends on the position operator for the atom $\hat{\mathbf{r}}$. It has frequency ω_f and wavevector k, assumed to be along the fiber axis direction z. The field is assumed to be a traveling wave in the positive z direction:

$$E(\hat{\mathbf{r}}) = \frac{1}{2}C(\hat{r})(e^{-i(\omega_f t - k\hat{z})} + e^{i(\omega_f t - k\hat{z})}) \tag{9}$$

$\hat{z}$ is the position operator along the fiber axis. We assume that the field intensity depends only on the radial cylindrical coordinate, r, figure 1. Defining the detuning $\Delta = \omega_f - \omega_a$ and making the rotating wave approximation the Hamiltonian (8) becomes in the interaction picture

$$\hat{H}_I = \frac{\hat{\mathbf{p}}^2}{2m} - \hbar\Delta\hat{\sigma}_+\hat{\sigma}_- + \frac{d}{2}C(\hat{r})(\hat{\sigma}_+e^{ik\hat{z}} + \hat{\sigma}_-e^{-ik\hat{z}}) \tag{10}$$

The last term of this Hamiltonian may be interpreted as describing the absorption of a photon ($\hat{\sigma}_+$) accompanied by a longitudinal momentum shift of $\hbar k$ ($e^{ik\hat{z}}$), and the conjugate process of photon emission accompanied by a momentum shift of $-\hbar k$.

We have also analyzed a more realistic model in which the atomic transition is $J = 0$ to $J = 1$. With a circularly polarized field the results are similar to those presented below.

3.1   The Schrödinger equation

The Schrödinger equation resulting from the Hamiltonian eq.(10) is

$$i\hbar\partial_t \ket{\Psi(\mathbf{r}, t)} = \hat{H}_I \ket{\Psi(\mathbf{r}, t)} \tag{11}$$

It is separable in cylindrical coordinates. A suitable ansatz is

$$|\Psi(\mathbf{r},t)\rangle = (\,\psi_g(r,t)\,|g\rangle_a\,|q\rangle_z + \psi_e(r,t)\,|e\rangle_a\,|q+k\rangle_z\,)\,|n\rangle_\phi\,e^{(-i\hbar q^2 t/2m)} \quad (12)$$

Here $|g\rangle_a$ and $|e\rangle_a$ are the ground and excited atomic states; $|q\rangle_z$ and $|q+k\rangle_z$ are longitudinal momentum eigenstates; $|n\rangle_\phi$, $n=0,1,2,\ldots$ are eigenstates of the angular momentum; and ψ_g and ψ_e are radial wavefunctions for the ground and excited states.

Substituting this ansatz into the Schrödinger equation (11) we find

$$i\hbar\partial_t\psi_g = -\frac{\hbar^2}{2m}\left(\partial_r^2 + \frac{1}{r}\partial_r - \frac{n^2}{r}\right)\psi_g + \hbar\Omega(r)\psi_e \quad (13)$$

$$i\hbar\partial_t\psi_e = -\frac{\hbar^2}{2m}\left(\partial_r^2 + \frac{1}{r}\partial_r - \frac{n^2}{r}\right)\psi_e$$

$$+\hbar\left(-\Delta + \frac{\hbar}{m}kq + \frac{\hbar}{2m}k^2\right)\psi_e + \hbar\Omega(r)\psi_g \quad (14)$$

$$\Omega(r) \equiv \frac{d}{\hbar}C(r) \quad (15)$$

We have introduced the Rabi frequency $\Omega(r)$. The second set of terms in eq.(14) represent detunings due to the atom-laser frequency difference, the doppler shift, and the atomic recoil shift, respectively. We assume that the first of these dominates all other terms in eq.(14) except the last, proportional to ψ_g. Then eq.(14) reduces to

$$\psi_e(r) = \frac{\Omega(r)}{\Delta}\psi_g(r) \quad (16)$$

This simplification is known as adiabatic elimination of the excited state. It is valid provided the excited state population is negligible. We have also assumed that the transverse kinetic energy, the doppler shift, and the recoil shift are all negligible compared to the detuning energy $\hbar\Delta$. These assumptions are not essential for reflection. However they greatly simplify the eqs.(13) and (14) as can be seen by substituting eq.(16) into eq.(13)

$$i\hbar\partial_t\psi_g = -\frac{\hbar^2}{2m}\left(\partial_r^2 + \frac{1}{r}\partial_r - \frac{n^2}{r}\right)\psi_g + \hbar\frac{\Omega(r)^2}{\Delta}\psi_g \quad (17)$$

This equation is valid provided the detuning Δ is large enough to keep the excited state population small

$$\frac{\Omega(r)}{|\Delta|} \ll 1 \quad (18)$$

The Schrödinger equation (17) has the form of that for a particle in the radial potential $\hbar\Omega(r)^2/\Delta$. The potential is repulsive provided the laser is

Table 2: Largest deBroglie wavelengths for the eigenfunctions of a particle confined to a disk of radius R_h: $\lambda_{dB}/R_h = 2\pi/j_{n,k}$.

(n,k)	$(0,1)$	$(1,1)$	$(2,1)$	$(0,2)$	$(3,1)$	$(1,2)$	$(4,1)$	$(2,2)$
λ_{dB}/R_h	2.62	1.64	1.22	1.14	0.98	0.90	0.83	0.75

tuned above the transition, $\Delta > 0$. We expect that particles with insufficient (transverse) kinetic energy to overcome the potential barrier will be reflected. This is the mechanism for the guidance of atoms. Note that the potential depends on the square of the Rabi frequency Ω so it can be large while still fulfilling the inequality (18).

3.2 Quantization of the transverse motion

According to the ansatz eq.(12) the motion along the fiber, in the z direction, can have arbitrary velocity. However the transverse motion, represented by the wavefunction ψ_g, satisfies the Schrödinger equation (17), and is hence quantized. For large waveguides the evanescent field can be approximated as a delta function at the wall. The physical problem is then that of quantization of a free particle confined by a circular wall. The eigenfunctions are Bessel functions of the first kind, $J_n(2\pi r/\lambda_{dB})$, where λ_{dB} is the deBroglie wavelength related to the transverse part of the energy E_T by

$$\lambda_{dB}^2 = \frac{2\pi^2\hbar^2}{mE_T} \tag{19}$$

The eigenvalue equation for the transverse deBroglie wavelengths is that the wavefunction be zero on the wall, $J_n(2\pi R_h/\lambda_{dB}) = 0$. So the quantized deBroglie wavelengths are given by

$$\lambda_{dB} = \frac{2\pi R_h}{j_{n,k}} \tag{20}$$

where $j_{n,k}$ is the k th zero of the Bessel function J_n. Table 2 gives the largest values of $2\pi/j_{n,k}$. Not surprisingly the largest wavelength is approximately the hole diameter.

It is difficult to cool atoms much below the so called photon recoil limit, at which the deBroglie wavelength is approximately the wavelength of the cooling light. For the Sodium D line this is about 0.6 μm. This estimates the hole diameter required to get a significant fraction of atoms from such a cooled source into the lowest transverse state. More fundamentally it estimates the hole diameter for which quantization of the atomic motion is significant.

3.3 Atomic reflection and tunneling

We now use the theory of the large optical waveguide discussed in section 2.1 to estimate the maximum guided transverse velocity. Using the small waveguide would yield similar results. Let the intensity of the field in the glass be I. Then for the sodium D transition the Rabi frequency of the evanescent field at the interface is, from eqs.(1,3,15), (this and all following quantities have SI units)

$$\Omega = \frac{d}{\hbar}\alpha \left(\frac{2I}{n\epsilon_0 c}\right)^{\frac{1}{2}} = 4.5 \times 10^6 I^{\frac{1}{2}} \tag{21}$$

The maximum transverse velocity classically reflected by the evanescent field, v_m is given by

$$\frac{1}{2}mv_m^2 = \hbar\frac{\Omega^2}{\Delta} \Rightarrow v_m = \left(\frac{2\hbar}{m}\frac{\Omega^2}{\Delta}\right)^{\frac{1}{2}} \tag{22}$$

To satisfy the excited state population inequality (18) we assume $\Delta = 10\Omega$. Although this justifies adiabatic elimination of the excited state, spontaneous emission will still occur and must be considered for any application requiring atomic coherence. Using $\Delta = 10\Omega$ in eq.(22) gives

$$v_m = 0.05 I^{\frac{1}{4}} \tag{23}$$

For the waveguide under consideration (section 2.1), 10mW of light gave an evanescent field with intensity $I \approx 400$ kWm^{-2}. Substituting this into eq.(23) gives $v_m \approx 1.2$ ms^{-1}. The corresponding Rabi frequency, eq.(21) is $\Omega \approx 3 \times 10^9$ s^{-1}, and the detuning $\Delta \approx 30$GHz.

However there is a certain probability that an atom quantum mechanically tunnels through the barrier to the wall [4]. We assume that if this happens the atom is lost. So we need to estimate the tunneling probability. We do this using the WKB estimate for the tunneling probability, T, through the potential barrier $V(r)$ extending from the classical reflection point, $r = R_r$, to the wall, $r = R_h$ [4, 6]

$$T \approx \exp\left[-\frac{2}{\hbar}\int_{R_r}^{R_h}\sqrt{2m(V(r)-E)}dr\right] \tag{24}$$

This is a good approximation provided it is very small. Let the atom of interest have an energy which is the fraction f of the maximum energy of the barrier, $E = fV(R_h)$. Assuming the potential to be exponential, $V(r) = V(R_h)\exp[2\kappa(r-R_h)]$ the integral can be evaluated and

$$T \approx \exp\left[-(2mv_m/\kappa\hbar)\left(\sqrt{1-f}-\sqrt{f}\tan^{-1}\left[\sqrt{(1-f)/f}\right]\right)\right] \tag{25}$$

Substituting in v_m=1.2 ms^{-1} and $\kappa = k = 1.1 \times 10^7$m^{-1}, we find the first round bracketed term to be ≈ 83. The function of f in round brackets varies

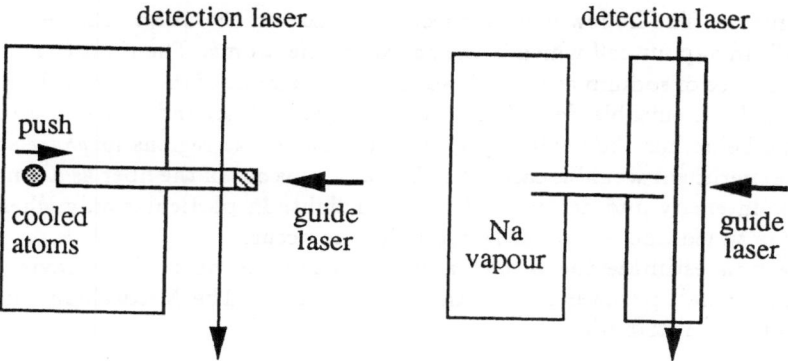

Figure 4: Schematic diagrams of "proof of principle" experiments for atomic waveguides. The left scheme uses cooled atoms, the right an atomic vapour.

between 1 for $f = 0$ and 0 for $f = 1$. For T to be less than 10^{-3} requires $f < 0.65$. So for an atom to have a probability of less than 10^{-3} of reaching the wall, per bounce, its transverse velocity must be less than $v_m\sqrt{f} \approx 1$ ms^{-1}.

Another loss mechanism is the non-adiabatic transition of the atom to the dressed state which is attracted, rather than repelled, by the evanescent field [7]. The probability of this is estimated by the ratio of the doppler shift to the detuning $kv/\Delta \approx 3 \times 10^{-4}$, which is of the same order as the tunneling probability.

4 Proof of principle experiments

In this section we outline experiments which could determine the practicality of atomic waveguides of the kind we have discussed. Only the large waveguide is considered since coupling in of both light and atoms should be easier than for the small waveguide. We first consider cooled atoms and then show that cooling may not be necessary.

Sodium atoms cooled to the recoil limit have velocities of about $\hbar k/m \approx 0.03$ ms^{-1}. Such atoms would be guided very well by the atomic waveguides we have discussed. A 50μm radius cooled ball could be launched at the large waveguide's hole by a laser push. A schematic diagram is given on the left of figure 4. Its arrival at the other end of the fiber could be detected with laser induced fluorescence, for example. Accumulation of sodium at the sealed end could be reduced by bending the fiber upwards so that gravity reflected the atoms back along the fiber.

Guidance may also be observable simply by exposing the fiber's hole to sodium vapour. A schematic of the proposed experiment is given on the right of figure 4. The hollow fiber connects two vacuum chambers. The left one is a sodium vapour cell which is the source of the atoms. The right chamber is kept as free of sodium as possible so that any flux out of the waveguide can be detected. A suitable detection scheme might be laser induced fluorescence. This scheme has the flexibility to detect various size regions in *phase* space. The experiment would measure the flux of atoms out of the fiber as a function of the intensity and detuning of the laser light. In particular only when the laser is tuned above resonance can guidance occur.

We now estimate the flux of atoms through the fiber's hole having sufficiently small transverse velocities to be guided. The Maxwellian velocity probability distribution is

$$f(\mathbf{v}) = N \exp(-\sigma \mathbf{v}^2), \quad N = (\sigma/\pi)^{\frac{3}{2}}, \quad \sigma = m/2k_b T \tag{26}$$

T is the temperature and $k_b \approx 1.4 \times 10^{-23} \mathrm{JK}^{-1}$ the Boltzmann constant. The relevant probability distribution for the atomic waveguide is that for having longitudinal velocity v_z, and transverse velocity less than some maximum v_m, $g(v_z, v_m)$. It is found by integrating over the transverse velocities. This can be accomplished by assuming that $\sigma v_m^2 \ll 1$, which is certainly true for $T = 500\mathrm{K}$ since $\sigma \approx 3 \times 10^{-6} \ (\mathrm{ms}^{-1})^{-2}$. Expanding the transverse exponentials we find,

$$g(v_z, v_m) \approx \pi N v_m^2 \exp[-\sigma v_z^2] \tag{27}$$

To determine the atomic flux we must find the spatial volume of atoms with a given longitudinal velocity v_z that enter the waveguide hole in a given time t. This is a cylinder of length $v_z t$ and cross sectional area A_h equal to that of the hole. The flux F is found by integrating over all longitudinal velocities and multiplying by the spatial atomic density ρ

$$F = \rho A_h \int_0^\infty v_z g(v_z, v_m) dv_z \approx \frac{\rho A_h \sqrt{\sigma}}{2\sqrt{\pi}} v_m^2 \tag{28}$$

We have assumed that the coupling between spatial position and transverse velocity can be ignored. The argument for this is based on the homogeneity of the gas: Although a transverse velocity will remove a velocity group from the cylinder centered on the fiber hole it will be replaced by a group with the same transverse velocity suitably displaced in the transverse direction.

For the temperature $T = 500\mathrm{K}$ eq.(28) gives

$$F \approx (5 \times 10^{-4}) \rho A_h v_m^2 \tag{29}$$

With a $50\mu\mathrm{m}$ radius hole and maximum transverse velocity $v_m = 1 \ \mathrm{ms}^{-1}$ this becomes

$$F \approx (4 \times 10^{-12}) \rho \tag{30}$$

So to obtain a flux of 1000 atoms s^{-1} requires the sodium density to be about 3×10^{14} atoms m^{-3}.

It is more difficult to estimate the flux due to ballistic flight through the hollow fiber. This is because the coupling between the transverse velocity and spatial position is important: Where an atom enters the hole determines the velocity vectors which will get through without hitting the walls. If we assume that the atoms which fly through ballistically have a transverse to longitudinal velocity ratio less than the fiber's radius to length ratio then the ballistic flux is

$$F_{ballistic} = \rho A_h \int_0^\infty v_z g(v_z, v_z(R_h/L)) dv_z \approx \frac{\rho A_h}{2\sqrt{\pi}\sigma} \left(\frac{R_h}{L}\right)^2 \quad (31)$$

Where L is the fiber's length and $g(v_z, v_z(R_h/L))$ is the probability distribution for having longitudinal velocity v_z and transverse velocity less than $v_z(R_h/L)$, eq.(27). Using the preceding parameters, and a fiber of length $L = 2$m, this becomes

$$F_{ballistic} \approx (10^{-15})\rho \quad (32)$$

With the previously assumed density of 3×10^{14} atoms m^{-3} the ballistic flux is less than one atom s^{-1}. However this estimate assumes that atoms are lost once they hit the walls. If they undergo an inelastic collision, continuing forward with reduced velocity, this estimate could be far too low.

So we consider the possibility of making a loop of fiber between the source and detection chambers. Then, at each bounce some of the longitudinal velocity is converted to transverse velocity due to the bend in the fiber. Our previous estimate of the wall tunneling probability suggests that 100 bounces should be possible with our usual parameters. Let the fiber bend angle be $\Delta\theta$. Then for 100 bounces to change the angle by 2π radians requires $\Delta\theta \approx 0.03$ radians, since each bounce deflects the atoms by $2\Delta\theta$. Assume that a maximum of 0.5 ms^{-1} of transverse velocity is due to the fiber bend angle $\Delta\theta$. Then the maximum allowable longitudinal velocity is $v_{zm} \approx 15$ ms^{-1}.

We can use $g(v_z, v_m)$, the probability distribution for having velocity v_z and transverse velocity less than v_m, eq.(27), to estimate the flux of atoms through such a looped fiber:

$$F_{2\pi} = \rho A_h \int_0^{v_{zm}} v_z g(v_z, v_m) dv_z \approx \rho A_h \frac{\sigma^{\frac{3}{2}}}{2\sqrt{\pi}} v_m^2 v_{zm}^2 \quad (33)$$

Using our usual parameters with $v_m = 0.5$ ms^{-1} and $v_{zm} = 15$ ms^{-1} this gives

$$F_{2\pi} \approx (6 \times 10^{-16})\rho \quad (34)$$

With a sodium density of 2×10^{17} atoms m^{-3} this flux is 100 atoms s^{-1}, which is detectable.

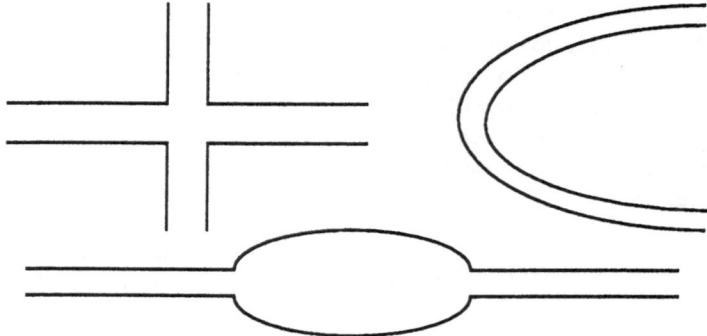

Figure 5: Examples of waveguide geometries which can produce bound states, when the ends are extended to infinity.

5 Atomic cavities

Atomic cavities using evanescent wave mirrors have been analyzed before [4, 8]. This section proposes a new type of cavity for atoms based on nonuniform waveguide geometry. We emphasize that most of this section is based on qualitative considerations and must remain speculation until verified by detailed calculation.

It has recently been shown that bending waveguides with Dirichlet boundary conditions, i.e. the wave zero on the waveguide boundary, can produce bound states with energies below the cutoff energy in the corresponding straight waveguide [9]. Similarly it has been shown that intersections [10] and bulges [11] in such waveguides can produce bound states, figure 5.

The latter case is particularly reasonable since in a wider region lower frequencies are expected to be supported. But if the wide region is bounded by narrower regions the corresponding modes will be localized in the wide region. Reference [11] gives conditions for the existence of a bound state. However the energy and wavefunction must be calculated numerically [10]. If the waveguide has finite length there will be some coupling of the state to the outside world, in which case we call it quasi-bound. There is a close analogy to optical cavity quasi-modes. Atoms injected into the waveguide with the quasi-bound state (QBS) energy will have some probability of transmission into the QBS. This probability will be a function of the overlap of the injected atomic wavefunction with the QBS wavefunction.

For definiteness let us consider the waveguide shown in figure 6, which also shows the qualitative form of the QBS wavefunction. It decays exponentially away from the bulge. The energy of the QBS below the continuum of modes (inset, figure 6) will be determined by the width of the waveguide as well as the details of the geometry. Its linewidth will be determined by the strength of the coupling to the outside world, which in turn is determined by the

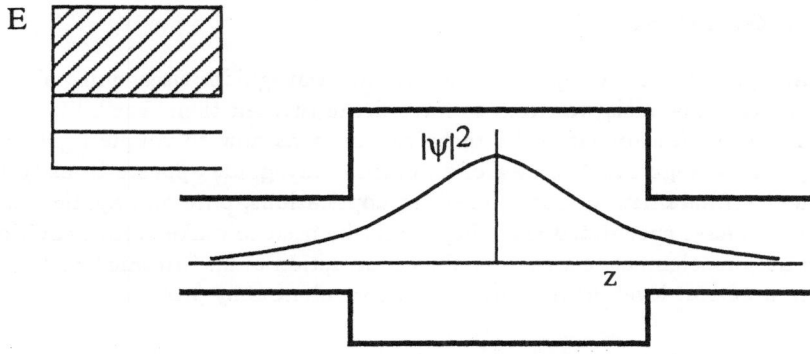

Figure 6: A waveguide geometry which can produce a quasi-bound state. The upper inset shows the eigenstate energies, the shading representing the continuum due to propagating modes. The graph suggests the localization of the quasi-bound state.

length of the narrow waveguide ends. For atoms cooled to the photon recoil limit the deBroglie wavelength is that of the cooling light and this sets the length scale of the waveguide with a QBS at the energy of the cooled atoms, as discussed in section 3.2. For visible light this is microns. Note that since the atoms are confined by light the effective width of the waveguide can be controlled via the strength of the evanescent field along the waveguide.

If the atoms are bosons then many can be accumulated in the QBS. However the density of atoms can be no higher than that in the injected beam if it is incoherent. This is because the probabilities to enter and escape the state are the same.

However if the injected beam were coherent the physical situation is analogous to the coherent excitation of an optical Fabry-Perot cavity. Then it is possible to build up much higher spectral densities in the quasi-mode (QBS) than in the injected beam. Unfortunately no such coherent atomic source is known.

If the state of the atoms changed appropriately within the waveguide, for example due to spontaneous emission, it might be possible to inject them at a higher rate than they can escape, even with an incoherent atomic beam. The accumulated atoms would then leak out of the QBS forming a coherent atomic beam of higher spectral density than the original atomic source. This would be the atomic device analogous to a laser, in which photons are similarly concentrated into a narrow quasi-mode of a cavity.

6   Conclusions

The analysis of hollow optical fiber atomic waveguides uses quite simple physics. On this basis we claim to have demonstrated their feasibility. The primary obstacles appear to be technical, such as how to couple light into the optical waveguide. The idea of an atomic waveguide appears to us to be sufficiently fundamental that interesting applications, pure and applied, will follow. We have speculated that they might be used to make atomic cavities, which are one element required for the construction of an "atomic laser". On a more practical level flexible mass transport alone may prove useful.

Acknowledgements

Bill Phillips and Steve Rolston suggested to us the possibility of making atomic waveguides from hollow optical fibers. Gerald Dunne told us about the bound states in nonuniform waveguides. Dana Anderson and Peter Marte helped clarify our ideas. JILA is supported in part by NSF. Craig Savage thanks JILA for a Visiting Fellowship which made this work possible.

References

[1] V.I. Balykin et al., Phys. Rev. Lett. **60**, 2137 (1988).

[2] J.V. Hajnal et al., Optics Commun. **73**, 331 (1989).

[3] A.W. Snyder and J.D. Love, *Optical Waveguide Theory* (Chapman and Hall, London, 1983).

[4] V.I. Balykin and V.S. Letokov, Appl. Phys. B **48**, 517 (1989).

[5] R.J. Longhurst, *Geometrical and Physical Optics*, (Wiley, N.Y., 1967).

[6] D. Bohm, *Quantum Theory*, (Prentice-Hall, New York, 1951).

[7] A.P. Kazantsev et al., *Mechanical Action of Light on Atoms*, (World Scientific, Singapore, 1990).

[8] H. Wallis et al., Appl. Phys. B **54**, 407 (1992).

[9] P. Exner et al., Phys. Lett. A **150**, 179 (1990); ibid **144**, 347 (1990); Czech. J. Phys. B **39**, 1181 (1989).

[10] R.L. Schult et al., Phys. Rev. B **39**, 5476 (1989).

[11] A.N. Popov, Sov. Phys. Tech. Phys. **31**, 1145 (1987).

PART II: Physics of Cooled and Trapped Particles

Towards higher densities of cold atoms: intense slow atom beams and dark light traps

Wolfgang Ketterle and David E. Pritchard

Department of Physics and Research Laboratory of Electronics
Massachusetts Institute of Technology, Cambridge, MA 02139, USA

Abstract. High density samples of cold atoms have been obtained using an increasing-field Zeeman slower and a dark light trap. The Zeeman slower produces more than 10^{12} slow atoms/s. In the dark spontaneous-force light trap repulsive forces between atoms due to rescattered radiation are reduced because the atoms are confined predominantly in a "dark" hyperfine level that does not interact with the trapping light. In such a trap, more than 10^{10} sodium atoms have been trapped at densities approaching 10^{12} atoms cm^{-3}.

1 Introduction

For many interesting experiments with cold atoms one needs both a large number of atoms and a high density . For example, for the study of cold collisions [1], the collision rate of an atom is proportional to the density n, and the observed signal therefore to n·N, where N is the number of atoms in the sample. In an atomic fountain [2], the density of atoms in the "near field" is equal to the density n before launching, and in the "far field", it is proportional to N divided by the solid angle of the fountain. For the observation of collective effects, one needs a density higher than the critical density at which the interatomic spacing becomes comparable to the thermal de Broglie wavelength, independently of the number of atoms in the sample. However, the most promising route to Bose-Einstein condensation is evaporative cooling [3], in which colder temperatures and higher densities are achieved by removing selectively the hot atoms from the trap. In that case, a high number of trapped atoms is crucial to reach eventually very high densities. And finally, many experiments with laser-cooled atoms start by collecting atoms in a magneto-optical trap [4] (MOT) and then transferring the atoms to other traps such as magnetic traps. In that case, a high density n readily achievable in a MOT might be lost by thermal spreading during the transfer unless a sufficiently large volume has been filled up to that density.

Unfortunately, as discussed in some detail below, the density in the standard magneto-optical trap was shown to decrease for a large number of trapped atoms due to absorption of the trapping light [5,6]. In this paper, we discuss our recent development of a dark spontaneous-force optical trap [7] (dark SPOT) in which the achievable density is the higher the more atoms are trapped. In such a trap, one can obtain both a high number and a high density of atoms.

Loading of such a trap requires an intense slow-atom source. We discuss our approach of employing an increasing-field Zeeman slower [8] along with potential improvements and practical limitations of this slower.

2 Production of intense slow beams by Zeeman slowing

We consider it desirable that a slowing method for producing a very high flux of slow atoms should be (a) continuous and (b) should ensure that all atoms reach the final velocity at the same location. Requirement (a) is essential to take full advantage of continuous atomic beam sources, (b) is important for very slow atoms, especially if they are of small mass and have a large recoil velocity. Such atoms have a large angular spread because of transverse heating. Special care is necessary to transfer them into a trap or to some other experiment (e.g. by a very small separation of the trap to the end point of slowing, by applying transverse collimation [8] or by imaging the end of the slower on the experiment [9]), and this is much simpler to accomplish if all atoms reach their final velocity at the same point.

Several methods for slowing atoms have been demonstrated. Chirped slowing [10] is a pulsed method. In this method, and also in the continuous methods of white light slowing [11], stimulated slowing [12], relay chirp slowing [13] and isotropic light slowing [14], atoms which enter the slower with a low velocity reach their final velocity already early. Only in Zeeman slowing [15], Stark shift slowing [16], isotropic light slowing with various frequencies [14] and the so far undemonstrated method of angle-tuned slowing [17] do all atoms reach the final velocity at the same point in space because slowly entering atoms are slowed down further only in the last part of the slower. This requires that the slower is not translationally invariant as is the case for the methods which don't posess this "spatial bunching" feature.

The most established technique which fulfills the above requirements is Zeeman slowing. However, in the past, difficulties were encountered in producing high fluxes of slow atoms with this kind of device [15,18]. In a conventional Zeeman slower which employs a decreasing magnetic field, there is considerable off-resonant scattering after the steep magnetic field gradient at the end of the slower, causing severe broadening of the resulting velocity distribution, and hindering the efficient extraction of very slow atoms. In addition, there are optical pumping effects near the end of the slower, when the magnetic field drops to zero. Both effects are much less pronounced in the recently demonstrated increasing-field Zeeman slower [19,20], where the atoms reach their final velocity near the maximum of the magnetic field. The fast drop-off of the field after this point rapidly detunes the atoms out of resonance with the slowing light, thereby minimizing subsequent off-resonant slowing.

In recent experiments we observed more than 10^{12} atoms/s in such an increasing-field Zeeman slower [8]. The density of slow atoms was high enough so that the velocity distribution could be determined with a transverse probe laser by single-pass absorption which was on the order of 10 % (Fig. 1). By imaging the cross-section of the slow beam with a light sheet technique, the spreading of the slow beam due to transverse heating could be directly observed. For velocities below ~60 m/s the slow beam started to fill the observation region (22 mm diameter) limited by the size of our vacuum chamber. The density of the slowed atomic beam showed an approximately linear decrease with final velocity (Fig. 2), as expected for a constant flux of atoms whose transverse velocity is independent of the final longitudinal velocity.

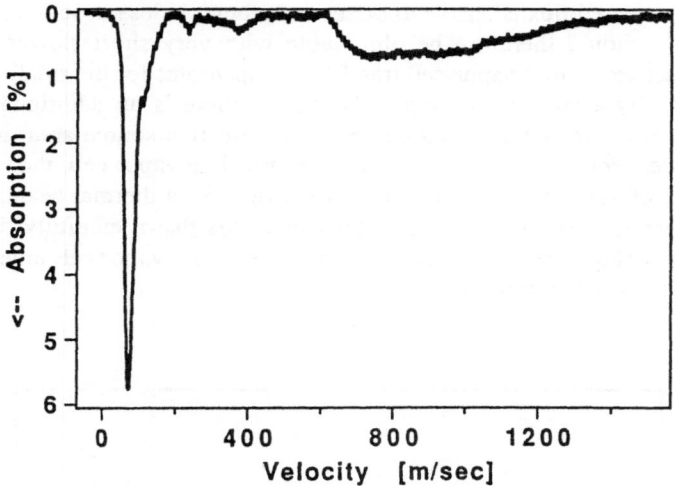

Figure 1. Absorption spectrum of a 70 m/s Na beam with $5 \cdot 10^{11}$ slow atoms/s.

For sodium atoms, one needs a final velocity below 20 to 30 m/s to load a magneto-optical trap. We obtained maximum loading rates of $\sim 10^{10}$ atoms/sec into a trap located ~20 cm downstream from the slower [7]. This implies that only ~1% of the slow atom flux was captured by the trap and 99 % missed the trap because of their large angular spread. We have recently demonstrated a way of reducing this spreading at the earliest stage possible by applying transverse collimation *inside* the Zeeman slower [8]. The observed density increase was sixfold, but another factor of ten could be achieved if the Doppler limit can be reached for the transverse temperature. For this, one has to avoid off-resonant slowing after the collimation by an improved geometry and a steeper magnetic field profile. In this case, the divergence of a 30 m/s beam would be only ±10 mrad, sufficiently small to transfer almost all slow atoms to the optical trap. Further improvements may be expected from a reduction of the distance between the slower and the trap which requires a more careful consideration of stray fields of the slower solenoid.

An important question is the optimum length or velocity capture range of such a slower. It is interesting to note that if the slowing starts right at the oven nozzle the highest atom flux is obtained with a slower of very small length l. For constant deceleration, the velocity capture range scales with $\sqrt{l}$ whereas the solid angle of an experiment with capture area A is A/l^2 (as long as $l^2 > A$). Since the thermal flux of atoms with a velocity smaller than v scales as v^4, one finds that the useful slow atom flux of a slower is independent of l! For larger l, the Boltzmann factor of the thermal velocity distribution causes the captured atom flux to increase less rapidly, the increased flux is unable to offset the geometric loss in solid angle; the optimum flux should therefore be obtainable with very short slowers. This argument illustrates how a vapor-cell trap [21] compensates for the small velocity capture range by a large solid angle. However, there is an additional factor proportional to the total flux of atoms or to the effective source area times the vapor pressure. For 1 cm size trapping laser beams in a vapor cell, the effective "nozzle" area of ~10 cm^2 is ~1000 times larger than for a thermal beam, but the vapor pressure is ~10^6 times smaller. This indicates that a carefully designed Zeeman slower should lead to a larger loading rate than a vapor cell unless large beams are used in such a cell [22].

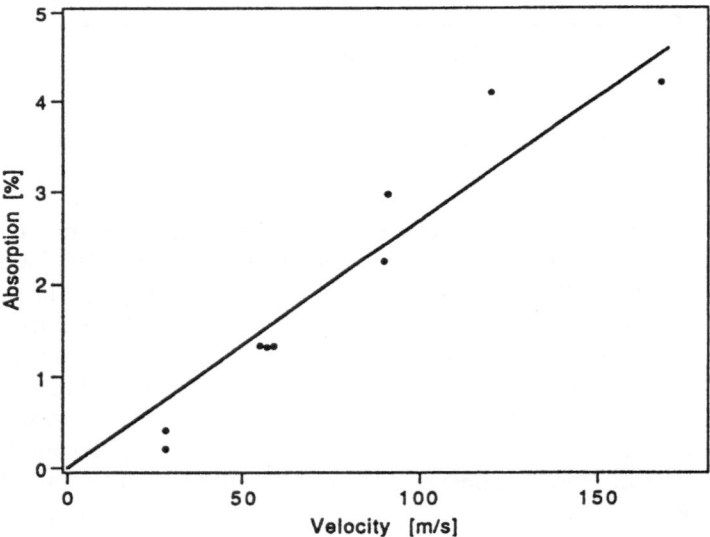

Figure 2. Maximum fractional absorption of a 1 μW probe laser beam as a function of the velocity of the slow atoms. A linear dependence is expected if the transverse velocity spread is independent of the longitudinal velocity.

The conclusion that the slow atom flux is maximized for a short slower is no longer valid if the slowing starts after a finite collimation distance. In this case the optimum length is on the order of the collimation distance. Furthermore, at low velocities deviations from a Maxwell-Boltzmann distribution become important. The slow part of the distribution is depleted by collisions in the nozzle ("jetting"). This effect, and to a lesser extent the shift of the Maxwell distribution to higher velocities set limits to the maximum useful oven temperature in our experiments. Although it is possible to achieve higher fluxes of thermal atoms by increasing the temperature, the absolute number of atoms in the velocity capture range of the slower will eventually even decrease. For our slower and oven geometry, the optimal oven temperature was between 300 and 330°C [8] corresponding to a mean free path of a few mm for sodium atoms in the oven.

By slightly increasing the velocity capture range of our slower we hope to achieve slow atom fluxes of $5 \cdot 10^{12}$ atoms/s. This would require 10^{17} photons/s or 30 mW of absorbed laser power, in practice even more because, due to imperfect collimation, there is additional light scattering by atoms which eventually hit the collimating apertures. In our recent experiments we found approximately 50 % longitudinal absorption of a slowing beam of ~150 mW. Another practical limit to the attainable fluxes might be set by repulsive forces due to rescattering of light. Such repulsive forces (see below) have so far only been discussed for atom traps [5], but they will eventually be a limiting process for intense slow beams.

If an atomic beam is slowed down, there is an outward radial flux of photons which leads to a transverse spreading of the slow beam. If f denotes the atomic flux density (which is regarded to be constant, i.e. transverse spreading during the slowing process is neglected), and the atomic velocity changes by Δv over a distance Δz, then the number of photons scattered within a radial distance r is $f \pi r^2 \Delta v / v_{rec}$, where v_{rec} is the recoil velocity. If one neglects geometric factors on the order of unity, the radial flux density of photons p(r) is obtained by dividing by the area $2 \pi r \Delta z$

$$p(r) = (f \, r \, / 2 \, v_{rec}) \, (\Delta v / \Delta z).$$

Again neglecting factors on the order of unity, the increase in transverse velocity Δv_t due to absorption of photons within the length Δz is

$$\Delta v_t = p(r) \, \sigma \, v_{rec} \, \Delta z / v = (f \, r \, \sigma / 2) \, (\Delta v / v),$$

where σ is the absorption cross section. By integration one obtains the final transverse velocity due to rescattering forces,

$$v_t = (f \, r \, \sigma / 2) \, \ln(v_{initial} / v_{final}).$$

The final result is independent of the details of the slowing process and depends only logarithmically on the final velocity. For a slow atom flux of 10^{12}/s·cm^2 in a 1 cm^2 cross section slowed down from 700 m/s to 50 m/s and a resonant cross section of $\approx 10^{-9}$ cm^2 one obtains the transverse velocity at r = 0.5 cm as $v_t \approx 6.5$

m/s. One actually expects a somewhat smaller value because the scattered photons are red detuned from the resonance by a detuning which depends on laser intensity and deceleration, and hence are absorbed with a smaller cross section. Further reduction results from the dipole pattern and polarization of the emitted photons.

We tried to observe a density-dependent transverse spreading of a slow sodium beam by comparing transverse beam profiles for high and low oven temperature. At the high temperature, we were close to the parameters discussed above, but found the dominant transverse velocity of ~2.5 m/s to be density-independent and therefore attributed it to transverse heating. In light of the discussion above, the expected additional spreading is on the order of 1 m/s and should add in quadrature, but was probably just below our detection threshold. Density-dependent spreading might become a limitation for compressing intense slow beams to sub-mm diameters in atomic funnels [23,24]. A two-dimensional version of the dark SPOT discussed below might avoid this problem.

It seems that longitudinal absorption and repulsive forces by rescattered light will set a practical limit at slow atom fluxes of ~10^{13}/s. The highest loading rate of an optical trap of $2\cdot10^{11}$ atoms/s was achieved in a vapor cell trap [22]. Our present loading rate of ~10^{10} atoms/sec [7] is an order of magnitude lower. However, the number of atoms trapped is similar (a few times 10^{10}) because in a vapor cell trap, such high loading rates were only achieved at fairly high vapor pressure limiting the loading time to 0.2 s whereas with a Zeeman slower the trap can be in ultrahigh vacuum with trapping times of at least a few seconds. We hope to improve the loading rate and number of trapped atoms by about an order of magnitude by achieving a more efficient transfer of atoms from the slower into the trap. An efficient coupling of a magnetic trap to a Zeeman slower was realized some time ago and resulted in loading rates of ~10^{11}/s [25].

3 Dark spontaneous-force optical trap

There are two kinds of processes which limit the density achievable in a conventional magneto-optical trap to about 10^{11} atoms/cm^3. The first process observed was collisions between ground- and excited-state atoms in which part of the excitation energy is transformed into kinetic energy, resulting in a trap loss rate per atom $\beta\cdot n$ with $\beta \approx (1\text{-}5)\cdot10^{11}$ cm^3/s [26,27]. For densities n approaching 10^{11} atoms/cm^3, the lifetime of the trap is limited to less than 1 sec, requiring extremely high loading rates to sustain a high density. This limit can be regarded as a dynamic limit due to loss mechanisms and could be avoided by loading many atoms into a weak trap and finally compressing them, but the large capture volume might be difficult to realize.

The second limit arises from repulsive forces between the atoms caused by reabsorption of scattered photons (radiation trapping) [5]. At a certain atomic density, the outward radiation pressure of the fluorescence light balances the confining forces of the trapping laser beams. Further increase of the number of trapped atoms leads to larger atom clouds, but not to higher densities. This limit is static and related to the balance of forces (see below). Finally, as a practical matter, the power of the rescattered light sets a limit to the *number* of atoms which can be confined in a magneto-optical trap: 10^{11} atoms scatter about 100 mW of near-resonant laser light.

In a dark SPontaneous-force Optical Trap (dark SPOT) all the above-mentioned limitations are mitigated by confining the atoms mainly in a "dark" hyperfine ground state - one which does not interact with the trapping light. The key idea is that optimum confinement of many atoms is not necessarily achieved with the maximum light force due to the limitations mentioned above. Light forces which are orders of magnitude smaller than the maximum saturated scattering force are still strong enough to confine atoms tightly, e.g. 1 m/s sodium atoms (corresponding to a temperature of 1 mK) can be stopped in a distance of 100 µm at 1 % saturation. All spontaneous-light-force traps realized so far have operated close to saturated excitation whereas our dark SPOT works at scattering rates two orders of magnitude smaller. This has allowed us to confine more than 10^{10} atoms at densities close to 10^{12} cm^{-3}, a density which has been achieved previously only with a much smaller number of trapped atoms, either in a conventional MOT [4,5,28], a dipole trap [29-31] or a stimulated magneto-optic trap [32].

The simple model used to explain the density limit in a MOT [5] is readily generalized to include a "dark" and a "bright" hyperfine ground state. The trapping force is

$$\vec{F}_T = - k\, p\, r\, \hat{r}, \tag{1}$$

where p denotes the probability that the atom is in the bright hyperfine state and k the spring constant of the normal MOT (i.e. for p=1). Attenuation of the trapping light and radiation trapping give rise to a density-dependent repulsive force which is quadratic in p because it involves two scattering events:

$$\vec{F}_R = k\, (n/n_0)\, p^2\, r\, \hat{r}, \tag{2}$$

where n_0 is a constant. From the stability criterion $|\vec{F}_T| > |\vec{F}_R|$ one obtains one limit for the maximum atom density in a MOT:

$$n < n_0/p. \tag{3}$$

For a very large number of atoms, the *column density* of atoms is limited by the fact that the atom cloud of diameter d has to be transparent for the trapping light [6], i.e. n d p < b_0, where b_0 is a constant. Substituting $d^3 = N/n$, one obtains a second limit for n:

$$n < (b_0/p)^{3/2}\, N^{-1/2}. \tag{4}$$

Finally, for small p, the density is limited by the confinement of a single trapped atom to a diameter $d = d_0\, p^{-1/2}$ as the spring constant of confinement is proportional to p (d_0 is the cloud diameter in a standard MOT for low N). This results in a third limit to the atomic density:

$$n < N\, p^{3/2} / d_0^3. \tag{5}$$

The constants n_0, b_0 and d_0 are the maximum density, maximum column density and minimum diameter of the atom cloud in a normal MOT. They depend on experimental parameters and are typically $5 \cdot 10^{10}$ cm^{-3}, $5 \cdot 10^9$ cm^{-2} and 200 μm [5].

In our simplified model, the atom density in a MOT is the largest value compatible with the three limits (3), (4), (5) as shown in Fig. 3. For low N, one has the ideal gas regime where independent atoms occupy a fixed confinement volume; the density is limited by the finite number of particles and given by (5). For higher N, the repulsive forces become dominant, and the density reaches the plateau value given by (3). Increasing the number of atoms leads to larger clouds with the same density. Finally, for very large N, the cloud expands even more in order to keep the optical density constant thus preventing too much light absorption. The density (4) decreases with increasing N in this regime. For very dark traps, the constant density limit which is governed by the repulsive rescattering forces is never reached because the (large) cloud becomes opaque before.

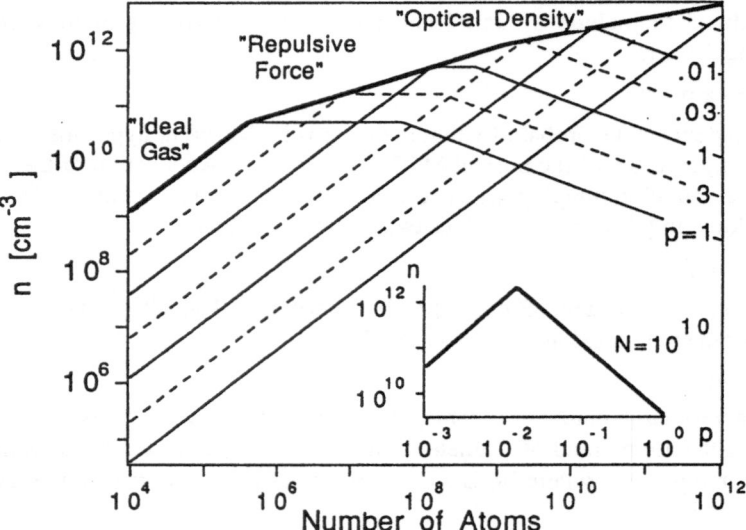

Figure 3. Atomic densities n vs. number N of trapped atoms for different values of the fractional population p of the "bright" hyperfine state. The thick line shows the highest density obtainable with an optimum value of p. For large N, this density is much higher than the one obtained in the normal MOT (p=1). For 10^{10} trapped atoms (inset), the optimum trap is a hundred times "darker" than the normal MOT, resulting in more than two orders of magnitude increase in density.

It should be noted that the bright atoms in dark traps are subject to the same density-limiting physical processes as in the bright trap: high optical density and repulsive forces. In our simple model, the overall improvement in number and density comes from the extra atoms in the dark state, which increases these quantities by a factor $1/p$ (one could call this a two-gas model of the trapped atom cloud). This improvement is achieved, however, by effectively weakening the trap which results in a larger size of the trap.

Fig. 3 shows that for a given number of atoms in the trap, there is an optimum value of p which maximizes density. If p is too large the density is small due to repulsive forces or due to the opacity limit, if p is too small the confinement volume is too large and cannot be filled up to a very high density. It is obvious from Fig. 3 that there are three different regimes for an optimum p:

For small N, the highest density is achieved in a bright trap ($p=1$). For intermediate N, the optimum density is obtained when the cloud is just at the cross-over from ideal-gas behavior to the constant-density regime. Finally, for large N, the density is maximized at the transition from ideal-gas behavior to the constant optical density regime. For the typical parameters given above, the boundaries between the regimes are at $N = 4 \cdot 10^5$ and $1.25 \cdot 10^9$, or more generally

$$N < n_0 \, d_0^3 \qquad p_{opt} = 1 \qquad n_{opt} = N/d_0^3$$
$$n_0 \, d_0^3 < N < b_0^5 \, n_0^{-4} \, d_0^{-2} \qquad p_{opt} = N^{-2/5} \, n_0^{2/5} \, d_0^{6/5} \qquad n_{opt} = N^{2/5} \, n_0^{3/5} \, d_0^{-6/5}$$
$$N > b_0^5 \, n_0^{-4} \, d_0^{-2} \qquad p_{opt} = (b_0/N)^{1/2} \, d_0 \qquad n_{opt} = N^{1/4} \, d_0^{-3/2} \, b_0^{3/4}.$$

The Stanford group [22] and our group have recently succeeded in trapping more than 10^{10} atoms in a normal MOT. For such N, the predicted optimum p of ~ 0.01 corresponds to a density increase of more than two orders of magnitude over the normal MOT (Fig. 3). For larger N, which might be realized by more efficient loading, one expects even higher density, however, because $n_{opt} \propto N^{1/4}$, the increase in density would be rather small. Further improvements might be accomplished by optimizing the parameters n_0, b_0 and d_0 with respect to detuning, laser intensity and magnetic field gradient. For a conventional MOT, this subject is discussed in Ref. 33.

The experimental demonstration of the dark SPOT has been published recently [7]. Briefly, a magneto-optical trap with a variable fraction of atoms in the dark hyperfine state has been realized. By using repumping beams with dark spots in the center, we obtained a "bright" capture region and a "dark" trapping region. The density and number of trapped atoms was determined by measuring the cloud diameter both in fluorescence and absorption imaging using a CCD camera, taking absorption spectra, and "counting" the atoms in a pulsed optical pumping experiment. Since the first two techniques for diagnostics are rather obvious, we will discuss only the latter in more detail. The atom counting was done by switching off the trapping and repumping beams and rapidly switching on a strong probe laser beam (0.5 mW/cm^2) close to resonance with F=1 atoms, optically pumping them into the F=2 state. From the transient absorption signal, the number of absorbed photons was obtained that was independent of the probe laser power. This number, divided by the number q of photons needed to optically pump one F=1 atom to F=2, gives directly the number of atoms in the trap. From known matrix elements, one obtains q=2 for the 1→2 transition and q=6 for 1→1 (the 1→0 transition is cycling, but bleaches out and, in addition, overlaps with 1→1). These numbers might change due to radiation trapping. However, in a simple picture, the probe laser "eats" its way through the sample having a dense cloud of F=1 atoms ahead and of F=2 atoms behind, 50 % of both Rayleigh and Raman photons are reabsorbed and q is unchanged. The number of absorbed photons varied by a factor of ~ 3 when the probe laser frequency was scanned by 150 MHz through the 1→F' transitions. Assuming q=2 for the smallest number, the number of trapped atoms deduced should be accurate to within a factor of two.

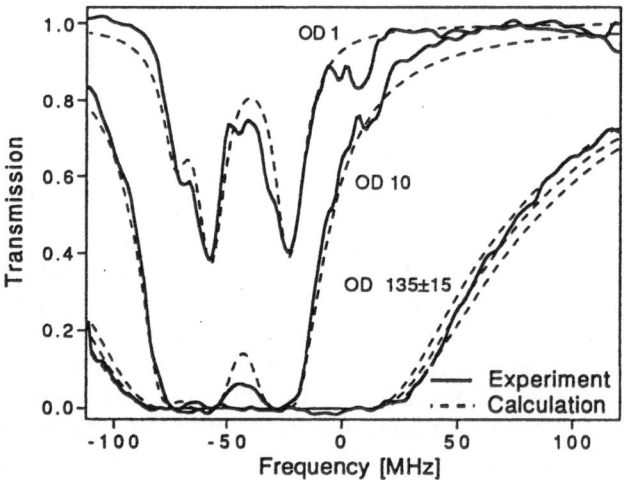

Figure 4. Absorption spectrum of a 4 mm diameter cloud of sodium atoms trapped in a dark SPOT. The best fit yields an optical density (OD) of 135 which corresponds to an atomic density of $7 \cdot 10^{11}$ cm^{-3} and $\sim 5 \cdot 10^{10}$ trapped atoms. Traces with lower OD were recorded with a reduced number of atoms. The dashed lines are calculated spectra for OD = 1, 10, 120, 135, and 150, respectively.

For a (3.0 ± 0.5) mm cloud of atoms, the optical density was 110 ± 10 implying a density $n = (8 \pm 2) \cdot 10^{11}$ cm^{-3}. Another less accurate value for n derived from the number of atoms $N = 1.5 \cdot 10^{10}$ and the observed size of the cloud agreed to within 30 %. In slightly larger clouds, optical densities up to 160 and $N = 5 \cdot 10^{10}$ trapped atoms were observed (Fig. 4). This combination of number and density is unprecedented in light traps: comparable densities have been reported to date only for at least 100 times fewer atoms [4,5,28-32]. A similar number of atoms has been trapped by light forces only at densities 20 times lower [22]. In some of the previous work with magneto-optical traps, large detunings were used [22,28,34] to reduce the reabsorption of scattered photons. This and other ways to affect n_0, b_0 and d_0 may eventually be combined with our approach of reduced repumping to achieve even higher densities.

The dark SPOT is the first cooling and trapping scheme in which the repumping light is intentionally reduced to "shelve" the atoms, i.e. cooling and trapping forces are only exerted on a small fraction of the atoms, while most of the atoms are kept "in the dark" thus avoiding strong interaction with the cooling light. This concept should avoid increased ultimate temperatures in polarization-gradient cooling observed at high atomic densities [28]. On the other hand, for very dark traps, one expects increased diffusion because the atom is moving without friction while it is in the dark state (in our trap this time was roughly 100 μs).

Another possibility for realizing a dark trap would be repumping on the $1 \rightarrow 1$ transition of the D_1 line with elliptically polarized light. This transition has a coherent dark state only for magnetic fields B=0 [35] which would inhibit repumping in the center of the trap. For finite B, the dwell time in such a dark state is limited by Zeeman precession into a bright state. This would result in a sufficiently dark state only in a very small region around the origin where magnetic fields are below ~0.1 G. With typical magnetic field gradients of 10 G/cm this would correspond to a trap diameter of only 100 μm. Assuming that the repumping rate and the population in the bright state are proportional to B which in turn is proportional to the displacement from the center r, one concludes from (1) that the effective potential varies like r^3.

The general idea behind the dark SPOT is to increase the trapping forces in relation to the repulsive forces due to radiation trapping. This was accomplished by optical pumping, but can, to a certain extent, also be achieved by changing the detuning δ and intensity of the trapping light. In a simple picture, assuming Doppler molasses and large detuning, the repulsive force is proportional to ω_R^4/δ^4 (see also Ref. 33), whereas the trapping forces scale with ω_R^2/δ^3 (ω_R denotes the Rabi frequency). One therefore expects larger densities for small intensities and large detunings. In addition, large detunings should increase b_0, the limiting column density. Indeed, in Cs, larger densities were achieved for larger δ [22,28,34].

The high densities achieved in a dark SPOT are promising for the study of cold collisions and for the observation of evaporative cooling after transferring the atoms into a magnetic trap. At densities of 10^{12} cm^{-3}, the estimated elastic collision rate exceeds 10 s^{-1}, much larger than the trap loss rate due to collisions with the background gas. Even if the density drops during the transfer of atoms by a factor of ten, there should still be 100 elastic collisions during a trapping time of 1 min (assuming a background pressure of 10^{-10} mbar) which should be large enough to see evaporative cooling. However, even at the highest density in a dark

SPOT and assuming that polarization-gradient cooling down to ~20 μK can be achieved, these densities are still more than three orders of magnitude below the critical density for Bose-Einstein condensation.

We would like to acknowledge the contributions of our coworkers Kendall B. Davis, Michael A. Joffe, Alex Martin and M. Mewes. This work was supported by ONR and AFOSR through contract N00014-90-J-1642, and NSF grant #8921769-PHY. W.K. would like to acknowledge a fellowship from the NATO Science Committee and DAAD, Germany.

4 References

[1]   see e.g. P.S. Julienne, in: "Laser Manipulation of Atoms and Ions," Proceedings of the Varenna Summer School, edited by E. Arimondo and W.D. Phillips (North-Holland, Amsterdam, in press).

[2]   M.A. Kasevich, E. Riis, S. Chu, and R.G. DeVoe, Phys. Rev. Lett. 63, 612 (1989).

[3]   N. Masuhara, J.M. Doyle, J.C. Sandberg, D. Kleppner, T.J. Greytak, H.F. Hess, and G.P. Kochanski, Phys. Rev. Lett. 61, 935 (1988).

[4]   E.L. Raab, M. Prentiss, A. Cable, J.E. Bjorkholm, S. Chu, and D.E. Pritchard, Phys. Rev. Lett. 59, 2631 (1987).

[5]   T. Walker, D. Sesko, and C. Wieman, Phys. Rev. Lett. 64, 408 (1990).

[6]   K. Lindquist, M. Stephens, and C. Wieman, Phys. Rev. A 46, 4082 (1992).

[7]   W. Ketterle, K.B. Davis, M. A. Joffe, A. Martin, and D.E. Pritchard, Phys. Rev. Lett. 70, 2253 (1993).

[8]   M.A. Joffe, W. Ketterle, A. Martin, and D.E. Pritchard, to be published.

[9]   H. Metcalf, W. Phillips, and J. Prodan, Bull. Am. Phys. Soc. 29, 795 (1984).

[10]   W. Ertmer, R. Blatt, J.L. Hall, and M. Zhu, Phys. Rev. Lett. 54, 996 (1985).

[11]   M. Zhu, C.W. Oates, and J.L. Hall, Phys. Rev. Lett. 67, 46 (1991).

[12]   M. Prentiss and A. Cable, Phys. Rev. Lett. 62, 1354 (1989).

[13]   C.C. Bradley, J.G. Story, J.J. Tollett, J. Chen, N.W.M. Ritchie, and R.G. Hulet, Opt. Lett. 17, 349 (1992); D.S. Weiss, Ph.D. Thesis, Stanford University (1993).

[14]   W. Ketterle, A. Martin, M. A. Joffe and D.E. Pritchard, Phys. Rev. Lett. 69, 2483 (1992).

[15]   W.D. Phillips, J.V. Prodan, and H.J. Metcalf, J. Opt. Soc. Am. B 2, 1751(1985).

[16]   L. Windholz, C. Neureiter, R. Gaggl, Book of Abstracts, Seminar on "Fundamentals of Quantum Optics III", Kühtai, Austria (1993).

[17]   J. Umezu and F. Shimizu, Jpn. J. Appl. Phys. 24, 1655 (1985).

[18]   V.S. Bagnato, C. Salomon, E. Marega, Jr., and S.C. Zilio, J. Opt. Soc. Am. B 8, 497 (1991).

[19]   T.E. Barrett, S.W. Dapore-Schwartz, M.D. Ray, and G.P. Lafyatis, Phys. Rev. Lett. 67, 3483 (1991).

[20] A. Witte, Th. Kisters, F. Riehle, and J. Helmcke, J. Opt. Soc. Am. B **9**, 1030 (1992).

[21] C. Monroe, W. Swann, H. Robinson, and C. Wieman, Phys. Rev. Lett. **65**, 1571 (1990).

[22] K.E. Gibble, S. Kasapi, and S. Chu, Opt. Lett. **17**, 526 (1992).

[23] E. Riis, D.S. Weiss, K.A. Moler, and S. Chu, Phys. Rev. Lett. **64**, 1658 (1990).

[24] J. Nellessen, J. Werner, and W. Ertmer, Opt. Comm. **78**, 300 (1990).

[25] K. Helmerson, A. Martin, and D.E. Pritchard, J. Opt. Soc. Am. B **9**, 483 (1992).

[26] M. Prentiss, A. Cable, J.E. Bjorkholm, S. Chu, E.L. Raab, and D.E. Pritchard, Opt. Lett. **13**, 452 (1988).

[27] L. Marcassa, V. Bagnato, Y. Wang, J. Weiner, P.S. Julienne, and Y.B. Band, to be published.

[28] A. Clairon, Ph. Laurent, A. Nadir, M. Drewsen, D. Grison, B. Lounis, and C. Salomon, Proc. of the 6th European Frequency and Time Forum, Noordwijk, Netherlands, 1992, J.J. Hunt ed. (to be published).

[29] S. Chu, J.E. Bjorkholm, A. Ashkin, and A. Cable, Phys. Rev. Lett. **57**, 314 (1986).

[30] W.D. Phillips, in: "Laser Manipulation of Atoms and Ions," Proceedings of the Varenna Summer School, edited by E. Arimondo and W.D. Phillips (North-Holland, Amsterdam, in press); K. Helmerson (private communication).

[31] D.J. Heinzen, J.D. Miller, and R.A. Cline, in: The Thirteenth International Conference on Atomic Physics, Munich, 1992, Book of Abstracts, Paper C4.

[32] O. Emile, F. Bardou, C. Salomon, Ph. Laurent, A. Nadir, and A. Clairon, Europhys. Lett. **20**, 687 (1992).

[33] A.M. Steane, M. Chowdhury, and C.J. Foot, J. Opt. Soc. Am. B **9**, 2142 (1992).

[34] E.A. Cornell and C.R. Monroe (private communication).

[35] A.M. Tumaikin and V.I. Yudin, Sov. Phys. JETP **71**, 43 (1990).

Two and Three-Dimensional Optical Crystals

Andreas Hemmerich[1] and Theodor Hänsch[1, 2]

[1] Sektion Physik der Universität München, Schellingstr. 4/III, W-8000
München 40, Germany
[2] Max-Planck-Institut für Quantenoptik, W-8046 Garching, Germany

1. Introduction

Atomic physicists have recently learned to spatially confine atoms in periodic
structures on the micron length scale by dipole forces inside optical standing
waves. One-dimensional arrays of planar de Broglie waveguides [1,2] , two-
dimensional arrays of linear de Broglie wave guides [3] or three-dimensional arrays
of microscopic light traps [4] have been demonstrated. Different two and three-
dimensional lattice geometries can be realized (e. g. cubic body centered) and the
atomic spins can exhibit ferromagnetic or antiferromagnetic order. The physics
describing such atomic samples resembles that of ultra cold and very dilute solids.
Such optical crystals may have a number of future applications. Unlike laser
cooled ions in a "Wigner crystal" [5], neutral atoms do not experience strong
Coulomb repulsion. Thus, it should be possible to study collective absorption
and emission processes and collisions in systems of two or more closely spaced
atoms. Interesting quantum statistical phenomena might occur if several atoms
are confined in the same microscopic light trap. On the other hand one may
investigate solid state properties, e. g. tunneling between adjacent potential wells
or spin waves. Since a full quantum treatment of optical crystals seems not to be
out of the scope they may serve as a model to increase our understanding of such
solid state phenomena. Theoretical investigations of the physics of optical
crystals have been presented for one and two dimensions in the case of a $J=1/2 \rightarrow$
$J=3/2$ atomic transition including phenomena like tunneling [6,7,8] (J = angular
momentum).

To achieve trapping and efficient cooling of the atoms at the same time,
polarization gradients of the light are essential. In one dimension the
configuration of two counterpropagating laser beams with orthogonal polarization
(i. e. the lin⊥lin scheme) is appropriate [1,2]. However, when two or more
standing waves are superposed, inherently multidimensional phenomena can arise
such as the action of vortical radiation pressure which may prevent the
localization of the atoms [9]. The presence of such complications depends on the
time phase differences between the employed standing waves and the choice of

polarizations. In section 2 we will discuss the physics of optical crystals by considering a two-dimensional example. In section 3 we consider a three-dimensional extension.

2. Two-Dimensional Optical Crystals

2.1 General Considerations

In this section we discuss two-dimensional (2D) quantized motion and confinement of atoms in the 2D-field geometry arising in the intersection region of two standing waves oriented along the x and y-axes and linearly polarized in the xy-plane which we call the xy:yx configuration in the following. As shown in fig. 1, this light field acquires two different types of spatial polarization gradients for the two characteristic values $\phi = 0°$ and $\phi = 90°$ of the time phase difference between the two standing waves. When $\phi = 0°$ the polarization is linear everywhere, only its direction varies. In the $\phi = 90°$ case there exists a 2D array of straight lines parallel to the z-axis where the light exhibits circular polarization with alternating sign. There is a continuous change to linear polarization when one moves away from those locations. The two cases shown in fig. 1 resemble the two one-dimensional optical fields which have been at the basis of the theory of polarization gradient cooling, namely the $\sigma+\sigma$-configuration and its counterpart the lin⊥lin configuration [10]. Thus, we expect our two-dimensional light field to contain the physics of both elementary 1D cases depending on the choice of the time phase difference. In particular, our 2D-field provides efficient sub-Doppler cooling for any value of ϕ [11]. At the same time it does not suffer from vortical radiation pressure [9].

A very similar 2D light field is realized if the polarization of the 1D standing wave along the y-axis is linearly polarized parallel to the z-axis instead of the x-axis as in the case discussed above. This 2D field (the xy:yz-configuration) exhibits the same spatial distribution of the polarizations as the xy:yx-configuration. The only difference is that the orientation of the polarization has been tilted by 90°. If we chose the quantization axis to be the z-axis for the xy:yx-configuration and the x-axis for xy:yz-configuration, the discussion of this section applies to both configurations.

Let us consider the case $\phi = 90°$ in some more detail for an atom with a $J \rightarrow J+1$ transition, e. g. $J = 3$. We choose the appropiate quantization axis as discussed above. The Zeeman-levels experience different spatially varying light shifts due to the different size of the Clebsch-Gordan coefficients and the spatially varying polarization and intensity of the light field. At negative detuning δ of the light frequency with respect to the atomic resonance the outermost Zeeman levels ($m = \pm 3$) are shifted by a maximum amount downward at the antinodes of the light field where circular polarization prevails. Thus, atoms in the $m = \pm 3$ states

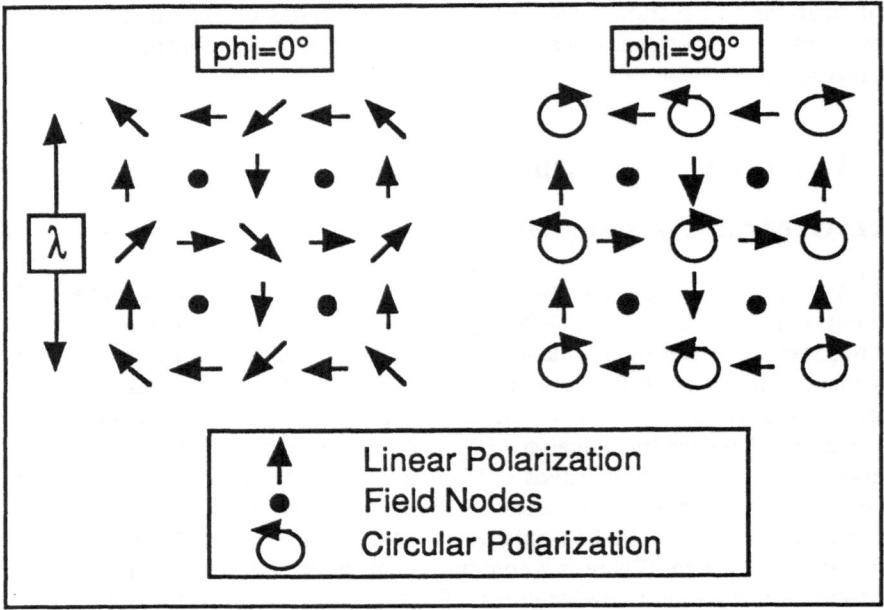

Fig. 1. The geometry of the electric field in the intersection region of two orthogonal standing waves propagating in the xy-plane (drawing plane) with linear polarizations parallel to that plane. At $\phi = 90°$ the antinodes are circularly polarized and form a two-dimensional array of linear de Broglie waveguides. The width of each section is an optical wavelength λ.

experience deep potential minima at those locations and we expect two-dimensional confinement of the $m = \pm 3$ -atoms. In particular, because of the circular polarization of the light at the antinodes, optical pumping prepares the trapped atoms as two-level systems. Due to the alternating sign of the circular polarization atoms trapped in neighboring waveguides should exhibit opposed magnetizations oriented along the z-axis similar to an antiferromagnetic medium. Because the atoms near an antinode are prepared as two-level systems we may treat their motion as due to an optical potential which at its bottom is approximately described by a two-dimensional harmonic oscillator potential. Therefore, we may expect discrete nearly equidistant vibrational quantum states at the bottom of that potential. A quantitative analysis has to take into account tunneling between different potential minima, which attributes a width to each vibrational state. At higher vibrational quantum numbers these widths significantly wash out the discrete energy structure. Therefore, a band-theory is more appropriate for the analysis of the problem. However, one-dimensional band calculations have shown that the widths of the low lying states are of the order of 10^{-6} times the recoil energy which is far below the energy separation of the vibrational levels [7]. When the atoms are sufficiently cold we may thus treat the potential minima independently.

In the following, we consider m=3 atoms near a potential minimum. We may, thus, study the atomic vibrational motion by considering the forces on a stationary two-level atom. This leads to the following simple expression for the trapping potential

$$U(x,y) = \frac{\hbar \delta}{2} \ln (1 + S),$$

$$\tag{1}$$

$$S(x,y) = S_0 \frac{1}{4} (\cos(kx) + \cos(ky))^2, \quad S_0 = \frac{\omega_{max}^2}{2(\delta^2 + (\frac{\Gamma}{2})^2)}.$$

Here, S_0 denotes the saturation parameter of the (J=3,m=3) $\rightarrow$ (J=4,m=4)-transition, ω_{max} is the antinode Rabi-frequency for this transition, Γ is the inverse life time of the excited state and we have parametrized our light field such that the locations (x,y) of positive circular polarization are given by $kx = \pm 2n \pi$, $ky = \pm 2n \pi$, where n is an integer. We may expand eq. (1) up to 2nd order obtaining:

$$\tag{2}$$

$$U(x,y) = \frac{\hbar \delta}{2} \left(\ln(1 + S_0) - \frac{1}{2} \frac{S_0}{1 + S_0} k^2 (x^2 + y^2) + O(4) \right).$$

For negative values of the detuning δ, the second term in eq. 2 provides a two-dimensional harmonic oscillator potential yielding a frequency separation of the equidistant vibrational levels of :

$$\Delta \nu_{osc} = 153 \text{ kHz} * \left(\frac{|\delta|}{\Gamma} \frac{S_0}{1 + S_0} \right)^{\frac{1}{2}}.$$

$$\tag{3}$$

The value of 153 kHz corresponds to the D_2 resonance line of [85]Rb. There are two sources of anharmonicity both leading to a lowering of the nth vibrational level and a splitting of its (n+1)-fold degeneracy which increase with the vibrational quantum number n. On the one hand there are higher order terms in eq. 2 which, however, introduce corrections of only a few kilohertz. On the other hand, the two-level model used in eq. 2 is correct only for an atom which is located exactly in a potential minimum. For a slightly displaced atom optical pumping partially populates the m=2 state and the atom thus experiences an effective potential which is lower than predicted by eq. 2. Both type of perturbations exhibit a spatial geometry such that the two-fold degeneracy of the first excited state is preserved and the three-fold degeneracy of the second state is lifted. The life times of the vibrational states should be determined by the optical pumping times which are strongly reduced for the well localized atoms due to the Lamb-Dicke effect.

We now briefly consider the case when $\phi = 0°$ in our 2D-field. The polarization remains linear in the plane of the light field (see fig. 1). Depending on their internal state and the corresponding Clebsch-Gordan coefficient the atoms now move in potentials of different size which arise from the spatial variation of the energy density of the light. The minima of these $0°$-potentials are at the same locations in contrast to the $90°$ case where two different optical potentials confine $m=-3$ and $m = 3$ atoms at separated locations. As in the $90°$ case optical pumping tends to prepare the atoms into the most light shifted Zeeman state; however this pumping is less complete for $\phi = 0°$. In addition the polarization gradient cooling at $0°$ is less efficient [11]. As a result we expect population of vibrational states up to higher order arising from various potentials of different depth. The life times of these vibrational states should be smaller than in the $90°$ case because of the rapid redistribution of the atoms among the different potentials by optical pumping. Thus, the discrete vibrational energy structure should wash out for $\phi = 0°$.

2.2 Raman Spectroscopy of Vibrational States

The vibrational quantum states can be experimentally observed by stimulated Raman spectroscopy. A weak probe laser beam is directed through the atomic sample and its transmission is recorded versus its frequency. When the probe frequency differs from that of the confining standing wave by the energy separation of two vibrational states, a Raman transition between these states can be excited, and Stokes or Anti-Stokes resonances are observed in the probe transmission spectrum. The excitation of Raman transitions between different vibrational states is connected with a change of the external vibrational energy of the atom by an amount much larger than the photon recoil energy. Thus, the photon language is not very appropriate to discuss the physics involved in these transitions. It appears to be more suitable to consider the oscillating distortion of the optical potential introduced by the interference between the probe beam and the confining 2D standing wave. The spatial geometry of this distortion depends strongly on the polarization of the probe beam. In particular, the excitation rates of Raman transitions including an even or odd number of vibrational quanta result from contributions to the distortion of different parity. This aspect plays an important role in two and three-dimensional experiments where different probe beam polarizations should lead to different relative strengths of the Raman sidebands. The probe beam can only interfere with the 2D-field and thus distort the optical potential if there exists a common polarization component. Thus, if we consider the xy:yx-configuration and a probe beam propagating within the xy-plane there is only one linear polarization component of the probe beam (namely parallel to the xy-plane) which allows for Raman transitions. A more interesting situation arises when we consider the xy:yz-configuration and a probe travelling along the x-axes. In this case we may expect Raman excitation for all probe polarizations.

In the following we would like to discuss how the dependence of the Raman sideband strengths on the probe polarization in case of the xy:yz-configuration can be understood within a rather oversimplified model. In analogy to refs. 6, 7, 8 we assume a $J = 1/2 \rightarrow J = 3/2$ atomic transition yielding an optical potential U for the atoms with magnetic quantum number $m_J = +1/2$

$$U \propto I_+ + \tfrac{1}{3} I_- \tag{4}$$

where $I_\pm$ denote the intensities of the positive and negative circular polarization components of the total optical field. In a first step we are interested in the energy level structure of our system and thus neglect the existence of the weak probe beam. We expand the optical potential obtained when only the 2D-field is considered in eq. (4) in the vicinity of a local minimum and keep only the harmonic part U_0 which gives rise to a set of equidistant vibrational levels similar to the prediction of eq. (3). In a second step the interference between the 2D-field and the probe beam is considered leading to contributions in the total intensities $I_\pm$ which oscillate at the difference Ω between the frequencies of the probe beam and the 2D-field. These interference terms contribute to the total optical potential according to eq. (4) and thus act to distort the 2D-field potential U_0. If we assume that the probe beam propagates along the x-axis and is linearly polarized at an angle ψ with respect to the y-axis, we obtain a perturbation

$$\Delta U \quad \propto \quad \cos(\psi)\,(A_1 \sin(\Omega t) + A_2 \cos(\Omega t))$$
$$+ \sin(\psi)\,(B_1 \cos(\Omega t) + B_2 \sin(\Omega t)) \tag{5}$$

where

$$A_1 \equiv \sin(x)\,(2\cos(x)+\cos(y))\ ,\quad A_2 \equiv \cos(x)\,(2\cos(x)+\cos(y))$$

$$B_1 \equiv -\sin(x)\,(2\cos(y)+\cos(x))\ ,\quad B_2 \equiv \cos(x)\,(2\cos(y)+\cos(x))$$

The contributions denoted by an A (i. e. A_1 and A_2) are connected with a polarization component of the probe beam parallel to the y-axis, i. e. they vanish when $\psi = 90°$. The B-terms correspond to a probe polarization orthogonal to the y-axis. The terms indicated by a 2 (i. e. A_2 and B_2) have even parity corresponding to transitions involving an even number of vibrational energy quanta, whereas the terms with index 1 (i. e. A_1 and B_1) are odd and thus give rise to transitions involving an odd number of vibrational quanta.

To evaluate the transition rate to first order between the vibrational states $|v_i\rangle$ and $|v_f\rangle$ introduced by the resonant perturbation ΔU we have to calculate the square of the matrix element $\langle v_i| \Delta U |v_f\rangle$. We assume all atoms are initially in the vibrational ground state $|0\rangle$ and we take into account only transitions to the first and second excited states $|1\rangle$ and $|2\rangle$. In fig. 2 we schematically show the spatial variation along the x-axis ($y = 0$) of the wave functions for the states $|0\rangle$, $|1\rangle$ and $|2\rangle$ (left column of fig. 2) and the distortions of the optical potential with odd and even parity (right column) resulting from the two kinds of probe polarizations. We immediately see that odd (even) perturbation terms can only mix the ground state $|0\rangle$ with the first (second) excited state $|1\rangle$ ($|2\rangle$). The solid lines indicate the perturbations B_1 and B_2, the dashed lines depict A_1 and A_2. We now consider the size of the matrix elements $\alpha_v \equiv |\langle 0|A_v |v\rangle|^2$ and $\beta_v \equiv |\langle 0| B_v |v\rangle|^2$ with $v = 1,2$ which determine the corresponding transition rates.

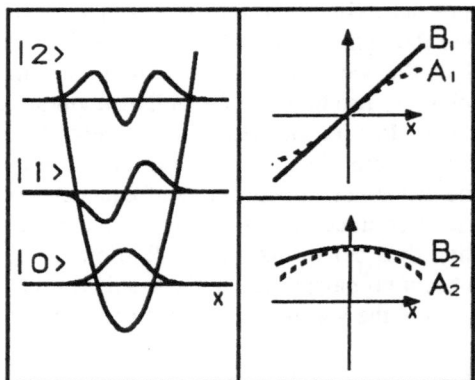

Fig. 2. Schematic spatial distribution of wave functions (left column) and probe induced perturbations of the optical potential (right column). The perturbation terms indicated by 1 (2) are responsible for the first (second) Raman sideband. The A-terms (B-terms) correspond to a probe polarization parallel (orthogonal) to the y-axis. The details are explained in the text.

We see that for transitions from $|0\rangle$ to $|1\rangle$ we have $|A_1| < |B_1|$ in the region where the wave function of state $|1\rangle$ is different from zero indicating that $\alpha_1 < \beta_1$. In the case of transitions from $|0\rangle$ to $|2\rangle$ we similarly have $|A_2| < |B_2|$ but this should lead to $\alpha_2 > \beta_2$ because the side lobes of the wave function $|2\rangle$ have a sign opposite to that of the center lobe which provides the main contribution to the transition matrix element. The evaluation of the matrix elements α_v and β_v in fact leads to the relations $\alpha_1 < \beta_1$, $\alpha_2 > \beta_2$ and thus $\alpha_1 / \alpha_2 < \beta_1 / \beta_2$. As a consequence, when the probe polarization is parallel to the y-axis, i. e. $\psi = 0$ in eq. (5), a larger second and a smaller first Raman sideband is predicted in the probe transmission spectrum in comparison to the case of a probe beam polarized parallel to the z-axis ($\psi = 90°$ in eq. (5)). We can slightly refine our model by assuming a Bose-Einstein distribution of the vibrational populations and taking into account all other transitions starting from higher lying vibrational levels. The same relations for α_v and β_v occur in this case. We do not claim that the preceding model has much quantitative significance, however, we believe that it offers a basic qualitative understanding of the physical mechanism leading to our experimental observations. A more elaborate picture should include a more realistic optical potential obtained by diagonalizing the reactive part of the Hamiltonian for a $J = 3 \rightarrow J = 4$ atomic transition and a distribution of the vibrational populations as suggested by the optical Bloch equation.

2.3 Observation of Probe Transmission Spectra

In our experiment a cold (4 μK) dense (10^{10} atoms/cm^3) cloud of atoms is prepared by a magneto-optical trap operating in a Rubidium-vapor cell [12]. The cloud is about 0.3 mm in diameter and contains approximately $3*10^5$ atoms. The trap is active for 40 ms. During the last 0.5 ms the magnetic field is switched off and the frequency detuning of the trapping light field is increased for most efficient sub-Doppler cooling. Then, the trapping light field is switched off for a probing period of 2 ms, while the repumping laser (which counteracts hyperfine pumping) is kept active. During this time our 2D-field is activated at the position of the atomic cloud.

The 2D-field is produced by crossing the two branches of a Michelson interferometer which is fed by the spatially filtered output of a grating stabilized diode laser. The time phase difference between both branches can be servo controlled to any desired value by analyzing the light reflected from the interferometer. We couple 3 mW in a 6 mm diameter beam into the interferometer. This results in a antinode Rabi frequency $\omega_{max} = 5 \Gamma$. The frequency of the 2D-field is detuned with respect to the (F=3,m=3) $\rightarrow$ (F=4,m=4)-transition of ^{85}Rb by $\delta = - 8 \Gamma$ ($\Gamma = 38.5$ s^{-1}) . According to eq. (3) the harmonic part of the frequency splitting between adjacent vibrational states for ϕ = 90° is $\Delta\nu_{osc} = 161$ kHz.

A weak linearly polarized probe laser beam of 0.4 mm diameter is directed through the cloud. This beam travels within the xz-plane and is tilted by 5° with respect to the x-axis. Its frequency ν_p is tuned across the frequency of the 2D-field ν_{2D} during the probing period, while its absorption by the atoms is recorded. We have studied four different situations. In fig.3(a) the 2D-field was adjusted in the xy:yx-configuration and the probe polarization was parallel to the z-axis. As expected no signature of Raman transitions is observed in this case for both values 0° and 90° of the time phase difference ϕ. In fig.3(b) the probe polarization was rotated to be parallel to the y-axis. In this case we observe

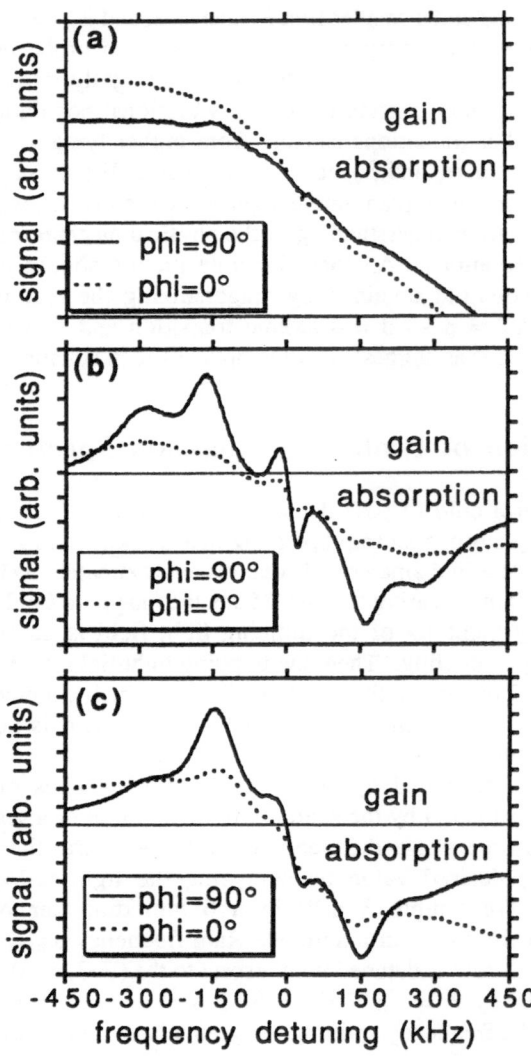

Fig. 3. Probe transmission spectra for a two-dimensional optical crystal. (a): xy:yx-configuration, probe polarization orthogonal to xy-plane. (b): xy:yx- or xy:yz-configuration, probe polarization parallel to xy-plane. (c): xy:yz-configuration, probe polarization orthogonal to xy-plane. Explanations in the text.

Stokes and anti-Stokes lines mainly due to the Raman transitions connecting the vibrational ground state with the first and second excited states. These resonances occur at center frequencies $\nu_{2D} \pm \Delta n_1$ and $\nu_{2D} \pm \Delta n_2$ symmetrically detuned from the frequency of the 2D-field ν_{2D}. We obtain the values $\Delta n_1 = 160$ kHz and $\Delta n_2 = 285$ kHz respectively. The value of Δn_1 is in good agreement with the value expected according to eq. 3. The spectrum remains unchanged if we switch to the xy:yz-configuration but keep the probe polarization fixed. This is expected because the two cases are equivalent and can be treated by the same formalism applied for different quantization axes. In fig.3(c) the probe polarization was rotated back to be parallel to the z-axes again as in (a). Again we observe Raman sidebands, however with different relative weights in accordance with our model discussed above. The line widths of the observed Raman transitions are limited by residual Doppler broadening due to the small angle of the probe beam with respect to the x-axis. This Doppler broadening occurs because of photon recoil heating along the z-axis, which is not counteracted by cooling forces in our 2D-field. The vibrational energy levels are not well resolved for $\phi = 0°$ (fig.3 (b) and (c)). In particular we observe more atoms at higher frequencies than in the 90° case corresponding to side bands of higher order.

Our interpretation of the probe transmission spectra is in accordance with the following considerations. The observation of well resolved side bands with significantly decreasing separation indicates that a large fraction of the observed transitions should start from the same state which should be the ground state if a thermal distribution of the populations is assumed. If transitions starting from highly excited states would strongly contribute to the spectrum we would expect to observe broader and more evenly spaced side bands because each of them would be due to a series of strongly decreasing energy separations. According to measurements similar as described by other authors [13] our atomic cloud is prepared at a temperature of about 4 microkelvin and due to the presence of efficient sub-Doppler cooling in the 2D-field [11] this temperature should be maintained. Assuming a 4 microkelvin thermal distribution of the vibrational populations leads to a ground state population which is about 4 times the population in the first excited state. The 1/e radius of the ground state probability density is only about $\lambda/30$. This indicates, that a large fraction of the observed atoms is localized along the axes of the light-induced waveguides.

The narrow structure in the center of the spectrum shown in fig. 3 (b) and (c) results from a Rayleigh scattering process arising from the periodic order in our atomic sample. A theoretical investigation in the context of a one-dimensional lin⊥lin-molasses in ref. 6 predicts Rayleigh resonances in the kHz range. The case of deep potential wells discussed there may serve as a model for our experiment. In this case the Rayleigh resonance should contain a significant contribution from the scattering of the pump waves into the probe by the time-modulated magnetization grating induced by the combined action of the pump waves and the probe. In the following we briefly sketch a physical model for this contribution to the Rayleigh resonance which is based on the one-dimensional considerations from ref. 1. In the following we assume the xy:yx-configuration for the 2D-field and the terms σ+ and σ- always refer to the corresponding quantization axis (i. e. the z-axis). The probe contributes equal amounts of σ+ and σ- polarization to the total local light field. This results in a modulation of the populations of the vibrational states within each waveguide with the frequency $|v_p - v_{2D}|$. Each of the four travelling (pump) waves which produce the 2D-field can be scattered by these modulated populations at each lattice site yielding a scattering component propagating on the same axis as the probe beam with the same frequency and polarization. To understand this, note that the oscillating electric field of each pump wave at a given lattice site can be split into σ+ and σ- portions of equal size. These are scattered by the oriented atoms with different phases, thus, giving rise to a small "Faraday rotation" of the linear pump polarization within the xy-plane. The scattering components from different lattice sites interfere constructively for the following reasons. In the case of the pump wave counterpropagating the probe the path length difference from components scattered by successive lattice sites is λ and the scattering phase is the same for each lattice site. For the pump waves propagating perpendicular to the probe there is an optical path length difference of λ/2 between components scattered by successive lattice sites. However, in this case we obtain a compensating phase difference of π from the opposed magnetizations. The total scattered wave interferes with the probe beam and gives a dispersively shaped resonance. The width of this resonance in our model is determined by the inverse relaxation time for the vibrational populations. However, similar as for the Raman sidebands our observations are limited by residual Doppler broadening.

3. A Three-Dimensional Spin-Polarized Optical Crystal

In this section we discuss a three-dimensional extension of the 2D considerations presented above. To obtain confinement of the atoms also in the third dimension we consider a one-dimensional standing wave extending along the z-axis with positive circular polarization superposed with the xy:yx-version of the 2D-field. The geometry of the total field (3D-field) depends on the time phase differences φ between the y- and x-wave and ψ between the z- and x-wave. We define ψ to be zero when the electric field vector at some antinode of the z-field rotates in phase with that in some antinode of the 2D-field. In fig. 4 we depict the two kinds of three-dimensional antinode structures arising for the two characteristic values 0° and 90° for ψ for fixed φ = 90°. The polarization is circular at these point-like antinodes thus giving rise to three-dimensional potential wells for the atoms in much the same way as discussed for two dimensions above. The microscopic traps in case of the body-centered cubic lattice of fig. 4 (a) (ψ = 0°) are deeper but only half as abundant compared to the case of fig. 4 (b) (ψ = 90°). All traps confine atoms with the same magnetization oriented parallel to the z-axis.

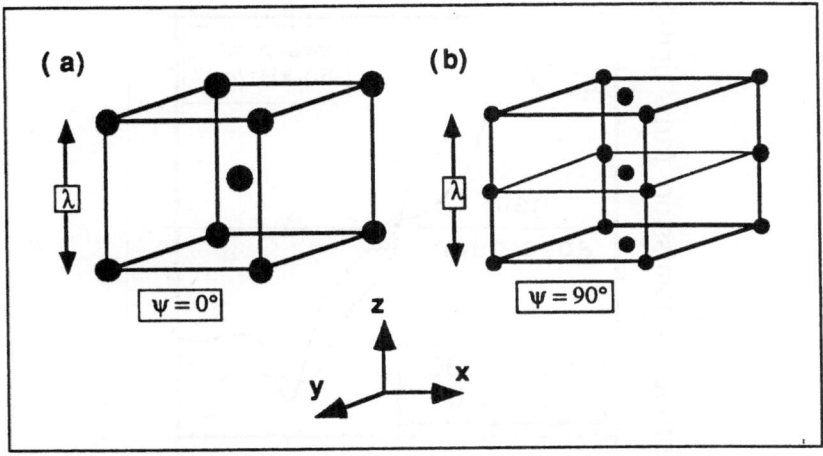

Fig. 4. Crystalline structures showing the spatial arrangement of the microscopic light traps for two characteristic values of the time-phase difference ψ between the light fields in the xy-plane and in the z-direction. The spins of the trapped atoms are oriented along the z-axis. Larger dots indicate deeper traps.

The vibrational atomic motion in the 3D potential wells may be observed as for the 2D case by Probe Transmission spectroscopy. In the following experiment the probe beam propagated along the same direction as before (within the xz-plane tilted 5° with respect to the x-axis) with linear polarization parallel to the y-axis. Probe transmission spectra are shown in fig. 5 (a) for the two cases $\phi = 90°$ (solid curve) and $\phi = 0°$ (dashed curve). The shape of these spectra does not change considerably when ψ is varied. Compared with the corresponding 2D-case presented in fig. 3 (b) we observe Raman sidebands significantly increased and with reduced line width (due to the absence of Doppler broadening and increased Lamb-Dicke narrowing) at a frequency of 115 kHz. We have confirmed that this frequency agrees with the value predicted by the shape of the optical potential that the atoms experience. Since the probe beam has only a small angle with respect to the xy-plane it should excite mainly vibrations in the xy-plane. The spring constant within the xy-plane remains practically unchanged by a variation of ψ leading to sideband frequencies independent of ψ in the observed spectra. For analogous reasons as discussed in the context of the 2D-case in fig. 3 (b) we do not observe discrete vibrational modes when $\phi = 0°$ (dashed curve in fig. 5 (a)).

The most striking change in the 3D case is a dramatic reduction of the line width of the central Rayleigh resonance (fig. 5 (a)). Even though our experiment is conducted in a cell trap containing a Rubidium atmosphere at a few 10^{-8} Torr the observed trapping times reach a few hundred milliseconds. In order to maintain the amount of trapped atoms constant on the percent level we adjusted the probing

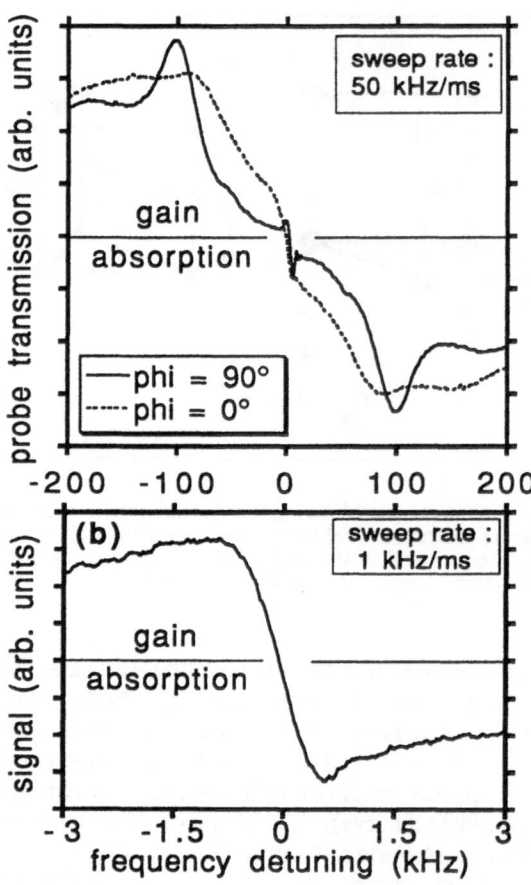

Fig. 5. Probe transmission versus the detuning between the probe and pump frequency in a 3D optical crystal. The signal is typically five to ten percent of the probe intensity. The frequency resolution in (a) is limited by the high sweep rate which causes the fringes on the right of the central resonance in the upper trace. In (b) the central Rayleigh resonance is shown on a extended frequency scale.

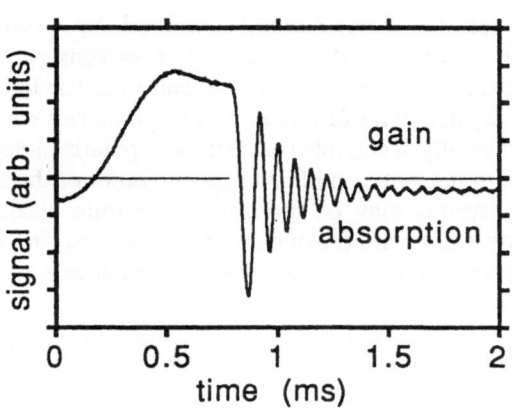

Fig.6. Time-resolved observation of the central resonance. Before t = 0 ms the probe frequency is far off resonance, at t = 0 ms it is switched into exact resonance for 0.8 ms, and at t = 0.8 ms it is rapidly detuned by 13 kHz.

time to be 10 ms. This limits the sweep rate of 50 kHz/ms in fig. 5(a) which does not allow to resolve the extremely narrow central resonance. This can be seen from the residual fringes on the right side of the central resonance in fig. 5 (a) which result from the heterodyning of a transient oscillation at the pump frequency with the constantly changing probe frequency. In fig. 5 (b) the central resonance was recorded at a much smaller sweep rate of 1 kHz/ms which reveals the extreme narrowness of this feature. Similarly as in the 2D case in fig. 3 (b) we do not observe well resolved Raman sidebands or Rayleigh resonances when ϕ is adjusted to 0°. In order to demonstrate the interference character of the central resonance we have used spectroscopy in the time domain (fig. 6). The probe frequency was adjusted to resonance for 0.8 ms to excite the Rayleigh resonance. After this period the probe frequency is detuned by about 13 kHz and thus acts as a local oscillator to produce a heterodyne signal with the decaying Rayleigh resonance. We then observe oscillation at 13 kHz with a decay time of about 0.3 ms which corresponds to a line width of the Rayleigh resonance below 1 kHz.

4. Conclusion

Dipole forces can form periodic structures of atoms inside optical standing waves on the micron length scale. Quantized motion and spatial order of atoms in such optical crystals can be experimentally studied by probe transmission spectroscopy. Our experiment allows direct comparisons of 2D and 3D field geometries. In particular, in the 3D case a sub-kHz Rayleigh resonance is observed. Optical crystals with different lattice geometries can be realized by controlling experimentally accessible parameters as polarizations and time phase differences. Antiferromagnetic and ferromagnetic order of the atomic spins are possible. Optical crystals may be interesting quantum systems which lend themselves to study solid state phenomena as tunneling. The investigation of collisions and quantum statistical phenomena also appears to be a promising perspective.

References

1. P. Verkerk, B. Lounis, C. Salomon, C. Cohen-Tannoudji, J. Courtois, and G. Grynberg, Phys. Rev. Lett. **68**, 3864 (1992).
2. P. Jessen, C. Gerz, P. Lett, W. Phillips, S. Rolston, R. Spreeuw, and C. Westbrook, Phys. Rev. Lett. **69**, 49 (1992).
3. A. Hemmerich and T. W. Hänsch, Phys. Rev. Lett. **70**, 410 (1993).
4. A. Hemmerich, C. Zimmermann, and T. Hänsch, Europhys. Lett, accepted for publication (Feb. 1993).
5. R. Blümel, J. Chen, E. Peik, W. Quint, W. Schleich, Y. Shen, and H. Walther, Nature **334**, 309 (1988).
6. J. Courtois and G. Grynberg Phys. Rev. A **46**, 7060 (1992).
7. Y. Castin and J. Dalibard, Europhys. Lett. **14**, 761 (1991).
8. K. Berg-Sörensen, Y. Castin, K. Möllmer, and J. Dalibard, preprint (March 1993).
9. A. Hemmerich and T. W. Hänsch, Phys. Rev. Lett. **68**, 1492 (1992).
10. J. Dalibard and C. Cohen-Tannoudji, J. Opt. Soc. Am. B **6**, 2023 (1989). P. Ungar, D. Weis, E. Riis, and S. Chu, J. Opt. Soc. Am. B **6**, 2058 (1989).
11. K. Möllmer, Phys. Rev. A. **44**, 5820 (1991).
12. C. Monroe, W. Swann, H. Robinson, and C. Wieman, Phys. Rev. Lett. **65**, 1571 (1990).
13. C. Salomon, J. Dalibard, W. Phillips, A. Clairon, and S. Guelatti, Europhys. Lett. **12**, 683 (1990).

Multidimensional Laser Cooling: Quantum Approaches

Yvan Castin[1], Kirstine Berg-Sørensen[2], Klaus Mølmer[2] and Jean Dalibard[1]

[1] Laboratoire de Spectroscopie Hertzienne de l'E.N.S.[3], 24 rue Lhomond, 75005 Paris, France.

[2] Institute of Physics and Astronomy, Aarhus University, DK-8000 Aarhus C, Denmark.

1   Introduction

The control of atomic motion by laser light is a field which has expanded very rapidly over the last few years. One of the most spectacular achievements in this domain is the possibility of reaching extremely low atomic kinetic temperatures, in the microKelvin range, by irradiating an atomic vapor with multiple quasi-resonant laser beams [1]. The limits of laser cooling in these so-called *optical molasses* correspond to *r.m.s.* atomic momenta $\bar{p}$ of only a few photon momenta $\hbar k$ [2, 3]. One can even pass beyond this *recoil limit* using some improved cooling schemes [4, 5].

The combination of these low temperatures with the possibility of trapping the atoms around a given point in space offers a new unique tool for atomic spectroscopy and quantum optics, and many fields of atomic physics can benefit from these new techniques: metrology using atomic fountains, collision physics, nonlinear optics, etc... These ultra-low temperatures also allow one to reach situations where the quantum nature of the atomic motion plays an important role. The atomic de Broglie wavelength $\Lambda_{dB} = h/\bar{p}$ is indeed quite large, of the order of a fraction of optical wavelength. This may be of great help for atomic interferometry experiments. For sufficiently high atomic densities, this might also offer a way of observing collective quantum effects in a sample of cold neutral atoms.

A direct experimental evidence of the quantum nature of the atomic motion in the molasses is made possible using spectroscopy techniques. In a one dimensional geometry (1D), the discrete energy levels related to the quantized atomic motion inside the potential wells have been observed experimentally [6, 7]. Recently experimental evidence for these quantum levels in a 2D [8] and a 3D [9] geometry has also been obtained.

---

[3] Unité de recherche de l'Ecole Normale Supérieure et de l'Université Paris 6, associé au C.N.R.S.

The purpose of this paper is first to discuss the possible theoretical approaches to study the atomic motion in this quantum regime. We then focus on the Sisyphus cooling mechanism, either in 1D and 2D. We discuss features specific of the 2D results, related to the existence of different types of bound motion in the potential wells created by the light-shifts.

2 Atomic motion in optical molasses

2.1 The Master Equation

We consider here a typical optical molasses geometry, where several plane running waves of equal frequency ω_L and equal electric field amplitude E_0, are present. The atom internal structure is modeled by a closed two-level system, with a stable ground state g and an excited state e with a lifetime Γ^{-1}. The ground state is supposed to have at least two Zeeman sublevels so that polarization gradient cooling can take place.

We restrict the discussion to laser fields with a saturation parameter $s_0 \ll 1$ which are known to lead to the lowest temperatures. The parameter $s_0 = 2\Omega^2/(\Gamma^2 + 4\delta^2)$ involves the Rabi frequency $\Omega = -dE_0/\hbar$ characterizing the coupling between the atomic dipole d and the field amplitude E_0 of each travelling wave, the natural width Γ of the atomic excited state and the detuning δ between the laser and atomic frequencies.

It is then possible to eliminate adiabatically the atomic excited state and to write down an equation of evolution for the restriction of the density operator to the atomic ground state g [10, 11]. The general form for this *master equation* is:

$$\frac{d\rho}{dt} = \frac{i}{\hbar}[\rho, H] + (\dot{\rho})_{relax}. \tag{1}$$

In (1), the Hamiltonian H describes the motion of the atom in the potential associated with the light-shifts. The light-shift amplitude scales as $\hbar\delta s_0$, and the spatial periodicity of the potential is the laser wavelength $\lambda = 2\pi/k$, so that H can be written:

$$H = \frac{p^2}{2M} + \frac{\hbar\delta s_0}{2}V(k\vec{r}). \tag{2}$$

where $\vec{r}$ and $\vec{p}$ are the position and momentum operators for the atomic center of mass. For a fixed value of $\vec{r}$, V is an operator acting in the $2J_g + 1$ Hilbert space of the internal atomic ground level. The relaxation part in (1) is linear and it corresponds to transitions between ground state sublevels due to *absorption– spontaneous emission* cycles. The rate of these optical pumping transitions scales as Γs_0, and we write therefore:

$$(\dot{\rho})_{relax} = \Gamma s_0 \mathcal{L}(\rho, k\vec{r}). \tag{3}$$

We emphasize that, due to the factorisation of δs_0 and Γs_0, the operators V and $\mathcal{L}$ appearing in (2) and (3) do not depend on the intensity of the light, nor its detuning to the atomic resonance. These operators are only functions of the geometry of the configuration which has been chosen (direction of the laser beams, polarizations) and of the angular momenta of the ground and excited atomic states. We note finally that we have neglected in the adiabatic elimination of the excited state, terms of the order of the Doppler shift kv when compared to Γ or δ. This amounts to neglecting the contribution of Doppler cooling, which is legitimate when the velocities achievable by polarization gradient cooling are much lower than the ones given by the Doppler cooling limit $M\bar{v}^2 \simeq \hbar\Gamma/2$.

2.2 Scaling laws for sub-Doppler cooling

In a given geometry and for a given atomic transition $J_g \leftrightarrow J_e$, several physical parameters are required to characterize a given laser cooling situation. Concerning the atom, one needs to know the atomic mass M, the wavelength λ and the natural width Γ. In addition, the atom-laser coupling is characterized by the Rabi frequency Ω and the detuning δ. Actually, it is possible to choose reduced units so that these 5 parameters are reduced to only 2 independent parameters. A possible choice is to take the reduced position $\vec{R}$ and momentum $\vec{P}$ operators and the reduced time scale τ given by:

$$\vec{R} = k\vec{r}, \qquad \vec{P} = \frac{\vec{p}}{\hbar k}, \qquad \tau = \frac{\hbar k^2}{M}t \qquad (4)$$

where the following commutation relation holds: $[R_k, P_l] = i\delta_{kl}$. The quantum equation of motion (1) then becomes:

$$\frac{d\rho}{d\tau} = i\,[\rho, \frac{P^2}{2} - u_0 V(\vec{R})] + \frac{u_0}{\Delta}\mathcal{L}(\rho, \vec{R}) \qquad (5)$$

where we introduced the two dimensionless parameters:

$$u_0 = \frac{-\hbar\delta s_0}{2E_R} \qquad \Delta = \frac{\delta}{\Gamma} \qquad (6)$$

and where $E_R = \hbar^2 k^2/2M$ stands for the recoil energy. The parameter u_0 gives the magnitude of the light-shift of the atomic ground level, measured in units of the recoil shift $E_R/\hbar$, and Δ is the detuning measured in units of the linewidth Γ. Note that this reduction of the problem to two independent parameters is useful not only for studying the steady state of the atom, but also transient regimes, spatial diffusion in the molasses, etc... Note also that if Doppler cooling had been kept, a third dimensionless parameter, such as $\epsilon = E_R/\hbar\Gamma$, would be required to characterize a given situation. The treatment presented here corresponds to the mathematical limit $\epsilon \rightarrow 0$, for given u_0 and Δ.

2.3   Solving the master equation

2.3.1   Direct numerical integration

The most direct method to solve the master equation describing the motion of the atom in the molasses is to perform a direct numerical integration of (1). From the steady-state density matrix, one can get all relevant quantities such as momentum or position atomic distributions. Unfortunately, this method is very cpu time-consuming. The number of density matrix elements which has to be evolved is very large, especially in multi-dimensional geometries. Therefore, this method is useful as a check for some particular parameter values in 1D or even 2D, but it cannot be used for realistic 3D configurations.

2.3.2   Monte-Carlo methods

The second approach that may be applied to solve (1) is the Monte-Carlo Wave Function (MCWF) approach [12, 13, 14, 15]. It consists in evolving wave functions $|\phi_{MC}(t)\rangle$ instead of the density matrix $\rho(t)$. This evolution involves a stochastic element which is chosen so that, in average over the various Monte–Carlo trajectories, one has:

$$\overline{|\phi_{MC}(t)\rangle\langle\phi_{MC}(t)|} = \rho(t) \tag{7}$$

Several types of stochastic evolution can be considered, which can be sub-divided in two main categories, depending whether they are continuous [12, 13, 16] or whether they involve quantum jumps [13, 14, 15]. Applying general statistical arguments [14] one obtains that this method is particularly valuable at providing *global* information such as the r.m.s. momentum $\bar{p}$, and it is less efficient at providing *local* information, such as the population of a single quantum state of the system. Our studies of the laser cooling problem, however, show that the momentum distribution, which is an example of such *local* information, is also calculated quite efficiently.

An example of Monte-Carlo result is plotted in fig.1, for $u_0 = 20$ and $\Delta = -3$ and for a $J_g = 1 \leftrightarrow J_e = 2$ transition. It corresponds to 3D cooling in a molasses formed by 3 mutually orthogonal standing waves. The 4 travelling waves forming the 2 standing waves in the xy plane are circularly polarized with a negative helicity, and the two travelling waves forming the standing wave along the z axis are circularly polarized with a positive helicity (magneto-optical trap configuration). The phases are respectively $0, \pi/3, 2\pi/3$ for the x, y, z standing waves. We have plotted in fig.1 the result for $\langle \vec{P}^2 \rangle$, averaged over 29 Monte-Carlo wave functions of the type:

$$|\psi(t)\rangle = \sum_{m_g=-J_g}^{J_g} \sum_{n_x=-N}^{N} \sum_{n_y=-N}^{N} \sum_{n_z=-N}^{N} \alpha(m_g, n_x, n_y, n_z, t)$$
$$|m_g, p_x = n_x\hbar k, p_y = n_y\hbar k, p_z = n_z\hbar k\rangle \tag{8}$$

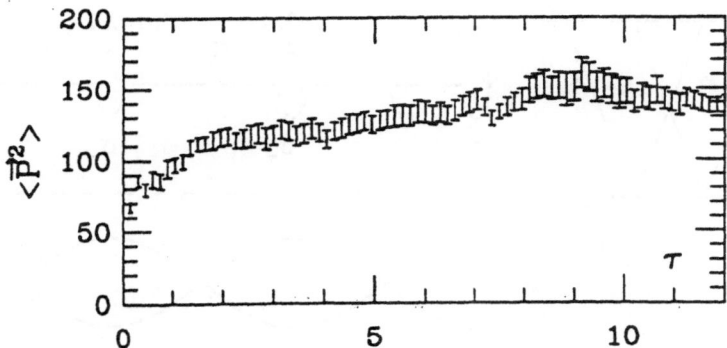

Figure 1: *Temporal evolution of $\langle \bar{P}^2 \rangle$ in 3D laser configuration formed by three mutually orthogonal standing waves. This result has been obtained by the average over 29 Monte-Carlo wave functions.*

We have chosen here a simplified spontaneous emission diagram in which photons are emitted only along axes x, y, z. The maximal momentum on the quantization grid is $N = 20$ photon momentum $\hbar k$, which corresponds to a wave function with $\sim 10^5$ components. The steady state result of fig.1, $\bar{p} \simeq 7.0 \pm 0.3 \hbar k$ along each axis, is in good agreement with the experimental results obtained for cesium atoms ($J_g = 4 \leftarrow J_e = 5$ transition) for these values of u_0 and Δ [17].

Note that this Monte-Carlo method has also been used successfuly to predict fluorescence spectra in a 1D geometry [18].

2.3.3 The secular method

The last method that we have used is based on an asymptotic expansion of (5) in the limit of large detunings Δ, for a given value of u_0. In this case, the reactive part of (5), i.e. the evolution with the Hamiltonian H, is much larger than the dissipative part, i.e. the relaxation. The procedure is then straightforward. One first look for the eigenstates $|\phi_i\rangle$ and the eigenenergies E_i of H, which depend only on u_0. One then calculates, using $(\dot{\rho})_{relax}$, the various transfer rates γ_{ij} from a given $|\phi_i\rangle$ to a given $|\phi_j\rangle$. Finally the steady-state, which is characterized in this limit by the set of populations $\{\Pi_i\}$ of the $|\phi_i\rangle$'s, is obtained by solving:

$$0 = -\sum_j \gamma_{ij} \Pi_i + \sum_j \gamma_{ji} \Pi_j \qquad (9)$$

Using the reduced units mentioned above, we see that the rates γ_{ij} are equal to $1/\Delta$ times a function of u_0. The term $1/\Delta$ then factorizes out of (9), and

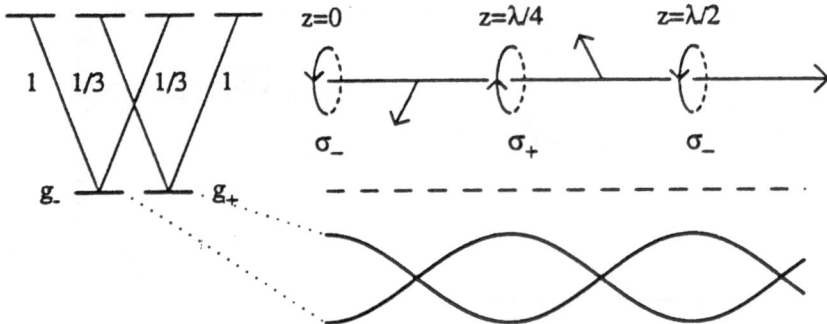

Figure 2: *Light-shifted ground state energy levels of a $J_g = 1/2 \leftarrow J_e = 3/2$ atom in a lin⊥lin laser field. Due to the gradient of ellipticity of the light, the two Zeeman sublevels oscillate in phase opposition with a period $\lambda/2$.*

the steady state populations depend only on u_0. Therefore the steady state in the secular regime is characterized by a single universal parameter u_0.

Contrary to the two previous methods, this approach has a restricted domain of validity. It consists in a secular expansion of (1), since we neglect any off-diagonal matrix element of ρ in the $|\phi_i\rangle$ basis. It is therefore valid only if all the Bohr frequencies $(E_i - E_j)/\hbar$, between two states $|\phi_i\rangle$ and $|\phi_j\rangle$ which could lead to a non-zero steady state density matrix element, are large compared to the relaxation rates appearing in $(\dot{\rho})_{relax}$. We will see that this criterion leads to very different conditions for 1D and 2D configurations.

3   Sisyphus cooling in 1D

3.1   The model of Sisyphus cooling

Sisyphus cooling is the simplest example of laser cooling with polarization gradients [19]. Consider for instance an atom with a $J_g = 1/2 \leftrightarrow J_e = 3/2$ transition, moving in the field of two plane waves counterpropagating along the z axis, with orthogonal linear polarizations. The axes of the resulting polarization are constant, oriented at 45° with respect to the polarization axis of the incoming beams. The ellipticity of the resulting polarization varies in space, going from circular to linear over a distance of $\lambda/8$. Since the electric field has only σ_+ and σ_- components when expanded on the standard basis associated with the z axis, the potential associated with the light shifts entering into (2) is diagonal in the basis $|g_\pm\rangle = |g, m_z = \pm 1/2\rangle$.

For a negative detuning, the two ground state Zeeman sublevels are shifted

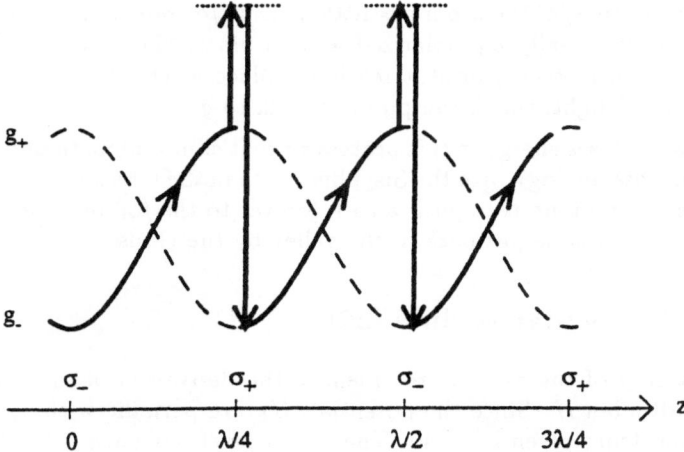

Figure 3: *Sisyphus effect: due to the spatial variation of the optical pumping rates, a moving atom climbs more than it goes down in its energy diagram. This causes a damping of its velocity in a much more efficient way than Doppler cooling.*

downwards. If the atom is located at a place where the light is σ_- polarized ($z = 0$ in fig.2), the shift of level $|g_-\rangle = |g, m = -1/2\rangle$ is three times bigger than the shift of level $|g_+\rangle = |g, m = 1/2\rangle$, because of the intensity factors (squares of Clebsh Gordan coefficients) of the $m_e - m_g = -1$ transitions. At a place where the light is σ_+ polarized ($z = \lambda/4$ in fig.2), the conclusion is reversed and the level $|g_+\rangle$ is shifted three times more than the level $|g_-\rangle$. In a place where the light is linear, one finds by symmetry that the two shifts are equal. The state dependent potential $\frac{1}{2}\hbar\delta s_0 V(k\vec{r})$ of (2) is found to be:

$$\frac{\hbar\delta s_0}{2} V(k\vec{r}) = \begin{pmatrix} U_+(z) & 0 \\ 0 & U_-(z) \end{pmatrix}, U_\pm(z) = \frac{-\hbar\delta s_0}{3}(-2 \pm \cos 2kz) \quad (10)$$

We consider now the transitions between g_+ and g_- caused by spontaneous emissions; these optical pumping processes also depend on the location of the atom. Suppose that the atom is moving towards the right and starts in $z = 0$ in level g_- (fig.3). At this place, since the light is σ_-, the atom cycles on the transition $|g, m = -1/2\rangle \leftrightarrow |e, m = -3/2\rangle$ and can never jump to the level g_+. Therefore it has to climb uphill until it reaches a place where there is a sufficient amount of σ_+ light so that the atom can be optically pumped to g_+ by a sequence $|g, m = -1/2\rangle \rightarrow |e, m = 1/2\rangle \rightarrow |g, m = 1/2\rangle$. This occurs preferentially around $z = \lambda/4$, at the top of $U_-(z)$, where the light is purely σ_+, and the atom is then put in a valley for $U_+(z)$. Once in level

g_+ in z close to $\lambda/4$, the atom has little chance to come back to g_-, because the light is essentially σ_+ polarized at this place. The atom has therefore to climb again in $U_+(z)$ until it reaches a place where there is a noticeable fraction of σ_- light, which can pump it back to g_-.

The atom loses energy in this process since it climbs more than it descends. This is in close analogy with the Sisyphus myth in the Greek mythology where Sisyphus was sentenced to push a rock forever to the top of a mountain, the rock being each time put back in the valley by the Gods.

3.2 The secular regime [20]

The first step of the secular treatment is the derivation of the spectrum of the Hamiltonian H. Since the potentials $U_\pm$ are periodic with z, we obtain a band spectrum given in fig.4a. The lowest levels coincide with those of an harmonic oscillator with a very narrow bandwidth induced by tunneling. The oscillation frequency in the bottom of the wells is given by $\hbar\Omega_{osc} \sim E_R\sqrt{u_0}$. The part of the spectrum corresponding to the above barrier motion is similar to the one of a free particle.

The steady state populations, obtained from (9) are plotted in fig.4b. They vary smoothly with u_0. The maximum of the ground state population, 0.34, is obtained around $u_0 = 45$.

The validity condition of the secular approximation requires that typical Bohr frequencies of the Hamiltonian, $i.e.$ Ω_{osc}, are much larger than the relaxation rates, $i.e.$ Γs_0, which can also be written:

$$|\Delta| \gg \sqrt{u_0} \qquad (11)$$

This method has also been extended to more complex transitions, still in a 1D geometry [11, 21].

Two experimental confirmations for this quantification of atomic motion in 1D molasses have been given recently . They are based on Raman absorption [6] or fluorescence [7] spectroscopy.

4 Sisyphus cooling in 2D

Very recently, experimental evidence has been given for the quantization of the motion in 2D and 3D [8, 9]. An important additional problem arises with respect to 1D. It concerns the stability of the interference pattern which leads to the optical potential wells. For instance in 2D, in a laser field formed by the superposition of two standing waves aligned respectively with the x and y axis, one has to stabilize the relative phase of the two standing waves to ensure this stability, which is essential for maintaining a fixed spectrum

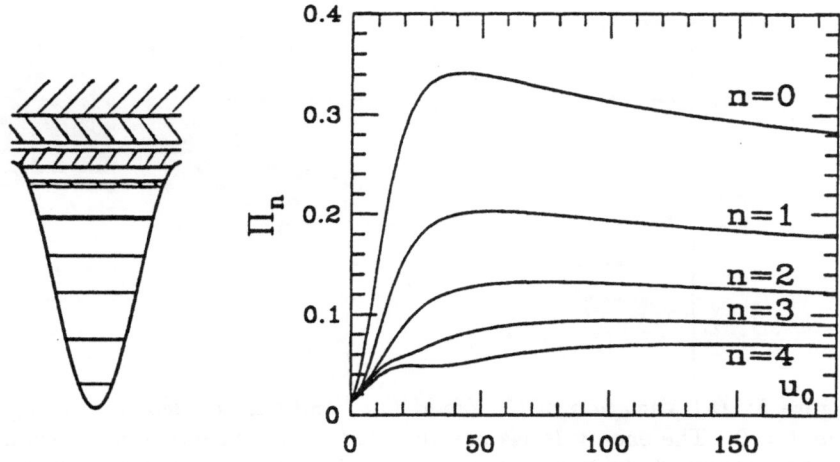

Figure 4: *(a) Band structure of the energy spectrum of H for $u_0 = 75$. (b) Variations of the steady state population of the first energy bands with u_0.*

for the Hamiltonian H. An alternative to this stabilization of the phase has been chosen in [9], using the minimal number of travelling waves to form the molasses: 3 coplanar beams in 2D and 4 non coplanar beams in 3D. In this case, a phase shift of a given beam implies a spatial translation of the interference pattern, which influences neither the structure of the optical wells nor the spectrum of the Hamiltonian H.

Our theoretical investigation has been done for the laser configuration consisting of two standing waves aligned with the x and y axes, and polarized along $\bar{u}_y$ and $\bar{u}_x$ [22]. As for the 1D configuration studied in the previous section, the electric field has only σ_+ and σ_- components along the z axis and the potential associated with the light shifts is diagonal in the basis $|g_{\pm}\rangle$:

$$U_{\pm}(x, y) = -\frac{4\hbar\delta s_0}{3}(\cos^2 kx + \cos^2 ky \pm \cos kx \cos ky \sin \alpha) \qquad (12)$$

In (12), α represents the phase between the two standing waves. As in 1D, Sisyphus cooling originates from the difference between U_+ and U_-. To maximize this difference we choose in the following $\alpha = \pi/2$.

In fig.5a, the variation of $U_+(x, y)$ is plotted along $y = 0$. Two types of minima are observed: the global minima $U_+ = 4\hbar\delta s_0$ in e.g. $(x, y) = (0, 0)$, and the local minima $U_+ = 4\hbar\delta s_0/3$ in e.g. $(\lambda/2, 0)$. The highest potential hills $U_+ = 0$ are located at $(n+1/2, n'+1/2)\lambda/2$ (field nodes). The topography of U_- is deduced from that of U_+ by a translation of $\lambda/2$ along x. Therefore the positions of the global and local minima are interchanged.

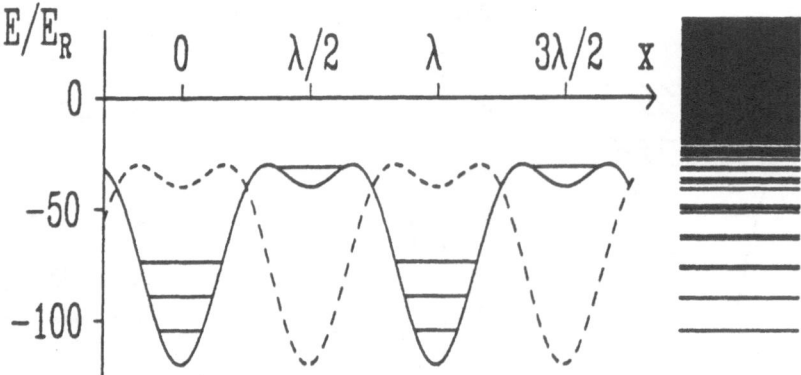

Figure 5: (a) Variation of U_+ (solid line) and U_- (dotted line) along the line $y = 0$. The energy levels are obtained in the harmonic approximation at the bottom of the potential wells. (b) Band energy spectrum calculated numerically for $-\hbar\delta s_0 = 30E_R$.

We first look for the eigenstates and eigenenergies for the motion of an atom of mass M in the periodic potentials $U_\pm(x, y)$. The eigenstates are labelled $|n, \vec{q}, \pm\rangle$, where the integer n is the band index, $\vec{q}$ is the Bloch vector chosen in the first Brillouin zone ($|q_x| + |q_y| \leq k$). The lowest levels of this spectrum can be approximated by the energy levels of a 2D harmonic oscillator located at the bottom of the well. It is also possible to identify in this spectrum the band corresponding to the "ground state" in the local minimum of $U_\pm$ (cf. fig.5b).

The steady-state populations of the various levels are obtained from the rate equations (9). We have indicated in fig.6a the variation of the band population Π_n summed over the internal state $g_\pm$, for the first ten bands, as a function of u_0. The maximum $\Pi_0 = 0.09$ is achieved for $u_0 = 11\,E_R$. In contrast to the result obtained in 1D, fig.6a shows several resonances for the populations of the lowest states as functions of u_0. We have checked that the main resonances appearing in fig.6a occur when the ground level $|\phi_a\rangle_\pm$ in the local minimum of $U_\pm$ is nearly degenerated with an excited bound level $|\phi_b\rangle_\pm$ in the main potential well. These two states may then be coupled by tunneling through the potential barrier separating the two wells, and the spectrum of H exhibits an avoided crossing as u_0 varies, the two corresponding eigenstates being linear combinations of $|\phi_a\rangle_\pm$ and $|\phi_b\rangle_\pm$. To explain the resonances in Π_0 for these values of u_0, suppose for instance that the atom reaches during the Sisyphus "cascade" the state $|\phi_b\rangle_+$ (fig.6b). The coherent tunneling coupling causes a precession of the atomic wave function towards $|\phi_a\rangle_+$. This state can then decay by optical pumping directly towards

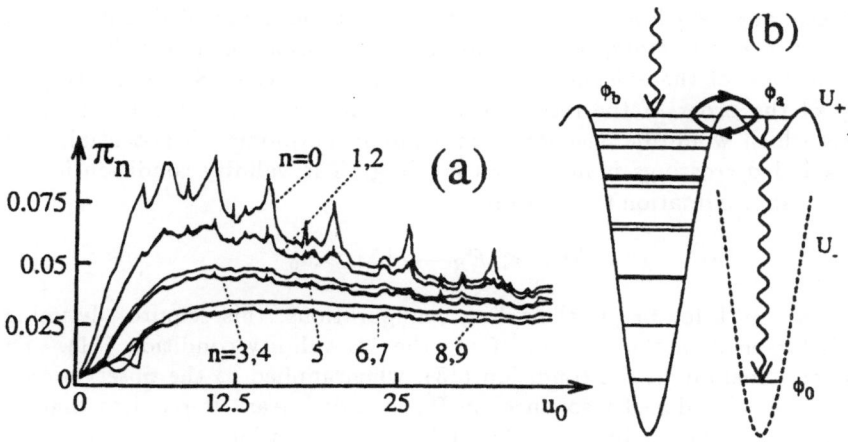

Figure 6: *(a) Steady-state populations Π_n of the lowest energy bands n as functions of u_0. (b) When the ground level in the local potential well of U_+ is close to an excited level of the main potential well, tunneling may open an extra cooling channel towards the lowest energy state of U_-, which enhances its steady state population.*

the ground state $|\phi_0\rangle_-$ of the main potential well of U_-. There is indeed a strong overlap between these two wave functions which are both localized in regions where the polarization of the light is σ_-. Outside of the avoided crossing, $|\phi_a\rangle_+$ is fed only by optical pumping from excited states of $U_\pm$. Due to the very different spatial dependence of those two types of states, the population of $|\phi_a\rangle_+$, and the corresponding feeding of $|\phi_0\rangle_-$ are smaller. To summarize, these resonances are due to the opening of a channel from highly excited states towards the ground state.

In addition to the resonances in Π_0, another important difference arises between 1D and 2D results. The maximum population of the ground state is strongly reduced, going from $\Pi_0 = 0.34$ in 1D to 0.09 in 2D. Most of this reduction is due to the degeneracy of excited levels in 2D: for a harmonic oscillator in thermal equilibrium with a frequency $\Omega_{osc} \ll k_B T/\hbar$, the population of the ground state changes from $\hbar\Omega_{osc}/k_B T$ in 1D to $(\hbar\Omega_{osc}/k_B T)^2$ in 2D. This decrease of Π_0 with increasing dimensionality is particularly important for the search of statistical effects related to quantum degeneracy. A natural "scale" for these effects is indeed the number N of atoms per potential well such that $N\Pi_0^{(3D)}$ is of the order of unity.

As mentioned above, the secular approach is valid if the relaxation rates of the levels, typically Γs_0, are small compared to the frequency splitting between any two levels leading to a non-zero off-diagonal density matrix element in steady state. This condition is much more restrictive in 2D than 1D, because of the existence of quasi-degenerate levels. Some splitting are only of the order of the recoil shift $E_R/\hbar$, for instance within a group of levels which would be degenerate in a purely harmonic 2D potential, or at the avoided crossings induced by tunneling. The validity condition for the secular approximation then becomes:

$$\hbar\Gamma s_0 \ll E_R \longrightarrow |\Delta| \gg u_0 \qquad (13)$$

Since $u_0 \gg 1$ for practical cooling configurations, this requires detunings much larger than the ones satisfying the 1D validity condition (11). The physical meaning of the condition (13), when applied to the observation of the tunneling induced resonances in Π_0, is quite clear. It requires that the precession frequency between $|\phi_b\rangle$ and $|\phi_a\rangle$ is much larger than the optical pumping rate, so that the transfer by tunneling has enough time to proceed before the atom is optically pumped away from $|\phi_b\rangle$.

For typical values of u_0 ($u_0 \sim 10E_R$), the detunings satisfying (13) are much larger than those normally used in laser cooling experiments. Therefore the observations of the resonances in Π_0 should require specifically designed experiments, for which alkali atoms may not be the best candidates, since the hyperfine splitting of their excited states limits the accessible detunings.

5   Conclusion

We have presented here practical methods to study laser cooling in multi-dimensional configurations. We have discussed in particular the secular approach, and we have formulated the results in terms of steady-state populations of the eigenstates of the atomic motion in the potential created by the laser light. Once these populations are known, it is possible to derive many other steady-state quantities, such as position or momentum distributions.

We have also predicted that the change of the light-shift potential from 1D to 2D leads to an interesting new phenomenon, due to tunneling of the atom between adjacent potential wells separated by a distance $\lambda/2$. The observation of the resonance in the ground state population induced by this tunneling would constitute a novel demonstration of the quantum nature of atomic motion in optical molasses. Similar phenomena should be expected in 3D but the complexity of the calculations makes a detailed theoretical study difficult. The MCWF seems to be the most realistic means for calculations in the general case. However the identification of population resonances such as the ones found in 2D by the secular approach might be hampered by the inherent statistical uncertainties of the MCWF method.

The authors are indebted to the ENS group of laser cooling and to E. Bonderup for many discussions. This work is partially supported by Collège de France and DRET. K. B.-S. acknowledges financial support from the Danish Natural Science Research Council and the Danish Research Academy and K. M. acknowledges financial support from the Alexander-von-Humbolt Foundation. Y. C. acknowledges numerical calculations on the Cray-2 of the French Centre de Calcul Vectoriel pour la Recherche.

References

[1] S. Chu, L. Hollberg, J. Bjorkholm, A. Cable and A. Ashkin, Phys. Rev.Lett. **55**, 48 (1985).

[2] P. Lett, R. Watts, C. Westbrook, W. Phillips, P. Gould and H. Metcalf, Phys. Rev. Lett. **61**, 169 (1988).

[3] C. Salomon, J. Dalibard, W.D. Phillips, A. Clairon and S. Guellati,Europhys. Lett. **12**, 683 (1990).

[4] A. Aspect, E. Arimondo, R. Kaiser, N. Vansteenkiste and C. Cohen-Tannoudji, Phys. Rev. Lett. **61**, 826 (1988).

[5] M. Kasevich and S. Chu, Phys. Rev. Lett. **69**, 1741 (1992).

[6] P. Verkerk, B. Lounis, C. Salomon, C. Cohen-Tannoudji, J.-Y. Courtois and G. Grynberg, Phys. Rev. Lett. **68**, 3861 (1992).

[7] P.S. Jessen, C. Gerz, P.D. Lett, W.D. Phillips, S.L. Rolston, R.J.C. Spreeuw and C.I. Westbrook, Phys. Rev. Lett. **69**, 49 (1992).

[8] A. Hemmerich and T.W. Hänsch, Phys. Rev. Lett. **70**, 410 (1993).

[9] G. Grynberg, B. Lounis, P. Verkerk, J.-Y. Courtois and C. Salomon, submitted for publication.

[10] C. Cohen-Tannoudji, in Fundamental systems in Quantum Optics, Les Houches Summer School 1990, J. Dalibard, J.-M. Raimond and J. Zinn-Justin Edts, North-Holland 1992 ; G. Nienhuis, P. van der Straten and S-Q. Chang, Phys. Rev. **A44**, 462 (1991). In these references the motion of the atomic center of mass is treated classically.

[11] Y. Castin, PhD Thesis, Université Paris 6, (Feb. 1992).

[12] N. Gisin, Phys. Rev. Lett. **52**, 1657 (1984), Helvetica Physica Acta **62**, 363 (1989); N. Gisin and I. Percival,

118

[13] H.J. Carmichael, Lectures notes at U.L.B., Fall 1991 (unpublished); see also H.J. Carmichael and L. Tian in OSA Annual Meeting Technical Digest 1990, p. 3.

[14] J. Dalibard, Y. Castin and K. Mølmer, Phys. Rev. Lett. **68**, 580 (1992). K. Mølmer, Y. Castin and J. Dalibard, J.O.S.A. **B10**, 524 (1993).

[15] R. Dum, P. Zoller and H. Ritsch, Phys. Rev. **A45**, 4879 (1992); R. Dum, A.S. Parkins, P. Zoller and C.W. Gardiner, preprint.

[16] Y. Castin, J. Dalibard and K. Mølmer, to appear in the proceedings of I.C.A.P. XIII, Münich, August 1992, T.W. Hänsch and H. Walther edts, A.I.P.

[17] A. Clairon, Ph. Laurent, A. Nadir, M. Drewsen, D. Grison, B. Lounis and C. Salomon, Proceedings of the 6th European Frequency and Time forum, J.J. Hund (1992) and private communication (March 1993).

[18] P. Marte, R. Dum, R. Taïeb and P. Zoller, Phys. Rev. **A47**, 1378 (1993)

[19] J. Dalibard and C. Cohen-Tannoudji, J.O.S.A. **B6**, 2023 (1989); P.J. Ungar, D.S. Weiss, E. Riis and S. Chu, J.O.S.A. **B6**, 2058 (1989).

[20] Y. Castin and J. Dalibard, Europhys. Lett. **14**, 761 (1991).

[21] P. Bergeman, in the poster abstracts of I.C.A.P XIII, Münich, August 1992 ; J.-Y. Courtois, PhD Thesis (Apr. 1993). For a derivation of the eigenstates of the Hamiltonian H, see also the contribution of P. Marte, R. Dum, R. Taïeb and P. Zoller in this volume.

[22] K. Berg-Sørensen, Y. Castin, K. Mølmer and J. Dalibard, submitted for publication. The same problem has been studied by the MCWF approach by P. Marte, R. Dum, R. Taïeb and P. Zoller, see their contribution to this volume and references therein.

Software Atom of Optical Physics

Juha Javanainen

Department of Physics, University of Connecticut, Storrs CT 06269–3046, USA
and
Department of Technical Physics, Helsinki University of Technology, FIN-02150 Espoo, Finland

Abstract. We have written a collection of computer programs to implement most of the semiclassical theory of light-matter interactions for an atom with an arbitrary level scheme and for arbitrary driving light. Both a direct numerical system and a symbolic-manipulation scheme employing *Mathematica* are discussed. Examples are given on resonance fluorescence of a sodium atom, laser cooling in three-dimensional optical molasses, and polarization gradient cooling of a trapped ion.

Keywords. laser cooling and trapping, resonance fluorescence, trapped ions, numerical experiments

1 Introduction

Theoretical physics ultimately is manipulation of symbols (mental images, or calculations on paper) representing objects and relationships that hopefully model Nature. The theme of our present work is to turn such a metatheoretical viewpoint into numbers and analytical expressions.

Let us build up an (exceedingly formal) example in optical physics. We assume that laser light is incident on an atom, and want to calculate the probability of finding the atom in an excited level after the laser has been on for a long time. To study the problem we introduce the density operator of the atom ρ, the atom-field interaction operator V that has been rendered independent of time by a rotating-wave approximation, the population operator for the excited level P_e, and the Liouville super-operator $\mathfrak{L}$ that includes spontaneous emission. The density operator is known to satisfy the equation of motion $\dot{\rho} = \mathfrak{L}\rho$.

In the usual case the Liouvillean $\mathfrak{L}$ has precisely one zero eigenvalue, so that the equation of motion of the density operator leads to a unique stationary state. However, the steady state, an eigenvector of the Liouvillean corresponding to the eigenvalue zero, is not uniquely determined by $\mathfrak{L}$ alone. We impose the auxiliary condition that the trace of the density operator be equal to one.

Our immediate goal is to incorporate the auxiliary condition directly into the formulation. To this end we define a projector $\mathcal{P}$ that picks up from any atomic operator O the part proportional to the unit operator 1. Denoting the total number of quantum states of the atom included in the model by N, such a projector is conveniently defined as $\mathcal{P}O \equiv 1/N \operatorname{Tr} O$. We furthermore define the conjugate of the operator $\mathcal{P}$, $\mathfrak{Q}$, by $\mathcal{Q}O \equiv O - \mathcal{P}O$. The projector $\mathcal{Q}$ renders its argument operator O traceless by subtracting a suitable multiple of the atomic unit operator 1 from it. Since an ordinary Liouville operator $\mathcal{L}$ preserves the trace of the density operator, $\mathrm{Tr}[\dot{\rho}] = 0$ and hence $\mathcal{P}\dot{\rho} = 0$ must hold for all density operators ρ. It follows that $\mathcal{P}\mathcal{L} = 0$. Multiplying the Liouville equation $\dot{\rho} = \mathcal{L}\rho$ from the left by first $\mathcal{P}$ then $\mathcal{Q}$ and using the decomposition $\rho = (\mathcal{P}+\mathcal{Q})\rho$ on the right-hand side, we are lead to the following equations for the projections $\mathcal{P}\rho$ and $\mathcal{Q}\rho$:

$$\mathcal{P}\dot{\rho} = 0, \; \mathcal{Q}\dot{\rho} = (\mathcal{Q}\mathcal{L}\mathcal{Q})\,\mathcal{Q}\rho + (\mathcal{Q}\mathcal{L}\mathcal{P})\,\mathcal{P}\rho \; . \tag{1}$$

Choosing the conserved projection as $\mathcal{P}\rho = 1/N$ (which has unit trace), we have the steady-state solution

$$\rho = \mathcal{P}\rho + \mathcal{Q}\rho; \mathcal{P}\rho = \frac{1}{N}, \; \mathcal{Q}\rho = -(\mathcal{Q}\mathcal{L}\mathcal{Q})^{-1}(\mathcal{Q}\mathcal{L}\mathcal{P})\frac{1}{N} \; . \tag{2}$$

Here $(\mathcal{Q}\mathcal{L}\mathcal{Q})^{-1}$ is the inverse of $\mathcal{L}$ taken in the projection space of $\mathcal{Q}$. Most often this projection space has only the vector 0 in common with the null space of the Liouvillean $\mathcal{L}$, hence the inverse is well defined. The desired excited level population finally is

$$p_e = \mathrm{Tr}[P_e\,\rho] \; . \tag{3}$$

While the mathematics of Eqs. (1)–(3) may seem like a case of overkill, it does have two distinct advantages. First, the form of the results does not depend on the atomic level system and light field configuration. Second, in principle Eqs. (2) and (3) constitute an explicit prescription for calculating the excited-level population: do the inverse of a superoperator, multiply operators and superoperators, take a trace.

The downside of Eqs. (1)–(3) is that they still are highly abstract. A lot more work needs to be done before numbers start coming out. The idea of the present work is to let the computer add flesh to the bare bones of the formal mathematics. We have implemented a numerical system that firstly constructs the required operators and superoperators from an elementary description of the atomic level scheme and light field, and secondly carries out the required operator inverses, multiplications and so forth. We basically create a "software atom", and subject it to a sequence of manipulations that model the interactions of an atom with light. A more modest scheme of the same kind has also been implemented symbolically with the aid of *Mathematica* [1]. In the simplest cases we may even extract analytical results.

In Sec. 2 we discuss the philosophy of the implementation of the numerical and analytical programs. Sections 3-5 are devoted to a few applications: resonance fluorescence from real sodium (Sec. 3), comparison of theory and experiment on laser cooling of atoms in 3D optical molasses (Sec. 4), and analysis of polarization gradient cooling of a trapped ion (Sec. 5). The remarks in Sec. 6 conclude the present paper.

2 Programming considerations

The concepts and symbols of the theory of light-matter interactions are supposed to model circumstances prevailing in the physical world. The next step is our computer programs. We implement the objects of the theory as software objects, and the manipulations required by the theory as subroutines that act on the software objects. The net result is that the software objects have a similar organization and even similar properties as attributed to Nature by the model of the physicist.

However, before the software world can be set up, a formulation of the theoretical concepts must be on hand that is precise enough to be expressed in a programming language. This was a main objective of Ref. [2]. There we promoted the viewpoint that a theory should not be formulated as a stream of algebra and analysis, but instead as a collection of algorithms.

We take up the general form of the rotating-wave approximation (RWA) as an example. The RWA for a given transition $1 \rightarrow 2$ ultimately boils down to subtracting a characteristic laser photon energy $\hbar\omega_{21}$ from the energy of the excited level. Thus, if the energy of the lower level is defined as zero, after transformation to the rotating frame and the RWA the energy of the excited level equals $\hbar$ times the detuning δ_{21}. For a cascade $1 \rightarrow 2 \rightarrow 3$ the same process is repeated for the second transition $2 \rightarrow 3$. The third level is then assigned an energy equal to $\hbar$ times the two-photon detuning $\delta_{31} = \delta_{32} + \delta_{21}$.

Let us now imagine two cascades $1 \rightarrow 2 \rightarrow 3$ and $1 \rightarrow 2' \rightarrow 3$ between the same end levels 1 and 3. Trouble may be on the way. There is no guarantee that the two-photon detunings calculated along the two branches are the same, hence the simple RWA cannot always be made. In this particular case the condition for a consistent RWA admittedly may be expressed in a simple algebraic equality. However, as the number of levels and laser couplings between them increases, the analysis of RWA level energies by algebraic methods rapidly becomes a daunting task.

The solution is an algorithm [2] that arbitrarily fixes an energy of an initial level, then proceeds to all levels coupled to the initial level and assigns a RWA energy to each of them, then branches to all levels coupled to the newly assigned levels, and so forth. The algorithm in effect executes a tree-like traversal through the levels of the system. Any level may be visited repeatedly, but as long as all visits assign the same RWA energy to the level, the RWA is consistent. Otherwise the user who posed the problem is sent back to the drawing board.

Next comes the actual software [3]. First, we define increasingly complex data objects, say an "atom", that consists of more elementary objects such as "levels" and "couplings" between the levels. Second, we define access functions that carry out the desired transformations on the objects. Special care is exercised to set up the access functions in such a way that the user does not have to know anything about the internal structure of the objects.

The numerical programs are illustrated in Fig. 1. This is functional C code, although the necessary declarations have been suppressed for brevity. The purpose of the program segment is to compute the excited-level population for an atom that has two levels g and e with angular momenta 1 and 2, respectively. The atom is driven by linearly polarized light with detuning D and Rabi frequency O, and the spontaneous decay rate from the excited level to the ground level is denoted by G. The first line creates an atom, and the subsequent two lines insert the levels g and e into it. The following three lines tell which level is the upper one, add spontaneous emission, and add the external field. The routine LEQ then forms the Liouvillean. The following two lines make the excited-level population operator Pe, and the next two lines form the steady-state density operator rho. The last line is a direct implementation of Eq. (3).

```
atom = initCoups();
g = putLevel(atom, "1", 0., 0.);
e = putLevel(atom, "2", 0., 0.);
addCp(atom, g, e, 1.);
addS0(atom, g, e, G, "default");
addI0(atom, g, e, O, 0., lin, D);

L = LEQ(atom);

Pe = LallocOper(atom);
LaddPopul(atom, e);
rho = LallocOper(atom);
LstatLU(L, rho);

pe = LtraceProd(rho, Pe);
```

Fig. 1. Program example to compute the excited-level population for a $j_g = 1 \rightarrow j_e = 2$ transition.

The same object structure applies to lower-level objects such as the atomic levels that were simply inserted in Fig. 1, and to higher-level objects such as those implementing the light pressure theory discussed in Sec. 4. We would like to emphasize once more that this kind of organization draws directly from the organization of experimental and theoretical objects carried out by physicists over many decades. As a result, design, coding and testing of the programs was remarkably fast and easy.

3 Resonance fluorescence from a multistate atom

It is nearly a quarter of a century since Mollow published his groundbreaking prediction of the spectrum of resonance fluorescence from a laser-driven two-state atom [4]. However, until very recently little has been said about resonance fluorescence and the closely related problem of weak-probe absorption taking into account the angular momentum degeneracy of the atoms [5-10]. The evident reason is the algebraic complexity of the analysis. We have now removed this bottleneck. Let us therefore consider resonance fluorescence from, say, a sodium atom.

We denote by $E^+(t)$ the positive-frequency part of the electric field operator in the Heisenberg picture at time t. It is well known [11] that an (idealized) apparatus set up to measure the spectrum of the scattered radiation at the field frequency ω and polarization $\hat{e}$ reports a signal proportional to

$$\mathcal{G}(\omega) = \text{Re} \int_0^\infty dt \, \langle [\hat{e} \cdot E^+(0)]^\dagger \hat{e} \cdot E^+(t) \rangle \, e^{i\omega t} . \tag{4}$$

It is equally well known that the field in the observation direction $\hat{n}$ is related to the Heisenberg picture dipole operator [rising and lowering parts $D^+(t)$ and $D^-(t)$] through the classical dipole radiation formula; neglecting various factors and effects immaterial for our purposes,

$$E^+(t) = \hat{n} \times [\hat{n} \times D^-(t)] . \tag{5}$$

The final link in the reasoning is the quantum regression theorem. It gives for the dipole correlation functions the result

$$\int_0^\infty dt \, \langle D_i^+(0) \, D_j^-(t) \rangle \, e^{i\omega t} = - \, \mathrm{Tr} \, [D_j^- \, (\mathfrak{L} + i\omega)^{-1} \, \rho_0 \, D_i^+] \,, \tag{6}$$

where $D_i^\pm$ are (for instance) the Cartesian components of the Schrödinger picture dipole operator, and ρ_0 is the steady-state density operator.

Equations (4)-(6) make an explicit prescription for calculating the spectrum of resonance fluorescence for an arbitrary atom and, for that matter, for any classical driving light for which a RWA can be made that renders the atom-field interaction time independent. Besides, an arbitrary direction of observation and an arbitrary polarizer in front of the spectrum analyzer are accommodated without problems.

As an introduction we consider resonance fluorescence from a $j_g = 2 \rightarrow j_e = 3$ transition [6]. The spectrum is reproduced in Fig. 2a. The unit for all our frequencies is the spontaneous decay rate of the excited level to the ground level, Γ. For this figure the Rabi frequency for a transition with a unit Clebsch-Gordan coefficient, e.g., $m_g = 2 \rightarrow m_e = 3$, is chosen as $\Omega = 20 \, \Gamma$, and the detuning of the laser from atomic resonance also is quite large, $\delta = 20 \, \Gamma$. The driving laser is taken to be linearly polarized. The detector is set in the plane containing the atom whose normal is parallel to the polarization of the driving light, and the detector only accepts linear polarization perpendicular to the polarization of the laser light. The spectrum is presented as a function of the difference of the detected frequency ω and the laser frequency ω_L.

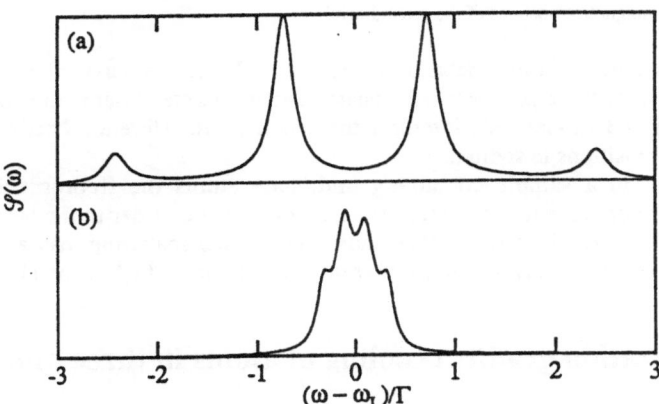

Fig. 2. Spectrum of resonance fluorescence for a $j_g = 2 \rightarrow j_e = 3$ transition (a), and the corresponding spectrum for the D_2 line of sodium (b).

The structure clearly cannot be Mollow sidebands. The number of peaks is not right. Besides, Mollow sidebands would be displaced from the atomic resonance by about Ω or δ, putting them out of the ω scale of the figure. Instead, we interpret the peaks as *quasielastic sidebands* derived from spontaneous Raman transitions between Stark shifted Zeeman substates of the ground level. Figure 3 shows the mechanism schematically. The differences of the Stark shifts of the states $m_g = -2$ and $m_g = -1$ on one hand and $m_g = -1$ and $m_g = 0$ on the other are different, so there will be two sets of peaks above and below the laser frequency. No unshifted peak is present because a detector whose polarization is set perpendicular to the quantization axis will not pick up $\Delta m = 0$ transitions at all.

Trace 2b shows the corresponding resonance fluorescence spectrum for the D_2 line of sodium [6]. In this case the ground level consists of two hyperfine components with $F_g = 1$ and $F_g = 2$, and the excited level has four hyperfine components, in ascending order in energy from $F_e = 0$ to $F_e = 3$. $F_g = 2 \rightarrow F_e = 3$ is the active transition driven with light whose parameters correspond to the values of Fig. 2a, namely laser intensity 5 W/cm^2 and laser detuning $2\pi \times 200$ MHz above resonance. To prevent optical pumping to the ground level $F_g = 1$ another laser beam is added whose intensity is 3 mW/cm^2 and which is tuned to resonance with the unshifted $F_g = 1 \rightarrow F_e = 2$ transition frequency. The linear polarization of the second laser beam is parallel to the polarization of the first beam.

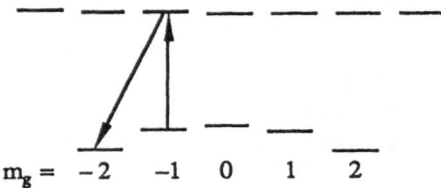

$$m_g = \quad -2 \quad -1 \quad 0 \quad 1 \quad 2$$

Fig. 3. Schematic explanation of the quasielastic peaks as spontaneous Raman transitions between Stark shifted magnetic substates of the ground level.

In real sodium the four sidebands have nearly collapsed into a nondescript structure, even though for these parameters the quasielastic peaks stand out in the corresponding $j_g = 2 \rightarrow j_e = 3$ model. We interpret the result as interference between different hyperfine transitions in sodium.

All told, in a sample containing multistate atoms the fluorescence contains quasielastic components whose frequencies typically are closer to the laser frequency than is expected of the Mollow sidebands. Quasielastic scattering may affect Doppler velocimetry of atoms [12,13], but at present it is not known by how much.

4 Polarization gradient cooling of atoms in three dimensions

A preponderance of evidence gleaned from experiments [14-17] and mostly one-dimensional theories [2,3,18-22] has shown that even in qualitative discussions of the three-dimensional optical molasses employed in laser cooling of atoms one has to take into account both the angular momentum degeneracy of atomic levels and the position dependence of the polarization of light.

Basically, in order to have cooling one wants to have a velocity dependent force. Given the atom-field interaction operator V(r) and the density operator of the internal degrees of freedom of the atom ρ, the force equals the negative of the gradient of the interaction energy,

$$\mathbf{F} = -\text{Tr}[\rho \nabla V(\mathbf{r})]. \tag{7}$$

The velocity dependence comes from the velocity dependence of the density operator. For an atom that travels with the velocity $\mathbf{v}$ to make it to the position $\mathbf{r}$ at time t, one may evidently find a velocity expansion in the form

$$\rho(\mathbf{r}, \mathbf{v}) = \rho^0(\mathbf{r}) + \rho^1(\mathbf{r},\mathbf{v}) + \dots \tag{8}$$

with $\rho^k \propto v^k$. The zeroth-order term would apply if the atom just stood still. Then the internal degrees of freedom simply settle to a steady state in the given field. However, once the atom moves, the density operator does not have time reach an equilibrium. Therefore ρ has a velocity-dependent nonequilibrium component.

From (7) and (8) the force may be expanded in velocity as

$$F_i(\mathbf{r},\mathbf{v}) = \phi_i(\mathbf{r}) - \sum_j \alpha_{ij}(\mathbf{r}) \, v_j. \tag{9}$$

Here $\phi(\mathbf{r})$ is a velocity-independent optical force that tends to confine the motion of the atom, and $\alpha_{ij}(\mathbf{r})$ is a damping tensor that conveys the first-order dependence of force on velocity.

To estimate the nonequilibrium it is useful to introduce the "memory time" of the atom τ, the time it takes the internal state to relax to equilibrium. The ρ^1 component of the density operator clearly is proportional to the change of the field as seen by the moving atom during the past memory time. We may construct the time scale of the change from the wavelength λ and velocity v as λ/v, and arrive at an estimate for the first-order component in the density operator

$$\rho^1(\mathbf{r},\mathbf{v}) \propto \frac{v\tau}{\lambda}. \tag{10}$$

This applies as long as the parameter on the right is small. For the present purposes our point is that the damping tensor will be proportional to the memory time, $\alpha_{ij}(\mathbf{r}) \propto \tau$.

In the time-honored two-state system only one evident memory time is available, namely the inverse of the spontaneous emission rate $\tau_s = \Gamma^{-1}$. However, in multistate atoms one also encounters optical pumping between the Zeeman substates of the ground level. At low intensity the rate of optical pumping is proportional to the intensity. By lowering the intensity one can make the optical pumping time arbitrarily long. If the polarization of light varies in space, the optically pumped equilibrium is different in different positions in space. The atom therefore has a true lag time equal to the optical pumping time τ_p, possibly with $\tau_p \gg \tau_s$. The associated damping of the motion $\alpha_{ij}(\mathbf{r})$ may consequently be much larger than the damping found with a two-state atom. Hence the cooling temperature of a multistate atom may be much lower than the temperature of a two-state atom.

While the basic idea of polarization gradient cooling thus turned out to be simple [18,19], only the completion of our numerical machinery [2,3,23] has facilitated the first direct comparisons of theory and 3D experiments. Some elaborate analysis is involved here, so below we give a strongly abbreviated summary of our development and its results.

We begin with the full quantum treatment of both the CM and internal degrees of freedom of the atom. We then eliminate adiabatically the internal degrees of freedom, and end up with a Fokker-Planck equation

$$\left(\frac{\partial}{\partial t} + \frac{\mathbf{p}}{M} \cdot \frac{\partial}{\partial \mathbf{r}}\right) f = \sum_{i\,=\,x,y,z} \frac{\partial}{\partial p_i}\,(F_i\,f) + \sum_{i,j} \frac{\partial^2}{\partial p_i\,\partial p_j}\,(D_{ij}\,f) \tag{11}$$

for the CM distribution function $f(\mathbf{r},\mathbf{p},t)$. To obtain the force $\mathbf{F}$ and the diffusion tensor D_{ij} at the phase space point $(\mathbf{r},\mathbf{p})$ at time t one imagines an atom traveling along a straight-line trajectory

$$\mathbf{r}(\theta) = \mathbf{r} + (\theta - t)\,\mathbf{p}/M\ . \tag{12}$$

Here θ is the parameter of the trajectory, "running time", belonging to the interval $\theta \in (-\infty, t]$. Next we find certain internal-state operators ρ and η_i $(i = x, y, z)$ by integrating the following equations of motion for an atom moving along this trajectory:

$$\frac{\partial}{\partial \theta}\,\rho = 2\Omega\Omega\,\rho + 2\Omega\mathscr{P}\,\frac{1}{N}\ , \tag{13a}$$

$$\frac{\partial}{\partial \theta}\,\eta_i = 2\Omega\Omega\,\eta_i + \frac{1}{N}\frac{\partial V}{\partial r_i} + \frac{1}{2}\,2\left(\frac{\partial V}{\partial r_i}\,\rho + \rho\,\frac{\partial V}{\partial r_i}\right) - \rho\,\mathrm{Tr}\!\left(\rho\,\frac{\partial V}{\partial r_i}\right). \tag{13b}$$

Given the solutions at the running time $\theta = t$, the force and the diffusion tensor are computed from

$$F_i = -\,\mathrm{Tr}\!\left(\frac{\partial V}{\partial r_i}\,\rho\right), \tag{14a}$$

$$D_{ij} = \frac{1}{2}\,\mathrm{Tr}\!\left(\frac{\partial V}{\partial r_i}\,\eta_j + \frac{\partial V}{\partial r_j}\,\eta_i\right) + \mathrm{Tr}\!\left[\mathscr{S}^{\,ij}\!\left(\rho + \frac{1}{N}\right)\right]. \tag{14b}$$

The force indeed is nothing but minus the gradient of atom-field interaction energy. The part of the diffusion tensor inside the parenthesis conveys the effect of photon statistics on the motion of the atom [22], and the last term reflects the angular distribution of spontaneous photons. $\mathscr{S}^{\,ij}$ are superoperators that govern the transfer of photon recoil in spontaneous emissions to the CM motion [2].

Of course, the force may be expanded in velocity so as to obtain the trapping force and the damping tensor. The explicit expressions are

$$\phi_i = \mathrm{Tr}\!\left(\frac{\partial V}{\partial r_i}\,[2\Omega\Omega]^{-1}\,2\Omega\mathscr{P}\,\frac{1}{N}\right), \tag{15a}$$

$$\alpha_{ij} = \mathrm{Tr}\!\left[\frac{\partial V}{\partial r_i}\,[2\Omega\Omega]^{-2}\left(2\frac{\partial\Omega}{\partial r_j}\,2\,[2\Omega\Omega]^{-1}\,2\Omega\mathscr{P} - 2\frac{\partial\Omega}{\partial r_j}\,\mathscr{P}\right)\frac{1}{N}\right]. \tag{15b}$$

As complicated as Eqs. (13)-(15) may seem, they are a completely explicit prescription to compute the desired quantities. Equations (13)-(15) may be implemented without problems on our software atom, for any atomic level scheme and three-dimensional light field.

Nevertheless, even though we have the force and the diffusion at our fingertips, the problem of solving the Fokker-Planck equation remains. In our work we have implemented a stochastic Langevin equation simulation. The idea is that the Fokker-Planck equation describes an ensemble of particles, each of which evolves according to a stochastic differential equation

$$M\ddot{r} = F(r,\dot{r}) + \xi(r,\dot{r},t) \ . \tag{16}$$

Here ξ is a random force whose statistics is related in a certain way to the diffusion tensor D_{ij} [24]. It is an easy matter to simulate the motion of an atom under the equation of motion (16) numerically. At every time step one simply adds to the momentum a random increment whose statistics depends on the statistics of the random force. Averaged over a long time, the results from such a one-particle simulation should reproduce the results obtained from the steady-state solution of the Fokker-Planck equation.

In order to reduce the CPU time we make an additional technical approximation. Every time the trajectory of an atom has taken it to a new position (r,p) in the phase space, in principle one should set up Eqs. (13) anew and integrate them over the interval $\theta \in (-\infty,t]$. Instead, we simply integrate Eqs. (13) continuously as the atom moves along its *stochastic* trajectory. The price of the approximation is a systematic error that gets the worse, the more the trajectory deviates from a straight line over an optical pumping time. To alleviate the error, we often increase the mass of the atom. This procedure may be justified from the empirical observation that in the simulations the time averaged kinetic energy of the atom tends to a mass independent limit as the mass goes to infinity.

Figure 4 presents a comparison between an experiment [16] and our numerical simulations [3,23]. In the experiment the D_2 line was used for cooling of sodium atoms. The light field constitutes a lin ∥ lin optical molasses: three pairs of counterpropagating beams set up in such a way that the beams in a pair share the same linear polarization, but the polarizations are orthogonal between different pairs. In the numerical runs we use a $j_g = 2 \rightarrow j_e = 3$ transition to model the sodium atom. Moreover, to alleviate the difficulties having to do with the surrogate trajectory, we take the atom to have ten times the true mass of a sodium atom. The figure gives the experimentally observed temperatures (squares) along with a fit (straight line) as a function of laser intensity. The temperature is expressed in units of the Doppler limit T_D, 240 μK for sodium, while the intensity is represented as the square of the ratio of the Rabi frequency of one of the six traveling waves to the spontaneous decay rate of the excited level. Time averaged kinetic energies of the atom from the simulations are expressed as temperatures, and are reported as filled circles. Our numerical experiments even come with statistical errors; in Fig. 4 we give the 1σ error bars. In these data the detuning of the laser is fixed at $\delta = -2\Gamma$.

Our theory and the experiments fully agree.

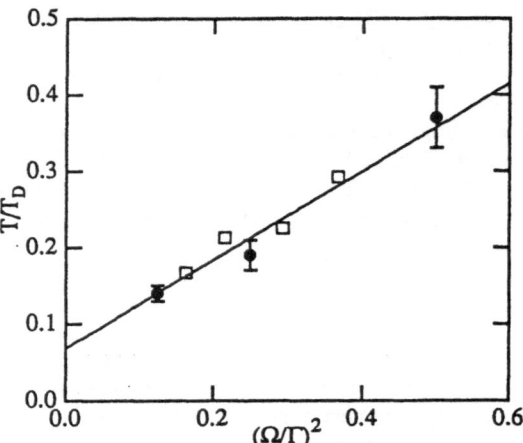

Fig. 4. Cooling temperature of sodium as a function of light intensity from experiments (squares, straight-line fit) and from our numerical simulations (filled circles).

In the hope of gaining more quantitative insights one might now reason as follows. We average away the variations of the damping tensor $\alpha_{ij}(\mathbf{r})$ and the diffusion tensor $D_{ij}(\mathbf{r},\mathbf{p}=0)$ on the scale of a wavelength. By the high symmetry of the light field the position averages have to be proportional to the unit tensor, $\langle\alpha_{ij}(\mathbf{r})\rangle = \alpha\delta_{ij}$ and $\langle D_{ij}(\mathbf{r},\mathbf{p}=0)\rangle = D\delta_{ij}$. One then has a prediction for the temperature, $T_0 = D/\alpha$. Unfortunately the temperature T_0 is low by a large factor, typically four. This observation demolishes an obvious compact characterization of laser cooling, but serves as a clue to novel physics as well [3]. The interplay of the position dependencies of the damping tensor and of the trapping force is responsible for the discrepancy. If the damping and diffusion tensors were of the form $\alpha\delta_{ij}$ and $D\delta_{ij}$, the thermal distribution function $f(\mathbf{r},\mathbf{p})$ uniquely determined by the temperature T_0 and the trapping potential $U(\mathbf{r})$ associated with the force $\phi(\mathbf{r})$ would ensue. In particular, the density of atoms would be largest at the minima of the potential $U(\mathbf{r})$. Now, the true position dependence of the damping tensor is such that the damping is at its weakest precisely at the minima of the potential. Around the minima the damping may even turn into *antidamping* with the wrong sign. The atoms therefore tend to boil out of the minima of the potential to extended trajectories. Kinetic energies of the atom should be comparable to the variations of the potential $U(\mathbf{r})$. This is precisely what we find in our numerical experiments.

Part of the agreement between our numerical experiments and real experiments is undoubtedly fortuitous, but the agreement still supports our numerical procedures. In the long run, though, qualitative insights gained from our calculations may be more important than quantitative numbers. Most notably, we have isolated a mechanism whereby the trapping potential rather than the damping and diffusion tensors primarily sets the temperature of the cooled atom. The result is a temperature comparable to the height of the trapping potential.

5 Polarization gradient cooling of a trapped ion

Laser cooling theories of a multistate atom have been the subject of an extensive recent literature. Nonetheless, the first successful laser cooling experiments on atom size particles were conducted with trapped ions [25,26], and vigorous trapped-ion studies continue up to this date. We therefore find it surprising that, in spite of analyses of some quite closely related topics [27,28], the basic concepts of polarization gradient cooling of a trapped ion apparently have not been examined in the literature. We present here some preliminary results [29] in this area.

Our semiclassical approach (quantized internal degrees of freedom, classical CM motion) to laser cooling [2] reviewed in the preceding Sec. 4. was actually developed from an earlier laser cooling theory of a trapped two-state ion [30]. We now revert back to the ion. Two major differences from a free atom are encountered. First, for a harmonically trapped ion the motion unperturbed by light no longer is free flight along a straight line, but instead harmonic oscillations. We write

$$r_i(\theta) - r_i^0 = (r_i(\theta) - r_i^0) \cos[\omega_i(\theta - t)] + \frac{p_i}{M\omega_i} \sin[\omega_i(\theta - t)] \qquad (17)$$

in lieu of Eq. (12). Here r^0 stands for the position of the center of the trap, and ω_i are the mechanical oscillation frequencies of the ion in the three principal axis directions of the trap $i = x, y, z$. Second, the equations for the operators η_i needed to compute the diffusion tensor are altered, and read

$$\frac{\partial}{\partial\theta}\eta_i = 2\Omega\Omega\,\eta_i + \cos[\omega_i\,(\theta - t)]\left[\frac{1}{N}\frac{\partial V}{\partial r_i} + \frac{1}{2}\,\Omega\left(\frac{\partial V}{\partial r_i}\rho + \rho\,\frac{\partial V}{\partial r_i}\right) - \rho\,\mathrm{Tr}\left(\rho\,\frac{\partial V}{\partial r_i}\right)\right]. \qquad (18)$$

The difference from Eq. (13b) is the cosine multiplier in front of the large square brackets.

While the cosine may look inconspicuous, it has quite dramatic implications both as a matter of principles, and in practice. To begin with, in Eqs. (13) the motion only enters through the classical trajectory of the atom. This no longer holds for Eq. (18). Even if we are computing the coefficient of diffusion for an ion that classically sits still at the center of the trap, $r_i(\theta) \equiv r_i^0$ and $p_i(\theta) \equiv 0$, through the cosine factor the ion still knows that it is trapped to a harmonic oscillator potential. All semiclassical theories of polarization gradient cooling advanced until now, except our rigorous expansion, would miss the cosine factor. In fact, we missed it, until the disastrous consequences such as laser cooling below the zero point energy of the harmonic oscillator made themselves felt.

To make further progress we seek an experimentally relevant expansion parameter analogous to velocity in Eq. (8). For a trapped ion the natural candidate is the ratio of the size of the quantum mechanical CM ground state to the wavelength. For notational simplicity we assume that the harmonic trap is one-dimensional (in the z direction) and has the trap frequency ω. Then the characteristic size of the trapped-ion cloud is $\ell = (\hbar/2M\omega)^{1/2}$, and the relevant expansion parameter is $\chi = \ell/\lambda$. An asymptotically small value of this parameter implies that a laser cooled ion resides in a region much smaller than the wavelength of light, a case known as the Lamb-Dicke limit.

We thus formally expand force up to first order and diffusion up to zeroth order in the parameter χ. While at that, we also expand in the intensity of the light, and take the limit when the mechanical oscillation frequency ω is much smaller than the spontaneous decay rate of the excited level Γ.

The ensuing formalism can make a rock cry, but *Mathematica* can turn it into simple analytical results. We look at a $j_g = 1/2 \rightarrow j_e = 3/2$ ion in 1D lin $\perp$ lin optical molasses consisting of two counterpropagating waves with orthogonal linear polarizations. For definiteness we choose the phases of the light waves in such a way that the net polarization is linear at $z = 0$. It is convenient to define the optical pumping rate

$$\gamma = \frac{4\Gamma\Omega^2}{9(4\delta^2 + \Gamma^2)}. \tag{19}$$

The inverse γ^{-1} is the memory time of the ion, the one and only long yet finite time scale of the internal evolution. A lengthy calculation shows that the position-velocity distribution of the cooled ion is thermal, and the expectation value of mechanical energy is

$$\frac{E}{\hbar\omega} = \frac{(37 + 4\xi)(\gamma^2 + \omega^2) + 45\,\phi^2\,\gamma^2(1 + \xi)^2}{60\,\phi\gamma\omega\,(1 + \xi)}. \tag{20}$$

Here $\phi = \delta/\Gamma$ is the scaled detuning, positive if the laser is tuned below the atomic resonance, and $\xi = \cos(8\pi z_0/\lambda)$ conveys the dependence on the position of the trap center z_0 with respect to the laser field.

For moderate detunings, $\phi \sim 1$, and keeping away from the field positions at which $\xi = -1$, the minimum of cooling energy with varying laser intensity is found at about $\gamma \sim \omega$. The minimum energy is of the order $E \sim 2\hbar\omega$. The lowest limit of energy, $E/\hbar\omega = (33/20)^{1/2}$, however, occurs when $\gamma \rightarrow 0$, $\phi \rightarrow \infty$, and $\xi \rightarrow -1$ in a certain interdependent way. A process with $\phi \rightarrow \infty$, etc., is experimentally infeasible, but the ion may still be cooled down until its average excitation $E/\hbar\omega - 1/2$ is less than one oscillator quantum above the zero-point energy.

The typical optimum $\gamma \sim \omega$ for the optical pumping rate is easy to understand in view of the mechanism of polarization gradient cooling. The internal state of the ion that has the velocity v at time t contains a non-stationary component, whose magnitude depends on the changes of the driving light sampled by the oscillating ion over its memory time γ^{-1}. In the case $\omega \ll \gamma$ the oscillations of the center of mass take place on a time scale much longer than the optical pumping time, so the ion with the given velocity v conveys little memory of its past velocities. The cooling takes place similarly to a free atom. In the contrary case $\omega \gg \gamma$ the ion oscillates back and forth many times during an optical pumping time, hence many oscillations are averaged over. The force then displays only a weak dependence on the *current* velocity of the ion, consequently the damping coefficient is small.

We have resorted to a specific example, $j_g = 1/2 \rightarrow j_e = 3/2$ ion in 1D lin $\perp$ lin molasses, but we expect that our findings apply qualitatively also to the other $j_g \rightarrow j_g + 1$ transitions with $j_g > 1/2$. In fact, extrapolating from the three-state Λ system [31] and from polarization gradient cooling of a free atom, we speculate that a judiciously chosen light field may lead to cooling of a trapped ion to an energy of the order of one

131

oscillator quantum above the zero-point limit $E = \hbar\omega/2$ whenever the level scheme of the ion is conducive to the existence of a long time scale.

6 Concluding remarks

Especially after the discovery of polarization gradient cooling, a theorist working on light-matter interactions has occasionally found himself/herself confronted with the full angular momentum degeneracy of the levels of a real atom. In such a case the algebraic complexity of the theoretical manipulations grows extremely rapidly with the number of states included in the model.

In this work we have discussed a way out. We suggest that, whatever the problem is, it is first formulated and analyzed in general terms such as Liouvilleans and abstract atomic operators. In the second step the abstract formulation is implemented either using an object-style program system to obtain the numbers, or symbolically on a computer to extract analytical results.

Our procedure is not the theory of everything in optical physics. Besides obvious constraints of computer time, our machinery cannot be set up before the basic concepts of the problem are well thought out and organized. This often is not the case in the initial stages of a new discovery. In exchange, the modular setup of our numerical and analytical schemes permits quantitative analyses to be made on the spot that would otherwise be major research projects.

Perhaps the most interesting potential outgrowth from our work might be the emergence of a new mode of thinking about light-matter interactions. Suppose, for instance, that a theorist or an experimentalist wants to find out how a given multistate atomic system should behave in a given light field. Instead of digging through the literature or carrying out painstaking analyses, the investigator would simply set up a numerical experiment and see what comes out!

Acknowledgment

This work is supported in part by the U.S. National Science Foundation and the Academy of Finland.

References

1. S. Wolfram, *Mathematica: A system for doing mathematics by computer* (Addison-Wesley, Reading, MA, 1988).
2. J. Javanainen, Phys. Rev. A **44**, 5857 (1991).
3. J. Javanainen, Phys. Rev. A **46**, 5819 (1992).
4. B. R. Mollow, Phys. Rev. **188**, 1969 (1969).
5. D. Polder and M. F. H. Schuurmans, Phys. Rev. A **14**, 1468 (1976).
6. J. Javanainen, Europhysics Lett. **20**, 395 (1992).
7. J. W. R. Tabosa, *et al.*, Phys. Rev. Lett. **66**, 3245 (1991).
8. D. Grison, *et al.*, Europhysics Lett. **15**, 149 (1991).
9. P. Verkerk, *et al.*, Phys. Rev. Lett. **68**, 3861 (1992).
10. M. Schubert, *et al.*, Phys. Rev. Lett. **68**, 3016 (1992).
11. P. Meystre and M. Sargent, *Elements of quantum optics* (Springer, Berlin, 1990).
12. C. I. Westbrook, *et al.*, Phys. Rev. Lett. **65**, 33 (1990).
13. P. S. Jessen, *et al.*, Phys. Rev. Lett. **69**, 49 (1992).
14. P. D. Lett, *et al.*, Phys. Rev. Lett. **61**, 169 (1988).
15. D. S. Weiss, *et al.*, J. Opt. Soc. Am. B **6**, 2072 (1989).
16. P. D. Lett, *et al.*, J. Opt. Soc. Am. B **6**, 2084 (1989).
17. C. Salomon, *et al.*, Europhysics Lett. **12**, 683 (1990).
18. J. Dalibard and C. Cohen-Tannoudji, J. Opt. Soc. Am. B **6**, 2023 (1989).
19. P. J. Ungar, D. S. Weiss, E. Riis, and S. Chu, J. Opt. Soc. Am. B **6**, 2058 (1989).
20. G. Nienhuis, P. van der Straten, and S.-Q. Shang, Phys. Rev. A **44**, 462 (1991).
21. V. Finkelstein, P. Berman, and J. Guo, Phys. Rev. A **45**, 1829 (1992).
22. K. Mølmer, Phys. Rev. Lett. **69**, 2301 (1991).
23. J. Javanainen, Optics Commun. **86**, 475 (1991).
24. C. W. Gardiner, *Handbook of Stochastic Methods for Physics, Chemistry, and Natural Sciences* (Springer Verlag, Berlin, 1989).
25. D. J. Wineland and W. M. Itano, Physics Today 34 (1987).
26. W. Neuhauser, M. Hohenstatt, P. E. Toschek, and H. G. Dehmelt, Phys. Rev. Lett. **41**, 233 (1978).
27. J. I. Cirac, R. Blatt, P. Zoller, and W. D. Phillips, Phys. Rev. A **46**, 2668 (1992).
28. D. J. Wineland, J. Dalibard, and C. Cohen-Tannoudji, J. Opt. Soc. Am. B **9**, 32 (1992).
29. S.-M. Yoo and J. Javanainen, unpublished.
30. J. Javanainen, J. Phys. B **18**, 1549 (1985).
31. M. Lindberg and J. Javanainen, J. Opt. Soc. Am. B **3**, 1008 (1986).

Quantum Optics with Cold Atoms

E. Giacobino, J.M. Courty, C. Fabre, L. Hilico, A. Lambrecht

Laboratoire de Spectroscopie Hertzienne de l'ENS et de l'UPMC, C74,
Université Pierre et Marie Curie,
4 place Jussieu, 75252 Paris Cedex 05, France

Abstract

Motionless two-level atoms constitute a very efficient nonlinear medium, which has been theoretically predicted to squeeze the quantum noise of light. Experiments made with cesium atoms, laser cooled and trapped in a magneto-optic trap, have demonstrated several nonlinear effects and give good prospect for quantum noise reduction.

Keywords. Bistability, Squeezing, Laser Cooled Atoms

1 Introduction

Laser cooled atoms have raised a lot of interest in the past few years. With the very low temperatures that can be achieved, the atoms can be considered as virtually motionless, i.e., the Doppler width of their optical resonances is much smaller than their natural width. Under such conditions, in the vicinity of the resonance lines, one can expect large nonlinear dispersive effects associated with very little absorption. We have actually observed several nonlinear effects, like Raman laser operation, bistability and instabilities with microwatt probe powers.

Moreover, media having a third order non-linearity are very good candidates to achieve quantum noise reduction. It has been theoretically shown by several authors[1-2-3] that a nonlinear optical cavity containing such a medium would be able to yield perfect quadrature squeezing in the output beam, under appropriate conditions. We outline a simple model that allows us to predict the squeezing expected from two-level atoms and we give preliminary results of an experiment using laser-cooled and trapped atoms for quantum noise reduction.

2 Nonlinear effects with cold atoms

The first studies of nonlinear effects in cold atoms showed large Raman-type gain on a probe beam passing through the cold atom cloud, at a frequency close to the cooling beam frequency[4-5]. Placing the cold atoms in an optical cavity, we have observed a laser effect linked to that gain feature.

On the other hand, when a probe laser beam is sent into the cavity, optical bistability is obtained with small light powers and on a large range of values of the detuning from resonance. When the number of atoms is large enough, instabilities appear.

In the experiments described in the following, we prepare a cloud of cold cesium atoms in a cell using the background pressure to fill the trap. The set-up includes three orthogonal trapping beams generated by a Ti:Sapphire laser and an inhomogeneous magnetic field (Figure 1).

The cell is an octogonal cylinder of electrolytically polished stainless steel, with good quality anti-reflection coated windows. It is connected to an ion pump and to a cesium reservoir, so that in the operating conditions, the cell contains a low pressure of cesium vapour of the order of 10^{-8} torr.

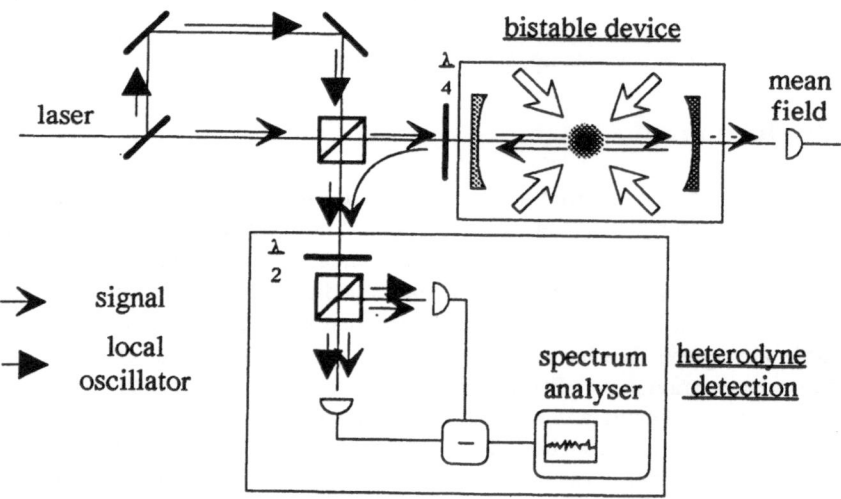

Fig. 1. Experimental set-up designed to study non-linearities and squeezing with cold atoms. Of the three cooling beams, only two, the horizontal ones, are shown. The magnetic coils are not shown. The bistability experiments (section 2.1) have been done using the photodiode placed behind the end mirror of the cavity to measure the mean signal field. The heterodyne detection shown here is the one used for the study of quantum noise (section 3). To investigate the Raman laser effect (section 2.1), the signal beam is suppessed, there are no λ/2 and λ/4 plates and the polarizing beam splitters are replaced by ordinary beamsplitters.

Each trapping beam contains a pair of counterpropagating waves with opposite circular polarisations (σ^+ and σ^-), having an intensity of about 10 mW/cm^2 and a diameter of 3 cm. The frequency of these beams, ω_T, is detuned by a few linewidths to the red side of the frequency ω_0 of the cesium transition $6\,S_{1/2}$ F=4 -> $6\,P_{3/2}$ F'=5.

In order to avoid optical pumping to the $6\,S_{1/2}$ F=3 ground state, we add to each trapping beam a beam coming from a semi-conductor laser tuned to the $6\,S_{1/2}$ F=3 -> $6\,P_{3/2}$ F'=4 frequency, which recycles the atoms.

The inhomogeneous magnetic field is produced by coils in the anti-Helmoltz position. Its gradient is 5 Gauss/cm in the symmetry plane of the coils and 10 Gauss/cm along their axis. We then obtain a cloud of about $2\,10^7$ cesium atoms of 2 mm in diameter. The typical temperature of the cloud is of the order of 1 mK, which gives a Doppler width much smaller than the natural width.

Because the cell has optical quality antireflecting windows, we can build a good finesse optical cavity around the atomic cloud. It is a 50 cm-long linear cavity, close to the hemi-confocal geometry, with a waist of 370 µm. The losses due to the two windows are of the order of 1%. Different mirrors have been used for the various experiments. The cavity is in the symmetry plane of the trap, making a 45° angle with the two trapping beams that propagate in this plane.

2.1 Cold atom laser

Recent experiments have measured the transmission of a probe beam through the sample of magneto-optically trapped atoms[4-5]. Very close to the frequency of the cooling laser (within a range of 1 MHz), they have shown that the spectrum exhibits new structures around the cooling laser ω_T : an absorption dip at a frequency slightly detuned to the blue side of ω_T and in a large gain peak (up to 20%) at a frequency slightly detuned to the red side of ω_T.

The width of these two structures, of the order of 200 kHz, is much smaller than the excited state natural width Γ, equal to 5.3 MHz. This very narrow structure has been interpreted as being due to a stimulated Raman effect between the Zeeman sublevels of the $6\,S_{1/2}$ F=4 ground state of the atoms, which are differently shifted and populated by the trapping beams.

Raman processes are then possible between the trapping beams and the probe beam. The Raman frequency shift is equal to the difference between the induced light shifts, and the width of the structures is of the order of the width of the ground state due to optical pumping among these Zeeman sub-levels.

Using a cavity of finesse about 70, we have observed a laser oscillation which has an unusual behaviour. When the cavity length is swept with a piezoelectric transducer, we observe a laser oscillation for several positions of the mirrors. The distance between two successive positions does not correspond to the free spectral range of the cavity, but rather to the transverse mode spacing of the cavity (which is about 2/7 of the free spectral range).

The characteristics of the Raman gain allows us to explain the behaviour of the laser oscillation. Because the gain width is very narrow compared to the width of the cavity, the laser can oscillate only when the frequency of one of the transverse modes of the cavity coincides with the Raman gain frequency. So, in contrast to ordinary lasers, the laser oscillation takes place in a definite transverse mode around definite values of the cavity length.

In order to check that the gain process of this laser corresponds to this stimulated Raman effect, we measured the beat frequency between the frequency of the trapping laser, ω_T and the frequency of the Raman emission, ω_l. At the output of the cavity, a 50/50 beam splitter mixes a local oscillator derived from the Ti:Sapphire laser (and shifted by 200 MHz from the trapping beams) with the cold atom generated laser. The beat note is detected by a photodiode and analyzed by a spectrum analyzer. Figure 2 shows the beat spectrum we obtained (curve b).

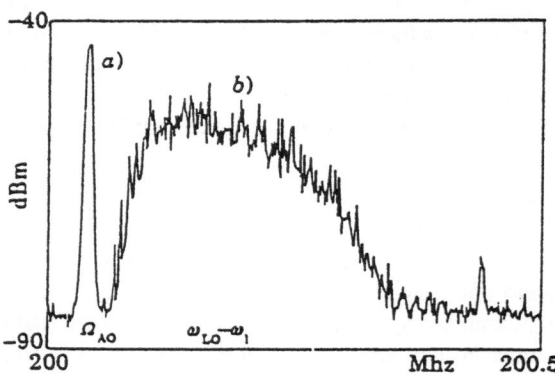

Fig. 2. Spectrum obtained by beating the Raman laser emission with a local oscillator. The narrow peak (a) corresponds to the trapping beams frequency, and the broad signal on the right hand side (b) to the cold atom laser. The full horizontal scale is 0,5 MHz.

The spectrum was recorded with the spectrum analyzer keeping the largest value obtained over a number of frequency scans while the cavity length, and consequently the cold atom laser frequency, was slowly swept. This procedure gives the gain profile of the amplifying medium. We observe a very narrow gain width of about 150 kHz, shifted by 150 kHz to the red side of ω_T. These values are in full agreement with the stimulated Raman gain characteristics.

2.1 Bistability and instabilities

To look for a bistable behaviour of the optical cavity containing cold atoms, we send a probe beam into the cavity. We scan either the length of the cavity or the

137

power of the input beam, and we monitor the output power transmitted by the cavity. Figure 3 shows the variation of the transmitted power as a function of the length of the cavity when the input power is kept fixed.

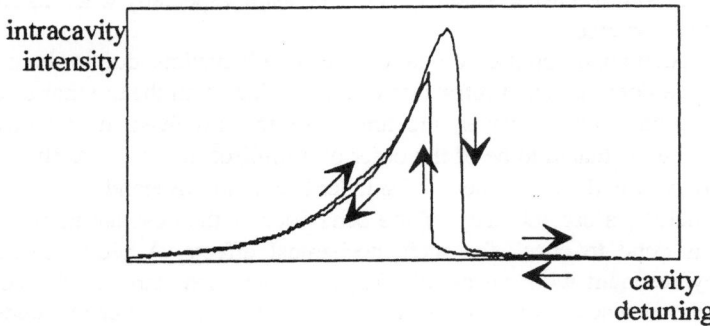

Fig. 3. Recording of the bistability hysteresis cycle in the intracavity intensity when the cavity length is scanned at a fixed input power. The intracavity intensity varies from 0 to 100 µW; C = 200.

The recording shows the characteristic hysteresis cycle due to the bistable behaviour, where the output power switches abruptly from high to low values for some value of the cavity length when the scan is in one direction, and switches from low to high for a different value of the cavity length when the scan is in the other direction.

Bistability has been observed in a very broad range of values of the relevant parameters of the experiment. The detuning of the probe laser from the atomic resonance has been varied up to 20 times the natural linewidth (120 MHz) on the blue side of the atomic transition frequency, and to 6 times the natural linewidth on the red side (on the red side, it is necessary to stay away from the resonance with the other hyperfine sublevels of the excited state). For detunings larger than 100 MHz, the absorption of the atomic medium becomes virtually negligible.

The finesse of the cavity for the experiments presented here was about 70, as in the previous experiment, but bistability has also been observed with a much lower finesse, of the order of 20. The number of atoms interacting with the probe beam has been varied by a factor of more than 10. The relevant parameter for the occurence of bistability is the so-called bistability or cooperativity parameter C, which is given by the ratio of the absorption of the atomic medium at resonance to the transmission of the coupling mirror of the cavity :

$$C = N g^2 / 2 \varkappa \gamma \tau$$

where N is the number of atoms, g is the coupling parameter of the atoms and the field, 2γ is the excited level width, $\varkappa$ is the amplitude transmission coefficient of the coupling mirror and τ is the cavity round trip time. We have obtained bistable behaviours for C parameters ranging from 20 to 300.

The probe laser input power has been varied over broad ranges of values. Bistability has been observed with powers as low as 5 μW. For high powers, the probe laser tends to push the atoms out of the beam, especially when its frequency is close to resonance.

This phenomenon could give rise to a bistable behaviour due to the change of the effective linear index of refraction of the medium with the number of atoms in the interaction zone. However, the time constants involved in this mechanical effect can be evaluated to be of the order of a millisecond, whereas the switching times observed in the experiment are smaller than a microsecond.

Nevertheless, a careful study of the behaviour of the cold atoms in the probe beam was done to check for such mechanical effects. A probe beam with a frequency resonant with the atomic frequency was sent through the cold atom cloud in the absence of the optical cavity and the transmission of the medium was measured for various values of the power. Above a value of 100 μW, the transmission starts to increase rapidly with increasing power. The cold atoms absorb photons from the probe beam and reemit fluorescence photons in all directions. This process heats the atoms, which tend to diffuse out of the interaction region. For large probe powers this effect can be seen directly on the image of the cold atom cloud given by an infrared camera.

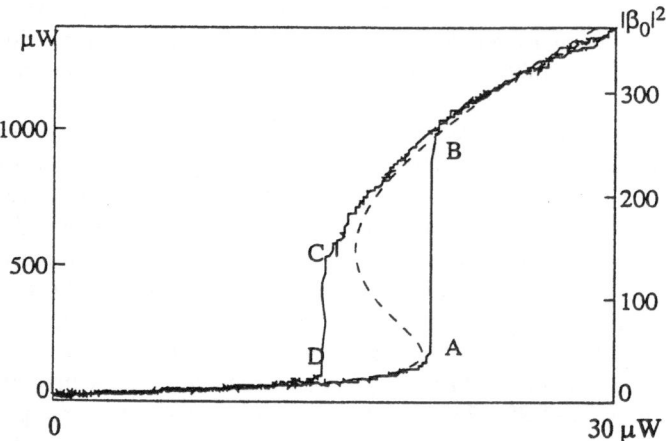

Fig. 4. Intracavity power recorded as a function of the input power for a detuning from resonance of 5,5 atomic linewidth on the red side, a cavity detuning equal to the cavity linewidth and C = 34. The broken curve is the theoretical fit. The scale on the right hand side gives the Rabi frequency normalized to the linewidth.

When the probe beam is out of resonance, this effect decreases like the square of the detuning, and is negligible for the parameters we have used. Another kind of force can be envisionned when the probe laser is out of resonance : the dipole force. Its signs depends on the sign of the detuning, and it should attract the atoms to the beam center for blue detunings and repell them for red detunings. We have seen no evidence for such an effect.

We have also studied the bistability cycle by scanning the input probe intensity for a fixed value of the cavity length, as shown in figure 5. The broken line corresponds to the theoretical transmission curve for two-level atoms interacting with a plane travelling wave in an optical cavity, while the values of the bistability parameter C and of the cavity detuning have been chosen to give the best fit. Introducing a Gaussian profile for the laser beam does not improve the fit very much. Other effects, like standing wave effects are under study.

When the bistability parameter C is increased, other phenomena appear. In figure 4, one can see an overshoot before the cavity reaches its steady state transmission on the rising side. Moreover, an oscillation can be seen on the right hand side of the curve. This oscillation is believed to be due to a single mode instability. Both the overshoot and the instabilities are predicted by the theory and have been observed with atomic beams[6]. Further checks are under investigation.

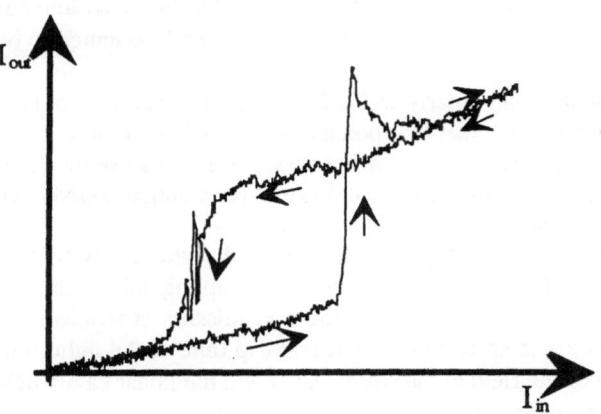

Fig. 5. Intracavity power recorded as a function of the input power for a detuning from resonance of 3,5 atomic linewidths on the blue side, a cavity detuning equal to the cavity linewidth and C = 100.

3 Quantum fluctuations in two-level atomic media

The experiments described in the previous section show that the cold atom cloud constitutes a medium having a high nonlinear index. Ideal squeezing in the output field of a nonlinear cavity has been predicted when the interaction of the nonlinear medium with the field is a purely parametric one [2-3].

When the nonlinear medium is made of near resonant two-level atoms, the quantum fluctuations in the output beam are affected by additional atomic fluctuations due to spontaneous emission and to other relaxation processes. Specific theories have to be used to take these effects into account.

3.1 Theory

The usual method consists in writing a Fokker-Planck equation for the coupled atomic and field variables and deriving an equivalent set of Langevin equations. This technique has been applied by several authors and yields analytical results[7-8-9-10].

An alternative method , which relies on the derivation of Langevin equations for the field only, has been introduced recently[11-12]. In this treatment, the atomic fluctuations are incorporated in a linearized input output theory similar to the semiclassical one[3], which applies to the parametric media only.

Schematically, the fluctuations in the output field are separated into two parts having different origins. The first part is linked to the parametric transformation of the fluctuations of the incoming field by the atomic medium, treated with a linear susceptibility calculated at the working point. The second part is associated with the resonance fluorescence of the atoms driven by the mean intracavity field and coupled with the empty modes of a bath field which is assumed to be independent of the cavity.

This treatment is particularly well adapted to large numbers of atoms in cavities having a geometry such that the spontaneous emission rate into the cavity mode is small compared to the total rate in free space. Here, we use this method to derive the spectrum of squeezing in the output field for a single ended cavity containing a collection of N motionless two-level atoms.

For the sake of simplicity, we will assume that the cavity has a good finesse. The amplitude reflection coefficient of the coupling mirror is $1-\varkappa$, where $\varkappa$ is small compared to 1. Subsequently, the transmission coefficient of the coupling mirror can be expressed as $\sqrt{2\varkappa}$. The round trip time of the light in the cavity is τ, the decay rate of the field in the cavity is $\varkappa/\tau$ and the linear cavity detuning is Φ.

The time dependent quantum field operators in the rotating frame (at the mean driving field frequency) are denoted $\hat{A}(t)$ for the field inside the cavity and $\hat{A}^{in}(t)$ and $\hat{A}^{out}(t)$ for the fields coming into and going out of the cavity. These fields are connected by input-output relations which are the classical reflection-transmission equation on a dielectric plate :

$$\hat{A}^{out}(t) = \sqrt{2\varkappa}\,\hat{A}(t) - \hat{A}^{in}(t) \tag{1}$$

On the other hand, the intracavity field $\hat{A}(t)$ verifies the differential equations :

$$\tau\,\partial_t\,\hat{A}(t) = -(\varkappa + i\Phi)\,\hat{A}(t) + \sqrt{2\varkappa}\,\hat{A}^{in}(t) + ig\,\hat{P}(t) \tag{2 a}$$

$$\tau\,\partial_t\,\hat{A}^{\dagger}(t) = -(\varkappa - i\Phi)\,\hat{A}^{\dagger}(t) + \sqrt{2\varkappa}\,\hat{A}^{\dagger in}(t) - ig\,\hat{P}^{\dagger}(t) \tag{2b}$$

The first term comes from the recycling of the field in the cavity and the loss through the coupling mirror, the second one comes from the input field A^{in}, the third one is the field emitted by the atomic polarization $\hat{P}(t)$, g being the coupling constant between the field and the atoms.

We will compute the quantum fluctuations of the fields in cases where they are small compared to the mean values. The equations for the fluctuation operators are obtained by linearizing the evolution equations around the steady state mean values. The mean values for the field and for the atomic polarization neglecting the effect of the fluctuations (mean field theory) can be calculated from the usual Maxwell-Bloch equations. To calculate the fluctuations in the intracavity field as a function of the input fluctuations and of the fluctuations of the atomic medium, we define the fluctuation operators as:

$$\delta \hat{A}(t) = \hat{A}(t) - <\hat{A}(t)> \qquad\qquad \delta \hat{P}(t) = \hat{P}(t) - <\hat{P}(t)> \qquad (3)$$

The equations for the fluctuations in the field are obtained in a straightforward manner from eqs.(1) and (2), since the latter are linear in the dynamic variables. In frequency space, the equations for the Fourier components of the field fluctuations operators $\delta A(\omega)$ write:

$$\delta A^{out}(\omega) = \sqrt{2\varkappa}\ \delta A(\omega) - \delta A^{in}(\omega) \qquad (4)$$

$$(\varkappa + i\Phi - i\omega\tau)\ \delta A(\omega) = \sqrt{2\varkappa}\ \delta A^{in}(\omega) + i\ g\ \delta P(\omega) \qquad (5)$$

$$(\varkappa - i\Phi - i\omega\tau)\ \delta A^{\dagger}(\omega) = \sqrt{2\varkappa}\ \delta A^{\dagger in}(\omega) - i\ g\ \delta P^{\dagger}(\omega) \qquad (6)$$

As mentionned earlier, the dipole fluctuations δP include two contributions, since the dipoles respond to the fluctuations of the cavity field and of the bath field. We will write:

$$\delta P(\omega) = \delta P_1(\omega) + \delta P_2(\omega) \qquad (7)$$

where δP_1 describes the response to the intracavity field fluctuations, and δP_2 is due to coupling of the dipoles with the vacuum fluctuations of the bath field and is associated with spontaneous emission.

For the calculation of the first term, we assume a linear response of the atoms around the working point of the system which is given by a susceptibility function $\chi(\omega)$. Since the atomic medium couples the frequency components $\delta A(\omega)$ and $\delta A^{\dagger}(\omega)$, it is convenient to introduce matrix notations, writing:

$$|\delta P_1(\omega)] = g\ N\ [\chi(\omega)]\ |\delta A(\omega)] \qquad (8)$$

where $|\delta P_1(\omega)]$ and $|\delta A(\omega)]$ are column vectors of the type $|X(\omega)]$:

$$|X(\omega)] = \begin{bmatrix} X(\omega) \\ X^{\dagger}(\omega) \end{bmatrix} \tag{9}$$

The matrix elements of the susceptibility $[\chi(\omega)]$ can be computed using the linear response theory[13-14]. In the time domain, $[\hat{\chi}(t)]$ can be shown to be a 4x4 matrix involving the commutators of the dipole operator[14].

The additional fluctuations δP_2 come from the coupling of the atomic dipoles to the vacuum bath field. They can then be calculated as in the resonance fluorescence theory[15], supposing that the fluctuations of the different dipoles are independent of one another. Their correlation matrix can be calculated from the Bloch equations using the quantum regression theorem.

To incorporate the susceptibility functions and the correlation functions of the dipole fluctuations in the calculation, we rewrite eqs.(5, 6) in matrix form as:

$$\{ \varkappa + i\,[\varepsilon]\,\Phi - i\omega\tau \}\ |\delta A(\omega)] = \sqrt{2\varkappa}\ [\delta A^{in}(\omega)] + i\,g\,[\varepsilon]\ |\delta P(\omega)] \tag{10}$$

The 2x2 matrix $[\varepsilon]$ introduced by complex conjugation is :

$$[\varepsilon] \quad = \quad \begin{bmatrix} 1 & 0 \\ 0 & -1 \end{bmatrix} \tag{11}$$

Using eq.(4), eq.(10) can be expressed as:

$$\{ \varkappa + i\,[\varepsilon]\,\Phi - i\omega\tau \}\ |\delta A^{out}(\omega)] =$$

$$\{ \varkappa - i\,[\varepsilon]\,\Phi + i\omega\tau \}\ |\delta A^{in}(\omega)] + i\,\sqrt{2\varkappa}\,g\,[\varepsilon]\ |\delta P(\omega)] \tag{12}$$

At this point, we have to take into account the fact that the fluctuating term $|\delta P(\omega)]$ is correlated with the fluctuations of the field, as can be seen from eqs.(7) and (8). Replacing $|\delta P(\omega)]$ by these expressions and using eq.(4) again, we obtain a new expression for the fluctuations of the output field:

$$\{ \varkappa + i\,[\varepsilon]\,\Phi - i\omega\tau - i\,N\,g^2\,[\varepsilon]\,[\chi(\omega)] \}\ |\delta A^{out}(\omega)] =$$

$$\{ \varkappa - i\,[\varepsilon]\,\Phi + i\omega\tau + i\,N\,g^2\,[\varepsilon]\,[\chi(\omega)] \}\ |\delta A^{in}(\omega)]$$

$$+ i\,\sqrt{2\varkappa}\,g\,[\varepsilon]\ |\delta P_2(\omega)] \tag{13}$$

The fluctuations of the output field are now expressed as a function of two uncorrelated sources: the fluctuations of the input field and the fluctuations of the atomic dipoles driven by the bath modes. The matrix $Ng^2[\varepsilon][\chi(\omega)]$ represents the modification of the phase and of the amplitude of the field fluctuations due to the atomic medium. The real part of this term causes dispersion in the same way as a transparent Kerr medium, while the imaginary part corresponds to absorption.

Comparison between eqs. (12) and (13) shows that these dispersion and absorption terms add to the linear losses and dephasing undergone by the field fluctuations in one round trip. When the driving field is very far from resonance, the imaginary part of $[\chi(\omega)]$ and the last term on the right hand side of eq.(13) can be ignored, and the effect of the atomic medium is represented by a nonlinear index, which corresponds to the Kerr limit.

We can now calculate the covariances of the output field as a function of the covariances of the input field, and the squeezing expected in the output field. A typical spectrum is shown in figure 5.

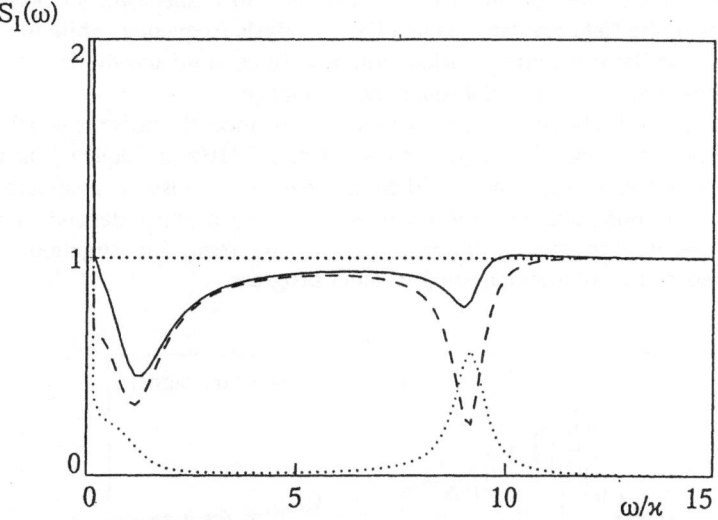

Fig. 5. Amplitude squeezing spectrum $S_I(\omega)$ for an atomic medium in the bad cavity case ($\gamma / \varkappa = 0.1$). The parameters are : $\delta / \gamma = 40$ (δ is the detuning of the field from atomic resonance), $C = 180$, $|\beta_0|^2 = 400$ (normalized Rabi frequency), $\Phi = 2.3$. The curve in broken line shows the contribution of the incoming fluctuations filtered by the atom-cavity system, and the dotted line is the contribution of the atomic fluorescence noise.

Investigation of the parameter space shows that large squeezing can be expected in the so-called bad cavity case ($\gamma / \varkappa \ll 1$), when the system is operated near the bistability turning points.

3.2 Experiments

Quantum noise reduction has been observed in an optical cavity traversed by an atomic beam of two-level atoms[16]. It is expected that the cold atoms should also yield good squeezing. To experimentally study the quantum fluctuations at the output of the cavity containing the cold atoms, we use a linear cavity having only one coupling mirror. The coupling mirror has a high transmission, so as to be in the bad cavity case, where the decay time of the cavity is larger than that of the atoms, and for which the theoretically predicted squeezing is the best. The other mirror is a highly reflecting one.

The field reflected from the cavity is separated by an optical isolator. The optical isolator is made of a polarizing beam splitter followed by a quarter wave plate (the field inside the cavity has a circular polarization).

The polarizing beamsplitter is also used to mix a local oscillator beam with the probe field going out of the cavity. the signal is detected with a balanced heterodyne detector (as shown in fig. 1), and sent to a spectrum analyser. The fluctuations of the field are detected at a fixed analysis frequency, while the phase of the local oscillator is rapidly varied with a piezo ceramic and the length of the optical cavity is slowly scanned through the resonance.

A typical curve is shown in figure 6. Out of resonance, the noise is equal to the standard quantum noise (the noise is recorded at 5.5 MHz, a frequency at which the classical noise is negligible). When the cavity is close to resonance, the presence of the nonlinear medium manifests itself by a phase dependent noise. Excess noise due to atomic fluorescence can be seen. Investigation of the parameter space to look for squeezing is under progress.

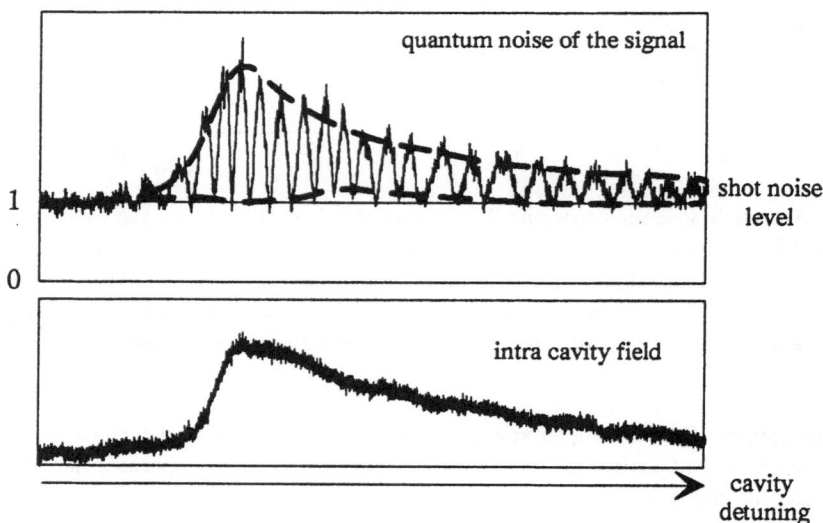

Fig. 6. Recording of the mean field inside the cavity (lower curve) and of the noise of the field reflected by the cavity (upper curve), measured with a heterodyne detection and a spectrum analyser. The frequency of the spectrum analyser is fixed at 5.5 MHz, and the cavity length is scanned through resonance. The large oscillations are due to the phase scan of the local oscillator. The dashed lines are the envelopes of these oscillations (i.e. the quadratures having minimum and maximum noise). The parameters are: $\delta/\gamma = 25$, $C = 150$.

4. Conclusion

The recent development of the magneto-optic trap, which enables to get clouds of motionless atoms with a density comparable to that of the atomic beams is of great interest for nonlinear optics and for squeezing. In such traps, one can get large non-linearities by setting the driving field close to resonance without having much absorption. The observation of dispersive bistability in such media makes it possible to do a detailed evaluation of the properties of the cold trapped atoms as a device to manipulate the quantum noise.

Acknowledgements :This work has been supported in part by the C.E.C. contracts ESPRIT BRA 3186 and 6934

References
1   L. Lugiato, G. Strini, Optics Commun. 41, 67 and 374 (1982)
2   M.J. Collett, D.F. Walls, Phys. Rev. A32, 2887 (1985)
3   S. Reynaud,C. Fabre, E. Giacobino, A. Heidmann Phys. Rev. A40, 1440 (1989)
4   D. Grison, B. Lounis, C. Salomon, J.Y. Courtois, G. Grynberg, Europhys. Lett. 15, 149 (1991)
5   J. Tabosa, G. Chen, Z. Hu, R. Lee, H.J. Kimble, Phys. Rev. Lett. 66, 3245 (1991)
6   L.A. Orozco, A.T. Rosenberger, H.J. Kimble, Phys. Rev. Lett. 53, 2547 (1984); L.A. Orozco, H.J. Kimble, A.T. Rosenberger, L.A. Lugiato, M.L. Asquini, M. Brambilla, L.M. Narducci, Phys. Rev. A39, 1235 (1989)
7   H.J. Carmichael, Phys. Rev. A33, 3262 (1986)
8   L.A. Orozco, M.G. Raizen, Min Xiao, R.J. Brecha, H.J. Kimble, J. Opt. Soc. Am. B 4, 1490 (1987)
9   F. Castelli, L.A. Lugiato, M. Vadacchino, Nuovo Cimento B 10, 183 (1988)
10 M.D. Reid, Phys. Rev. A37, 4792 (1988)
11 J.M. Courty, P. Grangier, L. Hilico, S. Reynaud, Opt. Commun. 83, 251 (1991)
12 L. Hilico, C. Fabre, S. Reynaud; E. Giacobino, Phys. Rev. A46, 4397 (1992)
13 H. Mori, Progress in Theoretical Physics, 33, 423 (1965); R. Kubo, Reports in Progress in Physics 29, 255 (1966)
14 J.M. Courty, Thèse de L'Université Pierre et Marie Curie (Paris 1990, unpublished); J.M. Courty and S. Reynaud, Phys. Rev. A46, 2766 (1992)
15 B.R. Mollow, Progress in Optics XIX ed. E.Wolf (North Holland, Amsterdam, 1981) p.1 and ref. therein; A. Heidmann and S. Reynaud, Journal de Physique 46, 1837 (1985) and ref. therein
16 M.G. Raizen, L.A. Orozco, Min Xiao, T.L. Boyd, H.J. Kimble, Phys. Rev. Lett. 59, 198 (1987)

Correlations of the resonance fluorescence from a single trapped ion

R. Blatt, M. Schubert, I. Siemers, W. Neuhauser
Institut für Laser–Physik, Universität Hamburg

W. Vogel
Fachbereich Physik, Universität Rostock

Abstract. Intensity correlation measurements of the fluorescence from a single trapped Ba$^+$–ion are presented. The feasibility of homodyne correlation measurements of non–classical effects such as squeezing and anomalous correlations is studied and potential signal–to–noise ratios are estimated.

Keywords. Intensity correlations, squeezing, resonance fluorescence

1 Introduction

Measuring photon correlations is of general interest for studying the quantum properties of light. For the first time it was demonstrated by Kimble, Dagenais and Mandel that the intensity correlations of the fluorescence from a low–density atomic beam reveal the non–classical effect of photon antibunching [1]. Later on, the techniques of cooling and trapping single ions made it possible to observe this effect from a single two–level atom [2]. However, these measurements were marred by the presence of the micromotion due the time–dependent trap potential. More recently, intensity correlations of the fluorescence from a single trapped multi–level ion have been observed [3].

Aside from the antibunching property, which more or less directly reflects the photon nature of light, other non–classical effects are of interest. An example is squeezing which is usually observed in a homodyne detection scheme [4, 5]. The squeezing property has also been predicted to appear in the resonance fluorescence of a single two–level [6] and three–level atom [7]. Mandel has shown that the observable effect due to squeezing in the photon statistics of the homodyne signal is extremely small [8]. Alternatively, the detection of homodyne intensity correlations is expected to offer an easier way to observe squeezing in resonance fluorescence [9], in particular for weak local oscillator intensities.

In the present paper we consider the intensity correlations of the resonance fluorescence observed from a single trapped Ba$^+$ ion (Section 2). In the context of squeezing in resonance fluorescence we consider homodyne correlations and estimate the size of the effect and the required conditions for such an experiment (Section 3). Some conclusions are given in Section 4.

2 Intensity correlations

In the experiment described below [3], we observe correlation functions of a single trapped Ba$^+$ ion which has an eight–level structure (see Fig. 1) and hence shows a manifold of dynamical features not appearing in two–level systems as, e.g. optical pumping and atomic coherence effects. For this purpose, single Ba$^+$ ions are confined in a miniature Paul trap of 1–mm ring diameter and optically cooled to a residual temperature of less than 3 mK. In order to minimize the residual micromotion, additional electrodes are, used to apply dc voltages in the ring plane and along the z–axis which allow to compensate offset potentials arising from surface charges on the electrodes.

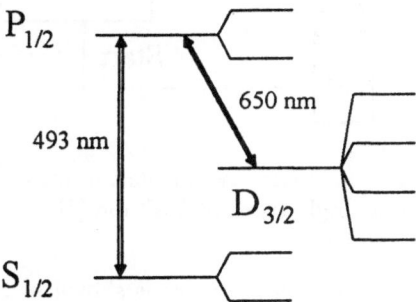

Figure 1: Level scheme of the Ba$^+$ ion. Due to the presence of a magnetic field, an eight–level system has to be considered.

Excitation spectra of the single Ba$^+$ ions are obtained by tuning the laser at 650 nm across the resonance while the laser at 493 nm is kept fixed in frequency, thus providing optical cooling. The rich sublevel structure of the Ba$^+$ ion leads to the observation of several dark resonances [10]. From these measurements the parameters required for the multi–level optical Bloch equations, i.e. the detunings, the Rabi frequencies and the magnetic field strength, can be uniquely determined. A measurement of the intensity correlation is then performed for the same parameters which are subsequently used to calculate the corresponding correlation functions.

Figure 2 shows the experimental setup for the intensity correlation mea-

148

surement. The lasers at 493 nm and 650 nm are collinear and focussed into the trap and excite the ion. The resonance fluorescence is observed via two detection channels which are perpendicular to the laser beams and opposite to each other. In front of the first detector (PM_1) a color glass filter is placed such that only radiation at 493 nm is recorded while the second detector is sensitive to the total fluorescence light. Correlation measurements are taken by means of a fast time–to–digital converter (TDC); the start pulses are derived from PM_1, and the stop pulses from PM_2 are electronically delayed. Thus we are able to measure different correlation functions for negative and positive time delays in a single measurement.

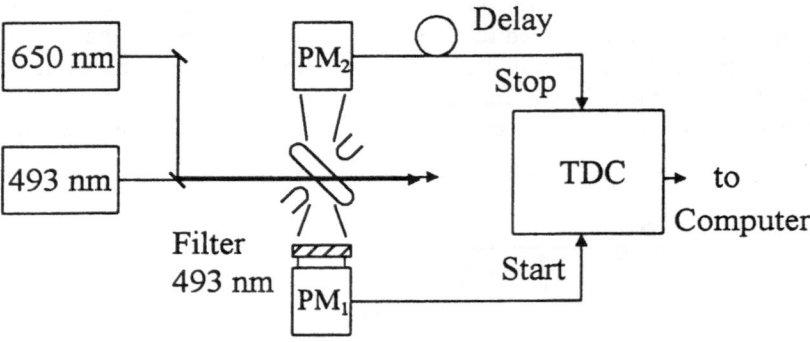

Figure 2: Experimental setup for intensity correlation measurements of the resonance fluorescence from a single trapped Ba$^+$ ion [3].

A typical intensity correlation measurement is shown in Fig. 3. The behavior of the curve around zero delay clearly reveals the antibunching nature of the fluorescence from the Ba$^+$ ion. The asymmetry of the correlation function with respect to zero delay time results from the different conditional probabilities effectively observed due to the presence of a filter in one of the detection channels. Note that the shapes of the observed correlations reflect the multi–level dynamics as can be seen from the ratio of the maximum value to the stationary value of the correlation function which increases the corresponding maximum value available for a two–level atom. The solid line in Fig. 3 shows the theoretical (normalized) correlation function

$$g^{(2)}(\tau) = \frac{G^{(2)}(\tau)}{\langle:\hat{I}:\rangle^2}, \qquad G^{(2)}(\tau) = \langle:\hat{I}(0)\hat{I}(\tau):\rangle \tag{1}$$

(with the intensity operator $\hat{I}$ and :: denoting normal and time ordering) which was calculated from the optical Bloch equations together with the quantum regression theorem using the parameters as derived from excitation spectra.

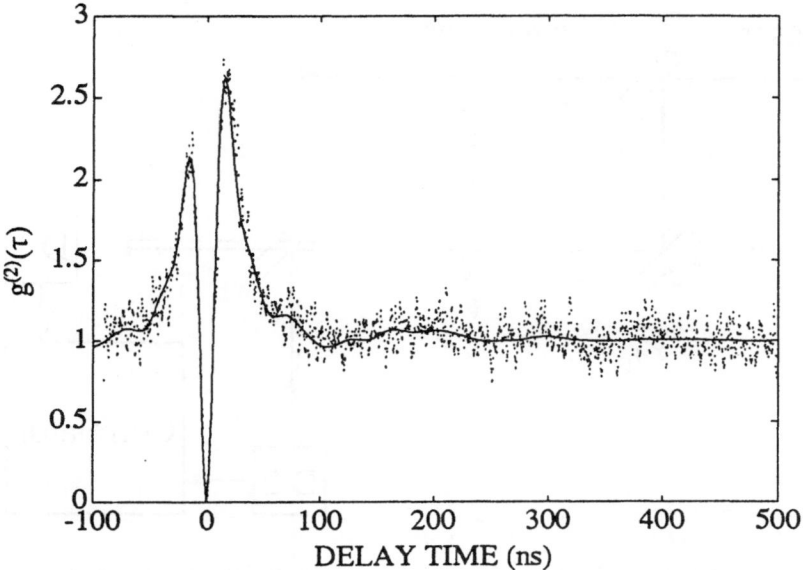

Figure 3: Measurement of an intensity correlation of the resonance fluorescence from a single trapped Ba$^+$ ion. The detuning of the laser at 493 nm is $\Delta_{493} = -13.5$ MHz and the Rabi frequency on the S–P transition was $\Omega_{493} = 0.81\ \Gamma_{493}$ with $\Gamma_{493} = 2\pi \cdot 15.1$ MHz. The laser at 650 nm was detuned to $\Delta_{650} = -7.7$ MHz, and the Rabi frequency on the P–D transition was $\Omega_{650} = 0.75\ \Gamma_{650}$, with $\Gamma_{650} = 2\pi \cdot 5.3$ MHz.

3 Homodyne intensity correlations

A measurement of the intensity correlations of the fluorescence superimposed by a local oscillator of the same frequency is of advantage for avoiding an explicit dependence of the squeezing effect on collection and quantum efficiencies. Moreover, one may record anomalous correlations of the fluorescence [9] which cannot be observed without using a homodyne scheme.

Figure 4 shows a typical setup for measuring homodyne intensity correlations. A laser beam is split into pump and local oscillator beam. The fluorescence light is superimposed with the local oscillator with a variable phase difference between both fields. The superimposed light is then investigated using standard intensity correlation techniques.

The intensity correlation function $G^{(2)}(\tau)$ can be decomposed as

$$G^{(2)}(\tau) = \sum_{i=0}^{4} G_i^{(2)}(\tau), \qquad (2)$$

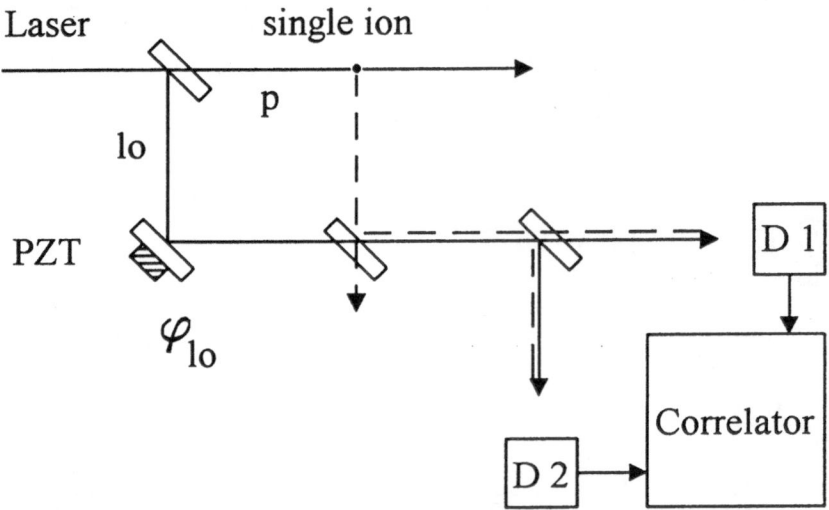

Figure 4: Experimental scheme for homodyne correlation measurements. The laser beam is split into pump beam (p) and local oscillator (lo) which is superimposed with the resonance fluorescence of the single trapped ion. One of the superimposed beams is analysed by an intensity correlation measurement with the detectors D_1, D_2 and the correlator.

where $G_i^{(2)}(\tau)$ denotes the ith order contribution with respect to the local oscillator field strength. When the value of the correlation function at short times is below the steady state value, i.e.

$$\Gamma^{(2)} = G^{(2)}(0) - G^{(2)}(\infty) < 0, \qquad (3)$$

the superimposed light shows a non–classical behavior, it exhibits a sub–Poissonian counting statistics. According to Eq. (2) various contributions to the overall non–classical effect appear. The zeroth order (in the local oscillator field) represents the intensity correlation of the fluorescence as studied in Section 2 and reveals the sub–Poisson nature of the fluorescence light. The first order contribution is of the type

$$\Gamma_1^{(2)} = (\mu^2/2)\mathcal{E}_{lo}\{\exp(-i\varphi_{lo})[\langle \hat{E}_{fl}^{(-)} \hat{E}_{fl}^{(+)} \hat{E}_{fl}^{(+)}\rangle - \langle \hat{E}_{fl}^{(-)} \hat{E}_{fl}^{(+)}\rangle\langle \hat{E}_{fl}^{(+)}\rangle] + \text{c.c.}\}, \qquad (4)$$

where μ is the product of detector efficiency and measurement time, $\mathcal{E}_{lo}$ is the local oscillator amplitude, φ_{lo} is the phase of the local oscillator and $\hat{E}_{fl}^{(\pm)}$ is the positive/negative frequency part of the fluorescence field (with $\hat{E} =$

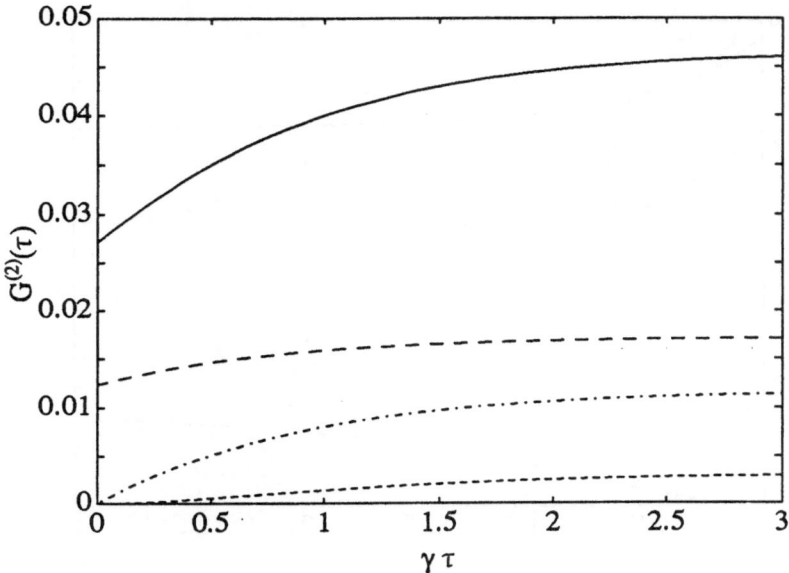

Figure 5: Correlation function $G^{(2)}(\tau)$ as a function of the delay time τ for a two–level atom, $2\gamma = \Gamma$ denotes the decay of the excited state population. The solid line represents the sum term, the long dashes show the squeezing term $G_2^{(2)}(\tau)$, the dashed dotted line represents the anomalous term $G_1^{(2)}(\tau)$ and the short dashes indicate the antibunching term $G_0^{(2)}(\tau)$. The parameters are: Rabi frequency $\Omega = 0.5 \Gamma$, detuning $\Delta = 0$, local oscillator phase $\varphi_{lo} = \pi/2$, and relative intensity $I_{rel} = I_{lo}/I_{fl} = 1$.

$\hat{E}^{(+)} + \hat{E}^{(-)}$). Since in usual photo–detection equal numbers of positive and negative frequency parts appear this contribution represents an anomalous term. The second–order term

$$\Gamma_2^{(2)} = (\mu^2/4)\mathcal{E}_{lo}^2 \langle : [\Delta \hat{E}_{fl}(\varphi_{lo})]^2 : \rangle, \tag{5}$$

contains the information about the squeezing property of the fluorescence light, which is defined by the condition that the normally ordered variance of the fluorescence field strength becomes negative, i.e.

$$\langle : [\Delta \hat{E}_{fl}(\varphi_{lo})]^2 : \rangle < 0, \tag{6}$$

with an appropriately chosen phase φ_{lo}. Note that the higher order contributions $\Gamma_i^{(2)}(i = 3, 4)$ vanish.

For a two–level atom, the intensity correlation $G^{(2)}(\tau)$ and the contributions $G_0^{(2)}$, $G_1^{(2)}$ and $G_2^{(2)}$ representing sub-Poisson statistics, anomalous

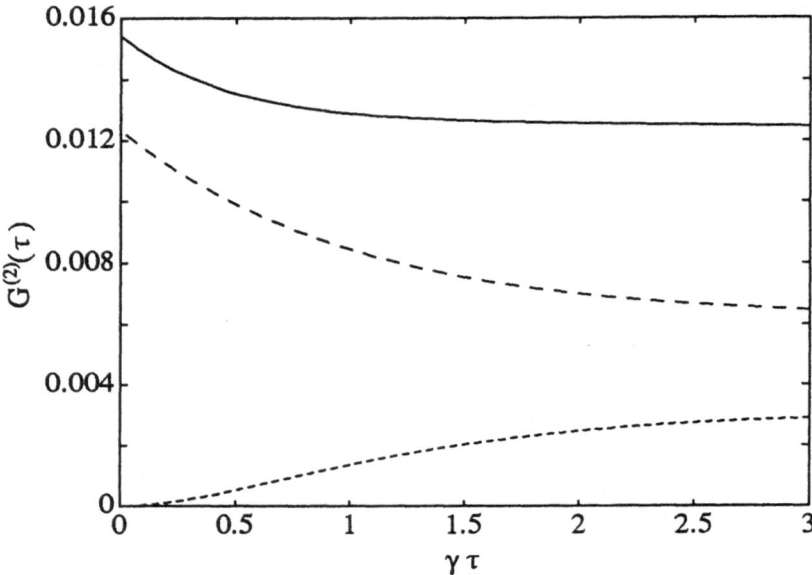

Figure 6: Correlation function $G^{(2)}(\tau)$ as a function of the delay time τ for a two–level atom, $2\gamma = \Gamma$ denotes the decay of the excited state population. The solid line represents the sum term, the long dashes show the squeezing term $G_2^{(2)}(\tau)$, the anomalous term vanishes and the short dashes indicate the antibunching term $G_0^{(2)}(\tau)$. The parameters are: Rabi frequency $\Omega = 0.5\,\Gamma$, detuning $\Delta = 0$, local oscillator phase $\varphi_{lo} = \pi$, and relative intensity $I_{rel} = I_{lo}/I_{fl} = 1$.

terms and squeezing, respectively, are shown in Fig. 5. It is seen, that for the chosen phase of the local oscillator all the terms contribute to the overall non–classical effect. Shifting the phase by $\pi/2$ (cf. Fig. 6), of course does not change the sub–Poisson term which is independent of the local oscillator phase. The anomalous term vanishes due to its periodicity given in Eq. (4). The squeezing term which has a periodicity of π with respect to φ_{lo} changes the sign of its initial slope. Note that the squeezing terms in Fig. 5 and Fig. 6 represent the field noise in the two field quadratures.

Figure 7 represents the observable non–classical contributions $\Gamma_i^{(2)}, i = 0, 1, 2$ for a two–level atom as a function of the detuning for equal intensities of fluorescence and local oscillator. In this figure the local–oscillator phases are adjusted to the detuning in such a way that the non–classical effects are maximum. It is seen that the non–classical character of the squeezing term and the anomalous term increases with detuning since both effects require a

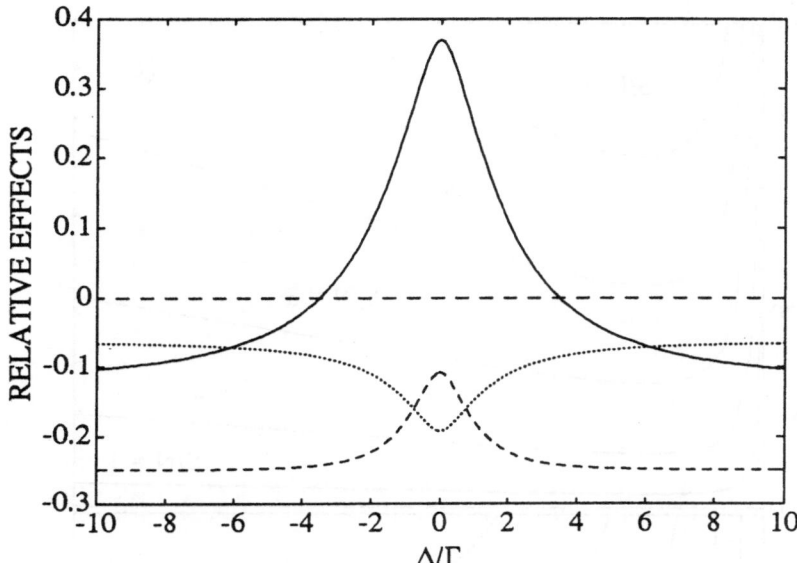

Figure 7: Relative effects as a function of the detuning Δ/Γ, Γ denotes the decay of the excited state population. The solid curve shows the squeezing term, the dotted curve represents the antibunching term and the dashed line indicates the anomalous effect. The parameters are: Rabi frequency $\Omega = 5\,\Gamma$, relative intensity $I_{rel} = I_{lo}/I_{fl} = 1$.

significant amount of coherence of the fluorescence.

In order to investigate the feasibility of such a homodyne correlation measurement, we estimate the achievable signal–to–noise ratio and the required measurement time. The total number of detected events N_c in a correlation measurement is related to the individual counting rates R_1 and R_2 of the two detectors, the total measurement time T and the time resolution ΔT of the correlator by

$$N_c = R_1 R_2 T \Delta T. \tag{7}$$

Assuming realistic counting rates $R_1 = R_2 = 10^4 \text{ s}^{-1}$ and $\Delta T = 10^{-9}$ s, we obtain the required measurement times as shown in Fig. 8 (a) and Fig. 8 (b) for the anomalous term and the squeezing term, respectively, for various signal–to–noise ratios S/N.

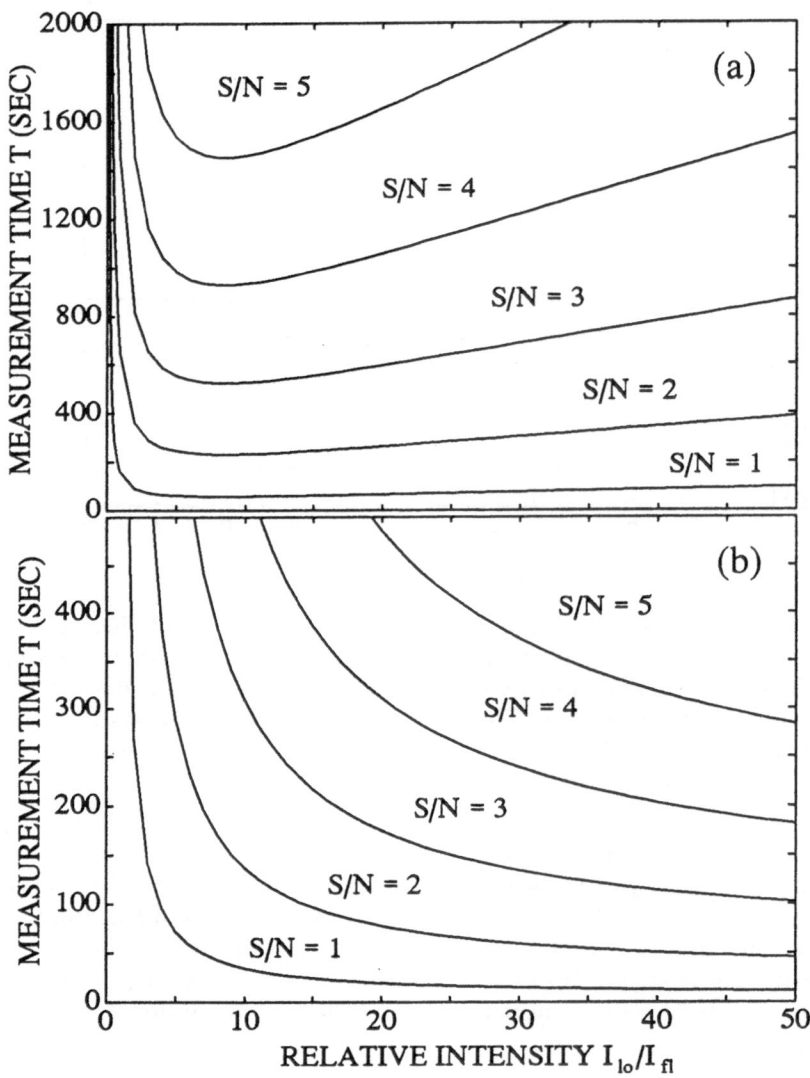

Figure 8: Measurement times required to reach the indicated signal–to–noise ratio S/N for the anomalous term (a) and the squeezing term (b) from a two-level atom as a function of the relative intensity $I_{rel} = I_{lo}/I_{fl}$. The parameters are: Rabi frequency $\Omega = 5\ \Gamma$, detuning $\Delta = -10\ \Gamma$, and local oscillator phase adjusted for maximum non–classical effects.

4 Conclusions

Intensity correlations observed in the fluorescence of single trapped ions are found to be in good agreement with calculations. Thus it can be expected that for the proposed measurements of homodyne correlations of the resonance fluorescence a quantitative comparison between experiment and theory should be possible. Both for the squeezing and the anomalous terms, the size of the effects of a few percent together with the required measurement time to achieve a sufficient signal–to–noise ratio is encouraging for performing homodyne correlation experiments in the resonance fluorescence of a single trapped ion.

Acknowledgements

We acknowledge continous support by Prof. P. E. Toschek. This research was supported by the Deutsche Forschungsgemeinschaft.

References

[1] H. J. Kimble, M. Dagenais, and L. Mandel, Phys. Rev. Lett. **39**, 691 (1977).

[2] F. Diedrich and H. Walther, Phys. Rev. Lett. **58**, 203 (1987).

[3] M. Schubert, I. Siemers, R. Blatt, W. Neuhauser, and P. E. Toschek, Phys. Rev. Lett. **68**, 3016 (1992).

[4] R. E. Slusher, L. W. Hollberg, B. Yurke, J. C. Mertz, and J. F. Valley, Phys. Rev. Lett. **55**, 2409 (1985).

[5] L.-A. Wu, H. J. Kimble, J. L. Hall, and H. Wu, Phys. Rev. Lett. **57**, 2520 (1986).

[6] D. F. Walls and P. Zoller, Phys. Rev. Lett. **47**, 709 (1981).

[7] W. Vogel and R. Blatt, Phys. Rev. A **45**, 3319 (1992).

[8] L. Mandel, Phys. Rev. Lett. **49**, 136 (1982).

[9] W. Vogel, Phys. Rev. Lett. **67**, 2450 (1991).

[10] I. Siemers, M. Schubert, R. Blatt, W. Neuhauser, and P. E. Toschek, Europhys. Lett. **18**, 139 (1992).

Non–classical states of motion in an ion trap

J.I. Cirac[1], R. Blatt[2], A.S. Parkins[3] and P. Zoller[3]

[1]Departamento de Fisica Aplicada, Universidad de Castilla–La Mancha, Campus Universitario s/n, E–13071 Ciudad Real, Spain.
[2]Institut für LaserPhysik, Universität Hamburg, Jungiusstrasse 9, D–2000 Hamburg, Germany.
[3]Joint Institute for Laboratory Astrophysics, and Department of Physics, University of Colorado, Boulder, Co 80309–0440, USA.

Abstract: We show that a single two–level ion trapped in a harmonic potential and laser cooled at the node of a standing wave can be described by the damped Jaynes–Cummings Model, in the appropriate limits. We propose a scheme for the preparation of Fock states of motion of the trapped ion based on the observation of quantum jumps. We also present a scheme for preparing squeezed states of motion based on the multichromatic excitation of the ion.

Keywords: Quantum Optics, Optical cooling of atoms; trapping

1 Introduction

The coupling of a harmonic oscillator with a two–level system represents one of the fundamental models in quantum mechanics. In a cavity QED context, this situation is described by the Jaynes–Cummings Model [1], in which a two–level atom interacts with a single mode of the radiation field in a cavity. In this problem, of particular interest is the *strong coupling* regime, in which the coupling constant g of the two–level system–harmonic oscillator interaction is much larger than the damping rates κ and Γ for the cavity mode and the atom, due to their coupling to independent heat baths. In this regime, interesting phenomena such as vacuum Rabi splitting [2,3] or the collapses and revivals of the population inversion [4,5] have been studied both theoretically [2,4] and experimentally [3,5]. With this success in cavity QED, new perspectives have arisen in the past two or three years, such as, for example, quantum nondemolition measurements of small photon numbers [6] or preparation of Fock states (or nonclassical superpositions of Fock states [7]) of the radiation field.

In this work, we discuss an alternative configuration for the realization of the strong coupling regime in the interaction of a two–level system with a harmonic oscillator, and we analyze two schemes for the preparation of Fock states [8] and squeezed states [9] of the harmonic oscillator. In particular, we consider a single trapped ion constrained to move in a harmonic oscillator potential of frequency ν and undergoing laser cooling at the node of a standing wave. In the ion trap configuration, preparation of nonclassical states of the harmonic oscillator corresponds to preparation of nonclassical states of motion of the trapped ion. Under conditions in which the vibrational amplitude of the ion is much less than the wavelength of the light (Lamb–Dicke limit), and when the trap frequency is much larger than the atomic spontaneous decay rate $\nu \gg \Gamma$ (strong confinement limit) we will show that this problem is mathematically equivalent to the cavity QED problem. However, in the ion trap problem there is negligible damping in the harmonic oscillator (i.e. $\kappa = 0$) and the coupling constant between the two–level system and the harmonic oscillator is proportional to the laser intensity, and therefore can be varied at will, which permits to reach the strong coupling limit by simply increasing the power of the laser field. This is in contrast with the cavity QED problem, where the presence of cavity losses is usually undesirable, and where the coupling constant is determined by atomic constants and cavity characteristics which cannot be modified easily. The significance of our use of the trapped ion configuration is that it demonstrates the potential of the well–established field of ion trapping for investigations of features of the Jaynes–Cummings Model which have thus far been studied only in the context of cavity QED. We also wish to point out that an analogy between an undamped trapped ion and the JCM has been noted by Blockly, Walls, and Risken [10] in the context of collapses and revivals of the population inversion. However, in contrast to their model, in which only traveling–wave laser field was considered, our model is based upon operation at the node of a standing–wave light field, which, as mentioned above, allows for a very direct connection with the JCM.

The scheme for the preparation of Fock states of the harmonic oscillator [8] is based on the observation of quantum jumps from the manifold of dressed levels $|n, \pm\rangle$ of the Jaynes–Cummings Hamiltonian to a third weakly coupled atomic level $|r\rangle$. Observation of the quantum jump in the sense of continuous measurement coincides with the preparation of a Fock state $|n\rangle$ of motion of the trapped ion. We will show that this scheme can also be used in cavity QED for the preparation of Fock states of the radiation field.

For the preparation of squeezed states of motion [9] we consider an ion to be located at a common node of two (different frequency) standing–wave laser fields. We show that, when the beat frequency between the light fields is equal to twice the trap frequency, the *steady state* of the system is a *pure* state, in which the motion of the ion is in a squeezed state and the (internal) two–level system is in its ground level. Consequently, the generation of the squeezed state of motion is indicated by the cessation of the fluorescence emitted by the ion, i.e., by a so–called "dark state" [11].

2 Model

We consider a single ion trapped in a one dimensional harmonic potential of frequency ν. The Hamiltonian for the ion in the trap is

$$H_{tp} = \frac{P^2}{2M} + \frac{1}{2}M\nu^2 R^2,$$ (1)

where R and P are the position and momentum operators, respectively. We will assume that the ion possesses *internal* structure, which can be modelled by a two–level system of ground level $|g\rangle$, excited level $|e\rangle$, and transition frequency ω_0. A complete basis for the whole system is $|n,\alpha\rangle = |n\rangle \otimes |\alpha\rangle$, where $\alpha = g,e$ and $|n\rangle$ $(n = 0,1,\ldots)$ denotes a Fock state of the harmonic oscillator with energy $\hbar\nu(n + \frac{1}{2})$. The ion interacts with laser light and is damped by its coupling to an electromagnetic reservoir. Absorption or emission of laser photons induce transition between these states, $|n,g\rangle \rightarrow |n',e\rangle$ and $|n,e\rangle \rightarrow |n',g\rangle$, respectively. Spontaneous emission induces transitions $|n,e\rangle \rightarrow |n',g\rangle$. Note that the absorption and emission of photons can be accompanied by a change in the quantum oscillator number n due to the recoil. The master equation describing this situation is ($\hbar = 1$)

$$\dot{\rho} = -i\left[\frac{1}{2}\omega_0\sigma_z + \nu a^\dagger a + H_{dip}, \rho\right] + \mathcal{L}\rho.$$ (2)

Here, the first and second terms in the commutator are the free Hamiltonian for the two–level system and the Hamiltonian for the harmonic oscillator, respectively, with $\sigma_z = |e\rangle\langle e| - |g\rangle\langle g|$, and a and $a^\dagger$ being the annihilation and creation operators for the harmonic oscillator, defined, as usual, according to

$$R = \sqrt{\frac{1}{2M\nu}}(a + a^\dagger), \qquad P = i\sqrt{\frac{M\nu}{2}}(a^\dagger - a);$$ (3)

H_{dip} is the Hamiltonian for the dipole interaction between the laser and the ion, and the Liouvillian $\mathcal{L}$ accounts for spontaneous emission, and it is given by

$$\mathcal{L}\rho = \frac{\Gamma}{2}(2\sigma_- \tilde{\rho}\sigma_+ - \sigma_+\sigma_-\rho - \rho\sigma_+\sigma_-).$$ (4)

Here, Γ is the spontaneous decay rate, $\sigma_+ = |e\rangle\langle g|$ and $\sigma_- = |g\rangle\langle g|$ are pseudospin operators for the two–level system, and $\tilde{\rho}$ accounts for the momentum transfer associated with the spontaneous emission of a photon, and is given by

$$\tilde{\rho} = \frac{1}{2}\int_{-1}^{1} du\, W(u)e^{ikRu}\rho\, e^{-ikRu},$$ (5)

where $W(u)$ is the angular distribution of spontaneous emission.

Master equation (2) can not be solved analytically, in general. Here, we will be interested in the regime in which the following conditions are fulfilled:

i) *Lamb–Dicke limit* (LDL): In this limit the ion is assumed to be well localized to a region much smaller than the laser(s) wavelength. Usually, in the LDL one expands master equation (2) in powers of a small parameter $\eta = k_L a_0 \ll 1$, where $k_L = \omega_L/c$ is the laser wavevector and a_0 is the spatial dimension of the ground state $|0\rangle$ of the harmonic Hamiltonian. In particular, the dissipative Liouvillian $\mathcal{L}$ takes the simple form

$$\mathcal{L}\rho = \frac{\Gamma}{2}(2\sigma_-\rho\sigma_+ - \sigma_+\sigma_-\rho - \rho\sigma_+\sigma_-) + o(\Gamma\eta^2), \tag{6}$$

where we have used that $W(u)$ is an even function of u.

ii) *Ion at the node of a standing wave*: We will assume that the ion interacts with laser standing wave fields. From the LDL it follows that the trapped ion is close to a fixed position with respect to the standing–wave field. In particular, we will consider the case in which the ion oscillates around the node of a laser standing wave. According to i) the LDL allows us to expand the electric field in powers of the small parameter η,

$$E = E_0 \sin(k_L R) \cos(\omega_L t) \simeq E_0\eta(a + a^\dagger) \cos(\omega_L t) + o(E_0\eta^3), \tag{7}$$

where we have used (3). The dipole interaction takes the form

$$H_{dip} = \eta\frac{\Omega}{2}(a + a^\dagger)(\sigma_+ e^{-i\omega_L t} + \sigma_- e^{i\omega_L t}) + o(\Omega\eta^3), \tag{8}$$

where Ω is the Rabi frequency. Thus the Hamiltonian part of Eq. (2) results in

$$H = \frac{1}{2}\omega_0\sigma_z + \nu a^\dagger a + \eta\frac{\Omega}{2}(a + a^\dagger)(\sigma_+ e^{-i\omega_L t} + \sigma_- e^{i\omega_L t}), \tag{9}$$

In Fig. 1 we have drawn the level scheme of the ion–trap system. The arrows indicate transitions induced by the laser standing wave.

iii) *Strong confinement limit*: This limit assumes that the trap frequency is much larger than the spontaneous decay rate ($\nu \gg \Gamma$). This fact allow us to perform further approximations in the dipole interaction. To do this, let us transform to an interaction picture defined by the unitary operator

$$U = e^{-i\frac{1}{2}\sigma_z\omega_0 t}e^{-ia^\dagger a\nu t}. \tag{10}$$

In this new picture the Hamiltonian becomes

$$\hat{H} \equiv U^\dagger H U + i\dot{U}^\dagger U = \eta\frac{\Omega}{2}[\sigma_+ a e^{-i(\omega_L-\omega_0-\nu)t} + \sigma_+ a^\dagger e^{-i(\omega_L-\omega_0+\nu)t} + h.c.]. \tag{11}$$

In the following, we will consider two situations:

160

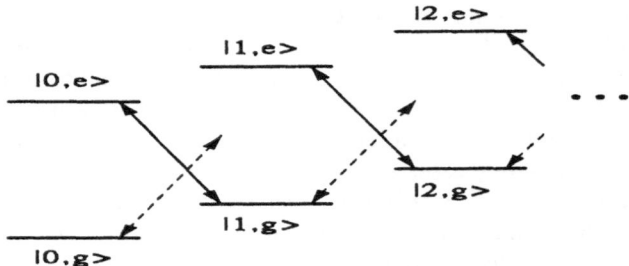

Figure 1: Level scheme for the ion–trap system. The laser–atom detuning is $\omega_L - \omega_0 = \nu$. Arrows pointing to the left indicate the resonant laser coupling with strength $\eta\Omega$; arrows pointing to the right correspond to non–rotating–wave terms.

a) $\omega_L = \omega_0 + \nu$: In this case, the second term of the Hamiltonian $\hat{H}$ oscillates with frequency 2ν. In the strong confinement limit, this oscillations will be much faster than any other dynamical evolution, and therefore one can neglect the terms proportional to $\sigma_+ a$ and $a^\dagger \sigma_-$ in the Hamiltonian. This corresponds to neglecting the arrows pointing to the right in Fig. 1, i.e., to neglect the off–resonance transitions $|n,g\rangle \leftrightarrow |n+1,e\rangle$. This approximation is equivalent to the usual rotating wave approximation in quantum optics. With this simplification, the Hamiltonian (11) results in

$$\hat{H}_a = \eta\frac{\Omega}{2}(\sigma_+ a + a^\dagger \sigma_-), \tag{12}$$

which is the so–called Jaynes–Cummings Hamiltonian.

b) $\omega_L = \omega_0 - \nu$: In this case, it is the first term of (11) the one that rotates very rapidly, and therefore one can neglect the terms proportional to $\sigma_+ a$ and $a^\dagger \sigma_-$ in $\hat{H}$. In Fig. 1, it amounts to neglecting the arrows that point to the left. Hence, the Hamiltonian (11) becomes

$$\hat{H}_b = \eta\frac{\Omega}{2}(\sigma_+ a^\dagger + a\sigma_-). \tag{13}$$

3 Preparation of Fock states

We consider here that the ion oscillates around the node of a laser standing [8] wave of frequency $\omega_L = \omega_0 + \nu$ [case a) above]. According to the above discussion, the master equation describing this situation is,

$$\dot{\rho} = -i\eta\frac{\Omega}{2}[\sigma_+ a + a^\dagger \sigma_-, \rho] + \frac{\Gamma}{2}(2\sigma_- \rho\sigma_+ - \sigma_+\sigma_-\rho - \rho\sigma_+\sigma_-). \tag{14}$$

This master equation corresponds to the damped Jaynes–Cummings Model. The Jaynes–Cummings Model describes a particular kind of interaction between a two–level system and a harmonic oscillator. This model usually appears in cavity QED problems, where the two–level system is a two–level atom and the harmonic oscillator is a single mode of the radiation field (a cavity mode). A significant difference between the cavity QED and the present problem is that in cavity QED, not only the two–level system but also the harmonic oscillator is damped via cavity losses. Further, in the ion problem, the effective coupling constant between the two–level system and the harmonic oscillator $\eta\Omega/2$ depends on the laser intensity, i.e., can be readily changed (note that an important consequence of having the ion at the node of the standing wave is that increasing the Rabi frequency Ω does not lead to heating of the ion [12]; this does not hold for a traveling–wave light field).

According to master equation (14), in steady state the ion remains in the *pure* state $|0, g\rangle$, i.e., both the internal and external degrees of freedom of the ion end up in the ground state. This cooling mechanism based on the Lamb–Dicke limit as well as on the strong confinement limit is the so–called sideband cooling [13]. In it, all the energy from the harmonic oscillator is removed via dissipation by spontaneous emission. Besides, since in steady state the two–level system is in its ground level, there is no fluorescence from the transition $|e\rangle$ to $|g\rangle$.

An interesting regime in the Jaynes–Cummings model is the strong coupling limit, where the coupling between the two–level system and the harmonic oscillator is stronger than the spontaneous decay rate. In this case, interesting features arise as a consequence of the quantum entanglements of both subsystems. In the trapped ion case, this regime corresponds to $\eta\Omega/2 > \Gamma$, which can be easily accomplished by increasing sufficiently the Rabi frequency. This control of the coupling strength is in contrast with current cavity QED experiments where the coupling is determined by atomic constants and cavity characteristics which cannot be modified easily. In the strong coupling regime, the spectroscopy of the system is best described in terms of transition between "dressed states" of the Jaynes–Cummings Hamiltonian H_a. The ground state is $|0, g\rangle$ and the excited dressed states are $|n, \pm\rangle = (|n - 1, e\rangle \pm |n, g\rangle)/\sqrt{2}$ ($n = 1, 2, \ldots$). We wish to point out that $|n\rangle$ are Fock states of the quantized motion, with excitation number $n = 0, 1, \ldots$. The energy corresponding to the excited states are $E_{n,\pm} = \pm\eta\Omega\sqrt{n}/2$, which shows ac Stark splitting proportional to Ω. The splitting between the levels $|n, \pm\rangle$ can be observed by measuring the spectrum of fluorescence corresponding to the two lowest transitions $|1, \pm\rangle \rightarrow |0, g\rangle$, which give rise to a doublet structure, the so–called vacuum Rabi splitting in cavity QED [2,3]. Note that in the conditions described above the ion does not emit light in steady state. Hence, in order to observe the fluorescence spectrum it is necessary to excite the dressed levels. This can be carried out, for example, by using broadband thermal light. Alternatively, one may avoid fulfilling completely the strong

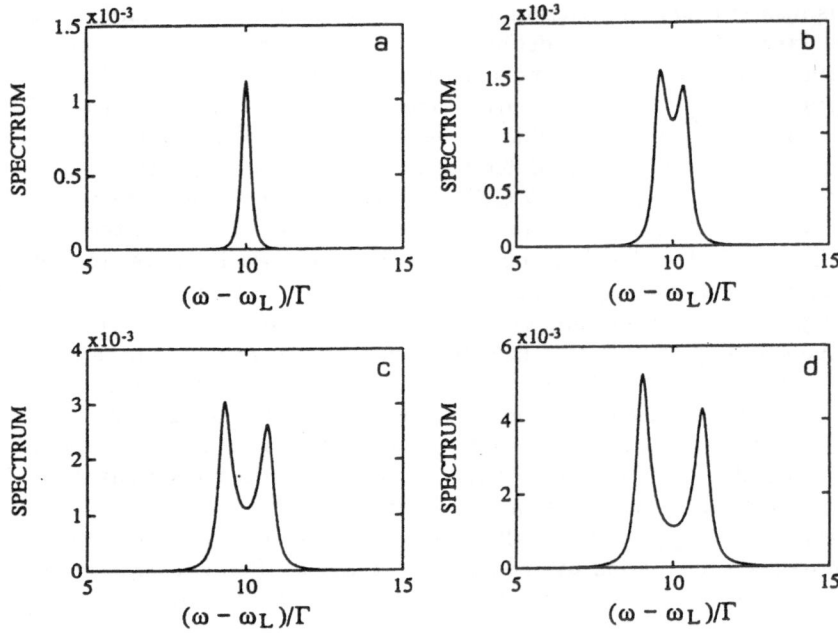

Figure 2: Spectrum of resonance fluorescence of a trapped ion: emission doublet $|1, \pm\rangle \rightarrow |0, g\rangle$ with $\eta = 0.01$, $\nu = 10\Gamma$ and Ω: (a) 50Γ, (b) 100Γ, (c) 150Γ, and (d) 200Γ.

confinement condition that we used to neglect the rapidly oscillating terms (eg, $\sigma_+ a^\dagger$) in (11), since what these non−rotating wave approximation terms do is to excite the system. Fig. 2 shows the spectrum computed from the numerical solution of the exact master equation (2) (using a finite basis set truncation at a suitable level) for $\nu = 10\Gamma$, $\eta = 0.01$ and $\Omega = 50\Gamma, 100\Gamma, 150\Gamma$, and 200Γ [(a), (b), (c) and (d), respectively]. As can be seen from this figure, the splitting between the two sidebands is proportional to Ω. The observation of splitting in the spectrum demonstrates the entanglement between the internal and external degrees of freedom, in analogy with what occurs in cavity QED. The asymmetry observed is caused by the non−rotating−wave−approximation terms, which also lead to enhanced pumping of the levels with increasing Ω.

Aside from this doublet structure, additional resonances, corresponding to transition involving higher excited states, may be observed in the fluorescence spectrum. This is shown in Fig. 3 where we have plotted the fluorescence spectrum derived from the numerical solution to master equation (2), including a broadband thermal noise beam with mean photon number $\overline{N} = 1$. In the figure it is also shown the dressed state level structure and the transitions

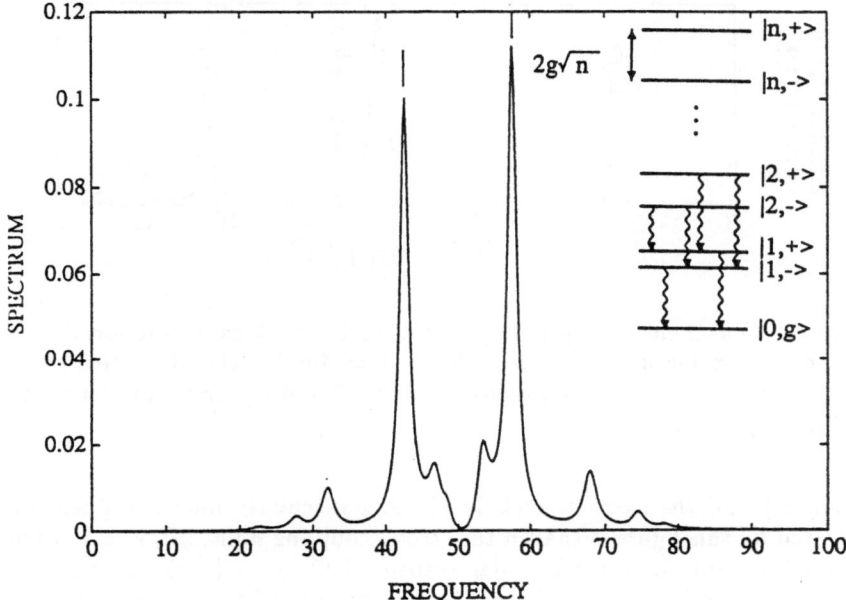

Figure 3: Spectrum of resonance fluorescence of a trapped ion excited by a broadband thermal reservoir of mean photon number $N = 1$. Parameters are: $\eta = 0.01$, $\nu = 50\Gamma$ and $\Omega = 150\Gamma$.

giving rise to the peaks in the spectrum. Similar spectra have been predicted in the context of cavity QED [14].

An alternative approach to the observation of the level structure is a measurement of the probe field absorption spectrum. In order to perform such a measurement without perturbing the cooling and strong coupling one can employ a third atomic level $|r\rangle$ very weakly coupled to the otherwise strongly coupled $|g\rangle - |e\rangle$ transition in "V" configuration, analogous to level schemes used for the observation of quantum jumps , where shelving of the electron in the third level $|r\rangle$ produces a dark period in the observed fluorescence. The absorption spectrum can be very sensitively measured by observing quantum jumps to and from the level $|r\rangle$ as a function of the detuning of the probe laser from the $|g\rangle \rightarrow |e\rangle$ transition frequency.

Figure 4 shows an absorption spectrum for such a three-level system. In order to facilitate excitation of higher levels $|n, \pm\rangle$ in the JCM ladder of states, we have assumed that additional thermal noise is pumping the ion, which could be easily achieved experimentally by adding a weak broadband noise to the laser. The central maximum corresponds to the transition $|0, g\rangle \rightarrow |0, r\rangle$, while the additional maxima correspond to transition between the dressed

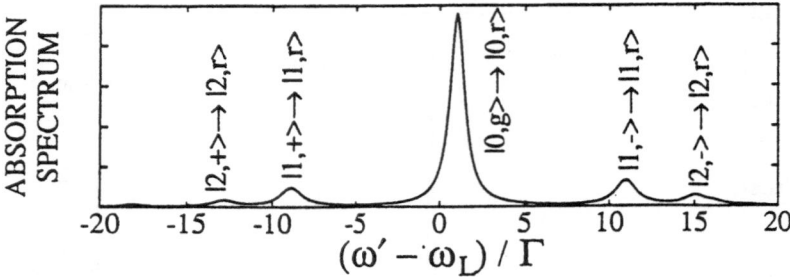

Figure 4: Weak field absorption spectrum to level $|r\rangle$ as a function of the probe detuning for $\eta = 0.05$, $\nu = 50\Gamma$, $\Omega = 400\Gamma$. The Rabi frequency corresponding to the $|g\rangle \leftrightarrow |r\rangle$ transition is 0.1Γ and the spontaneous decay rate from $|r\rangle$ is 0.01Γ.

states $|n, \pm\rangle$ and the excited levels $|n, r\rangle$. A particularly interesting feature illustrated by this figure is that in the strong coupling limit, $\eta\Omega/2 \gg \Gamma$, each spectral line represents a particular transition $|0, g\rangle \rightarrow |n, r\rangle$ or $|n, \pm\rangle \rightarrow |n, r\rangle$. This is consequence of the unequal spacing of the energy levels in the JCM, which means that at the frequency corresponding to a particular spectral line the probe laser is only resonant with the transition frequency between single pair of levels, and thus will only excite the system to a single state $|n, r\rangle$.

This ability to selectively excite a particular transition together with the state reduction associated with the observation of quantum jumps, offers the intriguing possibility of generating number states of the quantized trap motion. This follows from the fact that the probe laser exciting transitions to states $|n, r\rangle$ interacts only with the atomic ground state contribution to the particular dressed state $|n, \pm\rangle$ being excited. Given that we are able to distinguish spectroscopically between the different maxima characterizing the absorption spectrum (so that we can identify the dressed state being excited), observation of a quantum jump to the weakly coupled state $|n, r\rangle$ will tell us with certainty that the vibrational state of the ion is $|n\rangle$ and that we have produced a Fock state of the quantized trap motion. An obvious consequence of having the freedom to choose which transition is excited is the ability to choose the Fock state that is to be produce. This is in contrast to other schemes for the preparation of Fock states, where the Fock state to be produced is unknown (other method to produce selected Fock states based on the adiabatic passage of atoms through a cavity is given in Ref. 7).

Figure 5 shows the simulation of an experiment with a trapped three-level ion and the calculated fluorescence intensity in the strongly coupled two–level transition as a function of time. For this figure, we have taken the weak laser field on resonance with the transition $|2, -\rangle \rightarrow |2, r\rangle$ and we have also included a broadband field of mean photon number $N = 2$ in order to

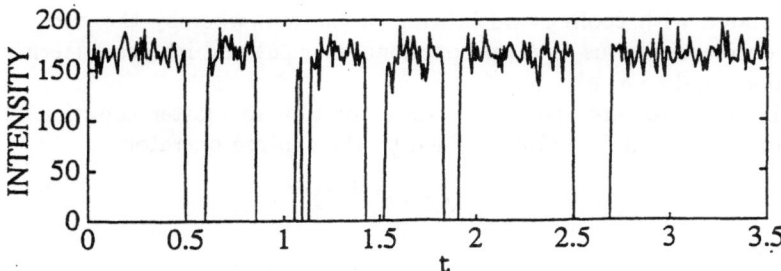

Figure 5: Simulation of quantum jumps as a function of time. The probe laser is tuned to the $|2, -\rangle \rightarrow |2, r\rangle$ transition.

excite the dressed levels. Quantum jumps to the state $|2, r\rangle$ are indicated when emission windows appear in the fluorescence intensity. Thus, a Fock state of the trap motion with $n = 2$ is prepared during these dark periods.

We would like to point out that all of these predictions are of equal relevance in the field of cavity QED, where Fock states of the radiation field could be produced. In this case one could send three–level atoms through the cavity, where they also interact with a weak laser field on resonance with the desired transition to produce the selected Fock state. Measurement of an atom in the level $|r\rangle$ at the output of the cavity would project the state of the cavity mode onto the selected Fock state, i.e., a state with n photons would be prepared.

4 Preparation of squeezed states

For the preparation of a squeezed state of motion [9] we consider here that the ion oscillates around a common node of two–standing waves: one of frequency $\omega_{L1} = \omega_0 - \nu$ and other of frequency $\omega_{L2} = \omega_0 + \nu$. The master equation which describes this situation is

$$\dot{\hat{\rho}} = -i[\hat{H}_a + \hat{H}_b, \hat{\rho}] + \mathcal{L}\hat{\rho}, \tag{15}$$

where

$$\begin{aligned}
\hat{H}_a &= g_1 \sigma_+ a + g_1^* a^\dagger \sigma_-, \\
\hat{H}_b &= g_2 \sigma_+ a^\dagger + g_2^* a \sigma_-,
\end{aligned} \tag{16}$$

and $\mathcal{L}$ is given by (6). Here, we have defined $g_{1,2} = \eta_{1,2}\Omega_{1,2}e^{-i\phi_{1,2}}/2$, with η_1 (η_2), Ω_1 (Ω_2) and ϕ_1 (ϕ_2) being the small Lamb–Dicke parameter, the Rabi frequency and the phase corresponding to the interaction of the ion with the first (second) laser, respectively. The coupling Hamiltonians $\hat{H}_a$ and $\hat{H}_b$ are associated with cooling and heating of the ion, whereby the energy of the ion decreases (terms of H_a proportional to $\sigma_+ a$) and increases (terms of H_b proportional to $\sigma_+ a^\dagger$).

In order to find the steady state solution to master equation (15) we perform the transformation defined by the squeeze operator

$$S(\epsilon) = e^{\epsilon^* a^2 - \epsilon a^{\dagger 2}}, \qquad (17)$$

where

$$\epsilon = r e^{i(\phi_1 - \phi_2)}, \qquad \tanh(r) = |g_2/g_1| < 1. \qquad (18)$$

(Note that in order for this transformation to be well defined $|g_1| > |g_2|$). Using the well known properties of the squeeze operator

$$\begin{aligned} S(\epsilon)^\dagger a S(\epsilon) &= a \cosh(r) - a^\dagger \sinh(r) e^{i(\phi_1 - \phi_2)}, \\ S(\epsilon)^\dagger a^\dagger S(\epsilon) &= a^\dagger \cosh(r) - a \sinh(r) e^{-i(\phi_1 - \phi_2)}, \end{aligned} \qquad (19)$$

it can be easily shown that the master equation for the density operator $\rho = S(\epsilon)^\dagger \rho S(\epsilon)$ is exactly (14), with $\Omega = |g_1| \cosh(r) - |g_2| \sinh(r)$. According to the discussion after (14) we have that in steady state, the ion will be in the state $|0, g\rangle$. Inverting the unitary transformation (17) we have that the steady state will be

$$|\epsilon, g\rangle = S(\epsilon)|0\rangle \otimes |g\rangle. \qquad (20)$$

Note that $S(\epsilon)|0\rangle$ is a squeezed vacuum state, i.e., in steady state this configuration yields a *pure* state given by the product of a squeezed state of the quantized motion with the ground internal state $|g\rangle$ of the ion. As we had in the case of an ion at the node of *one* standing wave, in this case the ion ends up in a state where it does not emit photons. Besides, in this state it can be readily checked that $(H_a + H_b)|\epsilon, g\rangle = 0$, i.e. ion does not see the laser once it has reached this state. In consequence, the generation of a squeezed state coincide with a dark state for the emitted fluorescence. An analogous dark state, coinciding with the production of a coherent field state, occurs in cavity QED when an external driven two–level atom is coupled to a lossless cavity mode [15].

Note that in order for the Lamb–Dicke limit to hold it is necessary that the mean quantum number $\langle a^\dagger a \rangle$ be not too large. Note also that if the frequency of the second laser differs from $\omega_0 - \nu$, the resulting Hamiltonian H_b is time–dependent, i.e., these results are not obtained. [1]

The important effect of squeezing is a reduction in the variance of one quadrature phase, which in our case corresponds to a reduction in the uncertainty of either the position or the momentum operators of the ion. In particular, defining the quadrature phase operators as usual

$$a_1 = \frac{1}{2}(a e^{-i\theta/2} + a^\dagger e^{i\theta/2}), \qquad a_2 = \frac{1}{2i}(a e^{-i\theta/2} - a^\dagger e^{i\theta/2}), \qquad (21)$$

---

[1] In general, it can be shown [9] that in order for the steady state to be $|\epsilon, g\rangle$ it is necessary for both lasers to have a frequency which difference coincides with 2ν.

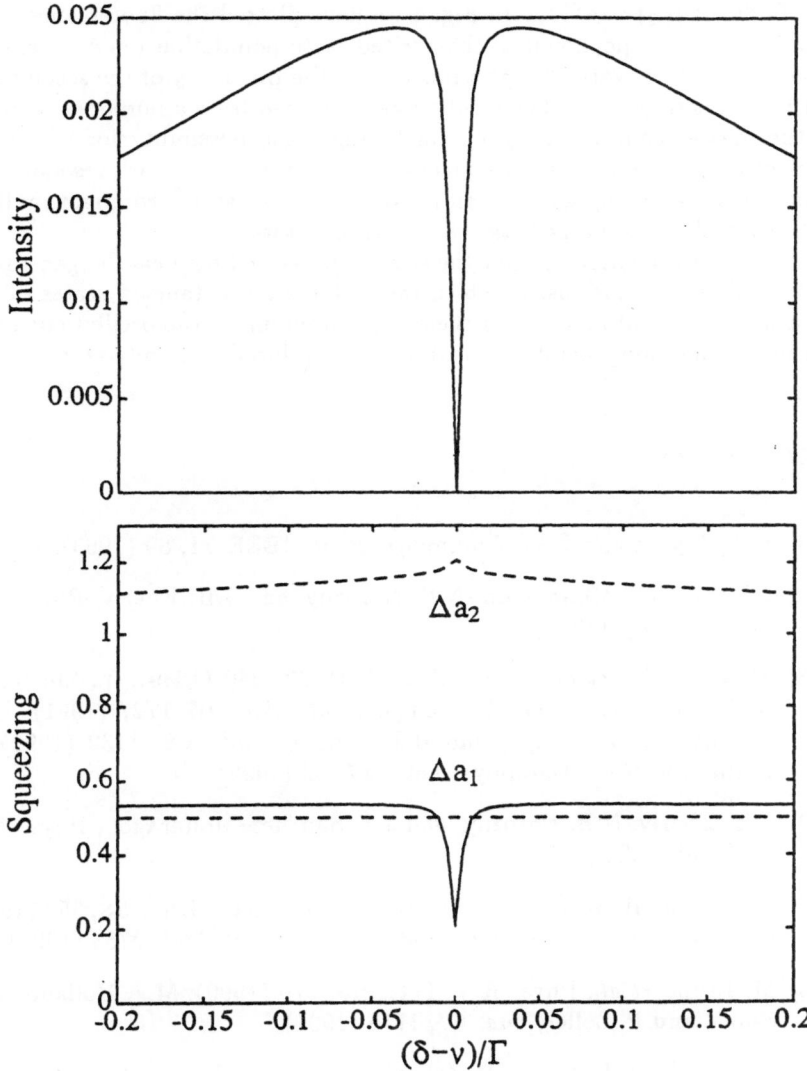

Figure 6: Fluorescence intensity and quadrature phase variances as a function of the frequency of the second laser standing wave (relative to $\omega_0 - \nu$). The dot–dashed line indicates the standard quantum limit.

one finds $(\Delta a_{1,2}) = e^{\mp r}/2$. In Fig. 6 we have plotted the fluorescence intensity (which is proportional to the excited state population $\langle \sigma_+ \sigma_- \rangle$) and the quadrature phase variances as a function of the frequency of the second laser with respect to $\omega_0 - \nu$. The results were obtained from a numerical solution of the master equation (15), but including a the possibility for the second laser of having an arbitrary frequency. At $\omega_{L2} = \omega_0 - \nu$ a dark resonance occurs, accompanied by a reduction in Δa_1 below the standard quantum limit, indicating that a squeezed state has been generated.

It is worth mentioning that the scheme presented here can be generalized to 2 or three dimensions, in which case interesting entangled states of 2 or 3 harmonic oscillators (each of them corresponding to the oscillations of the ion in 2 or 3 orthogonal directions) can be produced in steady state.

5 References

[1] E.T. Jaynes and F.W. Cummings, Proc. IEEE **51**, 89 (1963).

[2] J.J. Sanchez–Mondragon, N.B. Narozny, and J.H. Eberly, Phys. Rev. Lett. **51**, 550 (1983).

[3] M.G. Raizen *et al.*, Phys. Rev. Lett **63**, 240 (1989); Y. Zhu *et al.*, *ibid.* **64**, 2499 (1990); G. Rempe *et al.*, *ibid.* **67** 1727 (1991); R. J. Thompson, G. Rempe, and H.J. Kimble, *ibid.* **68**, 1132 (1992); F. Bernardot *et al.*, Europhys. Lett. **17**, 33 (1992).

[4] J.H. Eberly, N.B. Narozny, and J.J. Sanchez–Mondragon, Phys. Rev. Lett. **44**, 1323 (1980).

[5] G. Rempe, H. Walther, and N. Klein, Phys. Rev. Lett. **58**, 353 (1987); G. Rempe, F. Schmidtkaler, and H. Walther, *ibid* **64**, 2783 (1990).

[6] M. Brune *et al.*, Phys. Rev. Lett. **65**, 976 (1990); M.J. Holland, D.F. Walls, and P. Zoller, *ibid.* **67**, 1716 (1991).

[7] A.S. Parkins, P. Marte, P. Zoller, and H.J. Kimble, in this volume.

[8] J.I. Cirac *et al.*, Phys. Rev. Lett. **70**, 762 (1993).

[9] J.I. Cirac *et al.*, Phys. Rev. Lett. **70**, 556 (1993).

[10] C.A. Blockley, D.F. Walls, and H. Risken, Europhys. Lett. **17**, 509 (1992).

[11] G. Alzetta *et al.*, Nuovo Cimento Soc. Ital. Fis. **36B**, 5 (1976).

[12] J.I. Cirac *et al.*, Phys. Rev. A **46**, 2668 (1992).

169

[13] See, for example, S. Stenholm, Rev. Mod. Phys. **58**, 699 (1986) and references therein; F. Diedrich *et al.*, Phys. Rev. Lett. **62**, 403 (1989).

[14] J.I. Cirac, H. Ristch, and P. Zoller, Phys. Rev. A **44**, 4541 (1991).

[15] P.M. Alsing, D.A. Cardimona, and H.J. Carmichael, Phys. Rev. A **45**, 1793 (1992).

Limits in preparation of coherent population trapping states

E. Arimondo

Dipartimento di Fisica, Università di Pisa, Piazza Torricelli 2, 56216 Pisa, Italy

Abstract. The requirements to be imposed on the relaxation processes of atomic system for preparing atoms in coherent trapping superpositions, are discussed within the basis of coupled and noncoupled states.

1. Introduction

Coherent population trapping (CPT) represents the preparation of an atomic (or molecular) system, through the interaction with laser radiation, in a coherent superposition of states[1]. That coherent superposition of states results not coupled to the radiation field itself, with absorption (or emission) of the laser radiation not taking place because of the destructive interference in the absorption transition amplitudes. The states involved in the coherent superposition could be either the lower states, with the CPT wavefunction stable against spontaneous emission, or the excited state, with the CPT wavefunction decaying through spontaneous emission processes.

The applications of coherent trapping superpositions to different contexts of laser spectroscopy and quantum optics are quite spread, for instance high resolution spectroscopy[2,3], frequency stabilization[4], optical bistability[5], laser cooling[6], laser without inversion[7-9]. In all these applications a key factor is the amount of CPT prepared in the atomic system by the incident laser radiation. The laser detunings; the relaxation processes, the laser frequency fluctuations are between the factors limitinmg the preparation of atoms in CPT states. The laser detunings determine the frequency dependence of the CPT resonance. The laser frequency fluctuations are usually taken in care using a single laser source to drive both transitions of the three-level system. For what concerns the realaxation rates of the atomic coherence, most experimental investigations have been performed in conditions such that the interaction time with the laser radiation was the factor limiting CPT preparation.

However the damping rate of the coherent superposition of states will be the factor limiting the amount of CPT preparation by the incident laser radiation, in any experiment involving a very long interaction time. The experiments where the relaxation rate of the coherences have determined the final CPT result, have not being able to test its role on the CPT preparation, either because of the experimental difficulties, or because of the incompleteness of the theory.

The question of the limiting factors on the CPT preparation becomes very relevant in the extensions of laser cooling based on velocity-selective CPT. In the ^{4}He laser cooling experiment dealing with a Λ configuration[6], where the CPT state was an eigenstate of the kinetic energy, the CPT preparation was limited by the interaction time. When CPT laser cooling on higher angular momentum states is considered, the CPT state results to be not an eigenstate of the kinetic energy, and lasts only for a limited time[10]. For heavier atoms, this time is long enough to have an efficient preparation of CPT laser cooling. However for experiments exploring the limits of velocity-selective CPT laser cooling the role played by the relaxation should be carefully explored. From the theoretical point of view, the CPT Hamiltonian evolution can be framed in a very elegant way when the relaxation processes are neglected, except for the spontaneous emission decay required for the preparation in the CPT ground state superposition[11] The effect of fluctuations in the frequency of the incident laser radiation has been studied in refs.12 and the role played by the atomic coherence damping rate has been analyzed in refs.13.

In this contribution the role of atomic coherence relaxation rate in limiting the final CPT preparation is discussed within the basis of the coupled and noncoupled states, which results very useful in describing all the CPT behaviours. The role of other characteristics of both the atomic system and the laser radiation on the CPT preparation is also discussed and compared to published experimental investigations. The role of the relaxation rates on the laser cooling scheme based on CPT is investigated. Finally the propagation of radiation through a medium prepared in CPT states is examined. The analysis is focused on the Λ configuration in the steady state, because in such conditions CPT has been extensively explored experimentally. Thus the analysis here presented does not apply to the recent CPT experiments on upper states, where the transient evolution of the CPT states was explored by measuring the correlation function[14].

2. Density-matrix equations

The evolution of the three-level density matrix of an atomic system interacting with resonant laser radiation in the scheme of Fig.1a, with two laser fields acting on the $|1> \rightarrow |0>$ and $|2> \rightarrow |0>$ transitions, is investigated. The equation for the time evolution of the density matrix ρ subject to an atomic Hamiltonian and an Hamiltonian for the atomic interaction with the laser fields is:

$$i\hbar \, d\rho/dt = [\, H_{at} + V_{AL}\,, \rho\,] \tag{1}$$

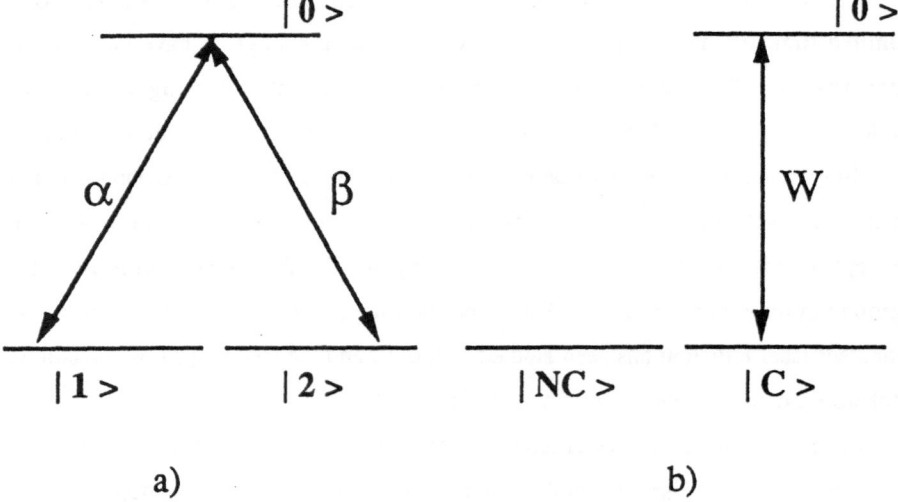

Fig. 1 Energy three-level scheme in the atomic basis, in a), and in the coupled and noncoupled basis, in b).

where

$$H_{at} = E_1 \,|1><1| + E_2 \,|2><2| + E_0 \,|0><0| \tag{2}$$

E_i (i=0,1,2) being the energy of the atomic state $|i>$, with E_0-E_1=$\hbar\omega_{01}$ and E_0-E_2 =$\hbar\omega_{02}$. The electromagnetic field is supposed to be composed by the following two modes at frequencies Ω_1 and Ω_2:

$$\vec{E} = \frac{1}{2}(\vec{E}_1 \exp(-i\Omega_1 t + ik_1 z) + \vec{E}_2 \exp(-i\Omega_2 t - ik_2 z) + c.c.) \quad (3)$$

Within the RWA the atom-laser interaction Hamiltonian is written as

$$V_{AL} = -\hbar\,\alpha\,|0><1|\,e^{-i\Omega_1 t} - \hbar\,\beta\,|0><2|\,e^{-i\Omega_2 t} + h.c. \quad (4)$$

where the Rabi frequency for each laser wave has been introduced:

$$\alpha = \left|\frac{\mu_{01} E_1}{2\hbar}\right| e^{-i\varphi_1} \quad \text{and} \quad \beta = \left|\frac{\mu_{02} E_2}{2\hbar}\right| e^{-i\varphi_2} \quad (5)$$

with μ_{0i} the dipole matrix element between excited level $|0>$ and ground level $|i>$, and φ_1 and φ_2 describing the phases associated to the propagation part of the radiation field.

The density matrix equations are more easily solved when the slowly varying off-diagonal elements of the density matrix are introduced:

$$\rho_{01} = \sigma_{01} \exp(-i\Omega_1 t) \exp(-i\varphi_1) \quad (6a)$$

$$\rho_{02} = \sigma_{02} \exp(-i\Omega_2 t) \exp(-i\varphi_2) \quad (6b)$$

$$\rho_{21} = \sigma_{21} \exp[-i(\Omega_1 - \Omega_2)t] \exp[-i(\varphi_1 - \varphi_2)] . \quad (6c)$$

The equations for the density matrix elements σ_{01}, σ_{02} and σ_{21} are written:

$$d\sigma_{01}/dt = -\sigma_{01}(\gamma_\perp + i\delta_1) + i|\alpha|(\rho_{11} - \rho_{00}) + i|\beta|\sigma_{21} \quad (7a)$$

$$d\sigma_{02}/dt = -\sigma_{02}(\gamma_\perp + i\delta_2) + i|\beta|(\rho_{22} - \rho_{00}) + i|\alpha|\sigma_{12} \quad (7b)$$

$$d\sigma_{21}/dt = -\sigma_{21}i(\delta_1 - \delta_2) + i|\beta|\sigma_{01} - i|\alpha|\sigma_{20} \quad (7c)$$

with $\delta_i = \Omega_i - \omega_{0i}$, and where we have introduced the transverse damping rate $\gamma_\perp$ for the optical coherences σ_{01} and σ_{02}. The transverse damping rate is determined by the spontaneous emission rate Γ from the upper level $|0>$ and by any other damping process of the optical coherences. The spontaneous emission from upper level $|0>$ contributes to the population evolution given by the following equations:

$$d(\rho_{11} - \rho_{00})dt = \frac{\Gamma}{2}\rho_{00} + 2i|\alpha|\sigma_{10} + i|\beta|\sigma_{20} + c.c. \quad (8a)$$

$$d(\rho_{22} - \rho_{00})dt = \frac{\Gamma}{2}\rho_{00} + i|\alpha|\sigma_{10} + 2i|\beta|\sigma_{20} + c.c. \quad (8b)$$

At this point, for the slowly varying off-diagonal and diagonal elements of the density matrix in Eqs. (7) and (8), the transformation to the following basis of the coupled/noncoupled states is applied[9]:

$$|C> = [|\alpha| \, |1> + |\beta| \, |2>]/W \qquad (9a)$$

$$|NC> = [\beta \, |1> - \alpha \, |2>]/W \qquad (9b)$$

with $W^2 = |\alpha|^2 + |\beta|^2$ proportional to the total intensity of the incident laser radiation. In the coupled and noncoupled basis the density matrix equations assume the following form:

$$\frac{d\sigma_{0,C}}{dt} = -\sigma_{0,C}(\gamma_\perp + i\frac{|\alpha|^2\delta_1 + |\beta|^2\delta_2}{W^2}) - i\frac{|\alpha||\beta|(\delta_1 - \delta_2)}{W^2}\sigma_{0,NC} + iW(\rho_{C,C} - \rho_{00})$$

$$\frac{d\sigma_{0,NC}}{dt} = -\sigma_{0,NC}(\gamma_\perp + i\frac{|\alpha|^2\delta_1 + |\beta|^2\delta_2}{W^2}) - i\frac{|\alpha||\beta|(\delta_1 - \delta_2)}{W^2}\sigma_{0,C}$$

$$\frac{d\rho_{C,C}}{dt} = \frac{\Gamma}{2}\rho_{00} + i\frac{|\alpha||\beta|(\delta_1 - \delta_2)}{W^2}(\rho_{NC,C} - \rho_{C,NC}) + i\,W(\sigma_{C,0} - \sigma_{0,C}) \qquad (10)$$

$$\frac{d\rho_{NC,NC}}{dt} = \frac{\Gamma}{2}\rho_{00} - i\frac{|\alpha||\beta|(\delta_1 - \delta_2)}{W^2}(\rho_{NC,C} - \rho_{C,NC})$$

$$\frac{d\rho_{C,NC}}{dt} = i\frac{|\alpha||\beta|(\delta_1 - \delta_2)}{W^2}\rho_{C,NC} + iW\sigma_{0,NC} - i\frac{|\alpha||\beta|(\delta_1 - \delta_2)}{W^2}(\rho_{NC,C} - \rho_{C,NC})$$

It may be noticed that in these equations the optical coherences depend on the population difference between coupled state and upper state and are decoupled from the low frequency coherence between coupled and noncoupled states. The $|NC>$ state is an eigenstate of the Hamiltonian $H_{at} + V_{AL}$ at the two-photon resonance, $\delta_1 = \delta_2$. In such condition, the solution for the density matrix equations becomes more straightforward with the population of the $|NC>$ state filled up through optical pumping, so that at the steady state $\rho_{NC,NC} = 1$. In the case $\delta_1 \neq \delta_2$ the use of the coupled/noncoupled basis for the slowly varying density matrix elements introduces a simplification. Hoewer the population on the noncoupled state depends on the low-frequency coherence, thus on other terms of the density matrix. Thus the complete pumping to the noncoupled state is not realized.

Relaxation terms for the ground state elements complete the above equations for the density matrix elements. An investigation of the relaxation processes affecting an atomic system prepared into coupled/noncoupled states was never performed. For such preparation it may be supposed that collisions, or any external perturbation, may produce either a disentangle of the wavefunction in the atomic states |1> and |2>, or introduce a random shift for the atomic states composing the superposition. For the case $\alpha=\beta$, the process of disentangle results described through a rate γ_p in the following equations:

$$\frac{d\rho_{C,C}}{dt} = -\frac{d\rho_{NC,NC}}{dt} = -\frac{\gamma_p}{2}(\rho_{C,C}-\rho_{NC,NC}) \tag{11a}$$

$$\frac{d\rho_{C,NC}}{dt} = -\gamma_p\,\rho_{C,NC} \tag{11b}$$

The random phase process leads to a damping of the atomic coherences with rate γ_c:

$$\frac{d\sigma_{21}}{dt} = -\gamma_c\,\sigma_{21} \tag{12}$$

so that the following rate equations are obtained in the coupled/noncoupled basis:

$$\frac{d\rho_{C,C}}{dt} = -\frac{d\rho_{NC,NC}}{dt} = -\frac{\gamma_c}{2}(\rho_{C,C}-\rho_{NC,NC}) \tag{13a}$$

$$\frac{d\rho_{C,NC}}{dt} = -\frac{d\rho_{NC,N}}{dt} = -\frac{\gamma_c}{2}(\rho_{C,NC}-\rho_{NC,C}) \tag{13b}$$

The total relaxation of the ground state density matrix is obtained by combining the contributions of Eqs. (10), (11) and (13).

It must be examined if the rates γ_p and γ_c depend on the spacing $\hbar\alpha$ and $\hbar\beta$ between the energy levels of the atom dressed by the photons at frequencies Ω_1 and Ω_2. In effect both relaxation processes introduced above, require an energy around $\hbar\alpha$ and $\hbar\beta$ to be exchanged in the collisional processes. For the typical values of the Rabi frequencies ($\alpha \cong \beta \cong \Gamma$) the energy is small as compared to the normal atomic interaction energy and no dependence of the relaxation processes on α and β is expected. However in a cold collision experiment, the spacing between the dressed atomic energy levels become comparable to the atomic thermal energy, and a dependence of the relaxation rates on the Rabi frequencies must be considered.

3. Density matrix solutions

The combination of Eqs. (10) , (11) and (13) may be solved for the relevant quantities. For instance in the case $\delta_1=\delta_2=0$, the populations of the upper level and of the coupled state are given by:

$$\rho_{00}=\frac{\frac{I}{I_s}}{1+\frac{I}{I_c}+\frac{3I}{I_s}} \qquad\qquad \rho_{C,C}=\frac{\frac{1}{2}+\frac{I}{I_s}}{1+\frac{I}{I_c}+\frac{3I}{I_s}} \qquad (14)$$

where the laser radiation intensity I, defined as $I = |\alpha|^2+|\beta|^2$, has been compared to the saturation of the optical transition, $I_s=\gamma_\perp\Gamma$, and to $I_c=\gamma_\perp(\gamma_c+\gamma_p)$ that has been defined as the coherence saturation intensity[8]. It may be noticed that if the radiation laser is switched off (I=0), there is no population in the upper level, because spontaneous emission repopulates the lower levels. In the case where the ground state relaxation is not present, i.e. $\gamma_p=\gamma_c=0$ and hence $I_c=0$, both populations in the upper level and in the coupled state are zero. In that regime, at the steady state the CPT mechanism operates with complete efficiency and all the population is concentrated in the noncoupled state |NC>.

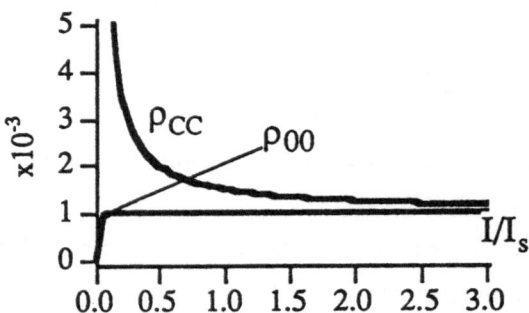

Fig. 2 Plots of the coupled state population $\rho_{C,C}$ and of the upper level population ρ_{00} versus the laser intensity I, in CPT conditions, as given by Eqs. 14, with $I_s = 1000I_c$.

In Fig. 2 the occupation of the upper level |0> and of the lower |C> state is reported as function of the laser intensity I for relaxation parameters appropriate for describing an experimental investigation on an alkali atom. It may be noticed that both occupations of the upper level and of the coupled lower state become $(\gamma_c+\gamma_p)/\Gamma$ for laser radiation intensity comparable to the saturation intensity of the optical transition. That limit value is reached by the upper level population already for a laser intensity comparable to I_c. Such dependence of the occupation probabilities modifies also the propagation of radiation through an atomic medium in CPT conditions, as shown in a following Section.

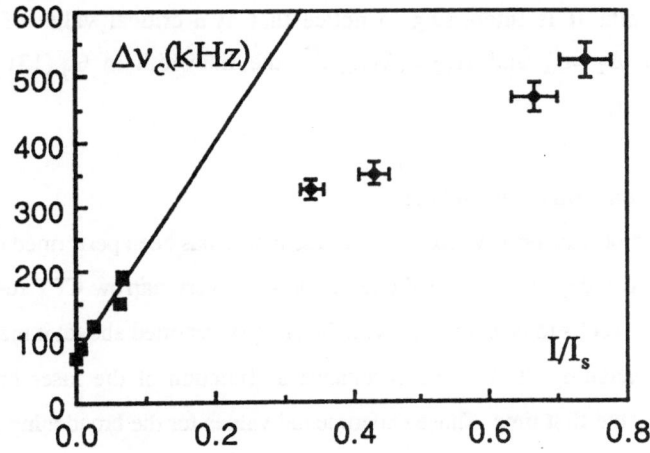

Fig. 3 Linewidth $\Delta \nu_c$ versus laser intensity I, in units of the saturation intensity I_s, for the Rb CPT resonance in the experiment of ref. 3 , and linear fit with the saturation broadening law of Eq.(15a).

From the solution for the density matrix equations, it is possible to derive the expression for the linewidth δ_c of the CPT resonance as function of the parameters introduced in the density matrix equations :

$$\delta_c = (\gamma_c + \gamma_p)\,(1+\frac{I}{I_c}) \qquad (15a)$$

with the coherence saturation intensity that in the general case depends on the laser detunings through:

$$I_c = (\gamma_c + \gamma_p)\, \gamma_\perp \left[\left(1 + \frac{\delta_1^2}{\gamma_\perp^2}\right)^{-1} + \left(1 + \frac{\delta_2^2}{\gamma_\perp^2}\right)^{-1}\right]^{-1} \qquad (15b)$$

The effect of stochastic laser field fluctuations on CPT has been described through a model where the Rabi frequencies are represented as zero mean Gaussian-Markoff processes delta-function correlated[12]. The final result of that model is that the I=0 linewitdh of the CPT resonance is modified and becomes:

$$\delta_c\,(I=0) = (\gamma_c + \gamma_p) + \Delta_1 + \Delta_2 - 2\Delta_{12} \qquad (16)$$

where Δ_1 and Δ_2 are the spectral widths of the laser fields, and Δ_{12} is the cross - spectral width. It is interesting to notice that at a critical value of the mutual correlation, $\Delta_1 = \Delta_2$ and $\Delta_{12} = (\Delta_1 \Delta_2^1)^{1/2}$, the linewidth in Eq.(13) reduces the collisional rate.

4. Comparison with an experiment

A recent and precise observation of CPT resonances has been performed on Rb atoms by using two independent laser diodes to produce very narrow CPT resonances[3]. Between the results to be compared with the analysis reported above, it may be noticed that the broadening of the CPT resonance as function of the laser intensity was observed for the first time. Those experimental values for the broadening are reported in Fig. 3 together with a linear fit dependence on the intensity I. Making use of the saturation intensity for the optical Rb transition ($I_s = 1.62$ mW/cm^2), the comparison with the dependence of the CPT resonance given by Eqs. (15) allows to derive a coherence intensity $I_c = 0.070$ mW/cm^2. The ratio of the saturation intensities allows to derive a ground state relaxation rate in the 200 kHz range; this value is not in agreement with the width of the CPT resonance at I=0 in Fig. 3. This suggests that homogeneous broadening did not produced the total contribution to the linewidth, and some inhomogeneous broadening source was also present. In effect an inhomogeneous broadening dependence leads to a better fit of the experimental points of Fig. 3. The authors of ref. 3 extimated that the limiting linewidth was affected by 10 percent by the flight time of rubidium atoms across the laser beams, and the remaining part by the spectral linewidths of the two diode lasers and optical misalignment.

179

5. Velocity-selective coherent population trapping.

In the laser cooling based on velocity selective coherent population trapping[6], the inclusion of the atomic kinetic energy into the density matrix equations leads to another term modifying the evolution of the noncoupled state. For laser fields E_1 and E_2 copropagating, the coupled/noncoupled states are eigenstates of the kinetic energy and the previous analysis remains substantially valid. The laser cooling is based on the counterpropagating configuration of Eq.(2), with density matrix equations including a kinetic energy term:

$$i\hbar\, d\rho/dt = [\, H_{at} + V_{AL} + p_z^2/2M, \rho\,] \qquad (17)$$

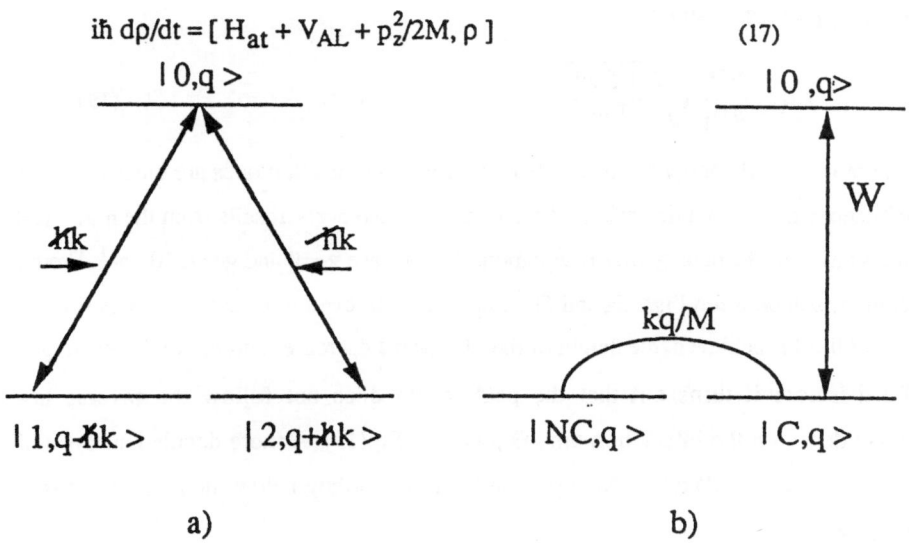

a)   b)

Fig. 4 Atom-laser interaction, with momentum exchanged, in the atomic basis and in the coupled/noncoupled basis

For the case $\alpha=\beta$, $\Omega_2=\Omega_2$ and $k_1=k_2=k$, considering the photon momentum $\pm\hbar k$ exchanged between the laser field and the atomic system (see Fig. 4a), the following basis of coupled/noncoupled states is obtained:

$$|C,q> = [\,|1,q-\hbar k> + |2, q +\hbar k >]/\sqrt{2} \qquad (18a)$$
$$|NC,q> = [\,|1,q-\hbar k> - |2, q +\hbar k >]/\sqrt{2} \qquad (18b)$$

having denoted q the momentum of the excited state $|0,q>$ connected by the

Hamiltonian V_{AL} to the atomic states $|1,q-\hbar k>$ and $|2,q+\hbar k>$. The kinetic energy term in the density matrix equation introduces a coupling term kq/M between the coupled/noncoupled states, as indicated in Fig. 4b.

The density matrix equations have been solved in refs. 6 as function of the interaction time Θ with the laser radiation, and, starting from a broad initial momentum distribution, atoms are concentrated in a distribution with two peaks centered at $\pm\hbar k$, as shown in the momentum distribution for ^{4}He atoms of Fig. 5. The width Δq of the momentum distribution, a parameter describing the efficiency of the cooling process, results:

$$\Delta q = \frac{\hbar W}{4E_R}\sqrt{\frac{1.594}{\Gamma\Theta}} \tag{19}$$

where the recoil energy has been introduced. When relaxation rates are introduced, the efficiency of the cooling process is reduced. Fig. 5b reports results from the numerical integration of the density matrix equations for the case $gp=0$ and $gc=1. \, 10^5$ s^{-1}. From a comparison between Figs. 5a and 5b, it appears that, even if a laser cooling below the recoil limit is achieved, the height of the $\pm\hbar k$ peaks decreases and their width is larger. Furthermore it turns out that the peak width does not follow the inverse law dependence on the interaction time Θ given by Eq.(19). A more detailed analysis of the influence of relaxation processes on the laser cooling below the recoil limit is in progress.

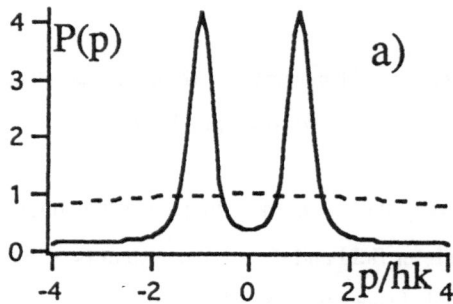

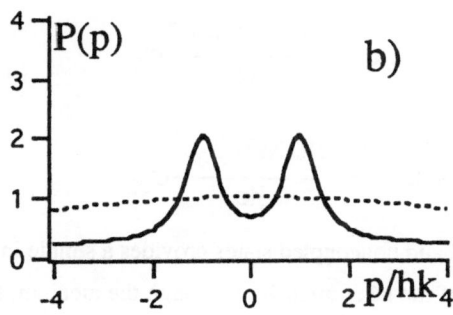

Fig. 5 Atomic momentum distributions P(p) for ^{4}He parameters at interaction time $\Theta=500\Gamma^{-1}$, Rabi frequencies $\alpha=\beta=0.3\Gamma$, lasers resonant with the atomic transition, Gaussian initial momentum distribution with width $\Delta p(t=0)=3\hbar k$. In a) all ground state relaxation rates are equal zero; in b) $\gamma_p=0$ and $\gamma_c=1.x10^5$ s^{-1}. The momentum distributions are scaled to the maximum value of the initial one.

6.Radiation propagation

Another interesting feature of a system in condition of CPT resonances is the attenuation for the laser radiation. To analyze the propagation equation in the case of electric fields interacting with the three-level system, the equations for the slowly varying amplitude of the electromagnetic field must be examined. Those wave equations for a copropagating two-mode degenerate field are[8]:

$$\frac{\delta E_1}{\delta z} + \frac{1}{c}\frac{\delta E_1}{\delta t} + \kappa_1 E_1 = \frac{4\pi i N\Omega_1 \mu_{01}}{c}\sigma_{01} \qquad (20a)$$

$$\frac{\delta E_2}{\delta z} + \frac{1}{c}\frac{\delta E_2}{\delta t} + \kappa_2 E_2 = \frac{4\pi i N\Omega_2 \mu_{02}}{c}\sigma_{02} \qquad (20b)$$

Here N is the atomic density, and κ_1 and κ_2 the damping rates for the electric fields, describing additional sources for attenuation of the laser radiation. In terms of the Rabi frequencies introduced above, the propagation equations become:

$$\frac{\delta\alpha}{\delta z} + \frac{1}{c}\frac{\delta\alpha}{\delta t} + \kappa_1\,\alpha = \frac{2\pi i N\Omega_1\,\mu_{01}{}^2}{\hbar c}\sigma_{01} \qquad (23a)$$

$$\frac{\delta\beta}{\delta z} + \frac{1}{c}\frac{\delta\beta}{\delta t} + \kappa_2\,\beta = \frac{2\pi i N\Omega_2\,\mu_{02}{}^2}{\hbar c}\sigma_{02} \qquad (23b)$$

The basis of coupled and noncoupled states provides a simple interpretation for the propagation equation of the laser intensity I through the medium, that for $\Omega_1=\Omega_2$ and $\mu_{01}=\mu_{02}=\mu_0$ is written as:

$$\frac{\delta I}{\delta z} + \frac{1}{c}\frac{\delta I}{\delta t} + (\kappa_1+\kappa_2)I + (\kappa_1-\kappa_2)(|\alpha|^2-|\beta|^2) = \frac{2 i\pi N\Omega\mu_0{}^2}{\hbar c} W(\sigma_{0c}-\sigma_{c0}) \quad (24)$$

Thus at $\kappa_1=\kappa_2$ the optical polarization corresponding to the coupled state only controls the laser propagation. Substituting the steady state solution of the density matrix equations into the steady state of Eq. (24) ($\delta I/\delta t=0$) leads to the equation for the spatial propagation of the laser radiation inside a CPT medium. By considering the case of $|a|^2=|b|^2$ and $\delta_1=\delta_2=0$, the following equation is obtained

$$\frac{\delta I}{\delta z} + (\kappa_1+\kappa_2)I = -\frac{gI}{1+I(\frac{1}{I_c}+\frac{3}{I_s})} \qquad (25)$$

with the parameter g given by:

$$g = \frac{2\pi N\Omega\mu_0{}^2}{\gamma_\perp \hbar c} \qquad (26)$$

Thus the CPT propagation is described by an equation equivalent to that of saturation bleaching except that the saturation intensity for the bleaching is $1/(1/I_c+3/I_s) \simeq I_c$, hence the bleaching intensity is given by the coherence intensity[8,13]· Integration of the propagation equation for the case of no losses ($\kappa_1=\kappa_2=0$) leads to the z dependence for the laser intensity I reported In Fig. 5. It may be noticed that at initial I_0 values small as compared to I_c, the attenuation results exponential, while at values large compared to I_c a linear decrease takes place.

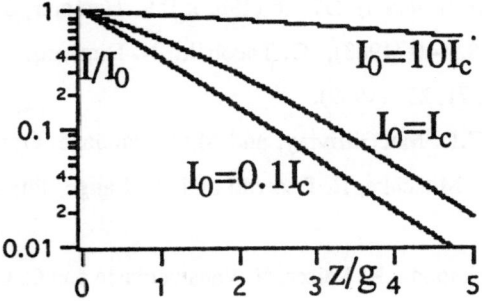

Fig. 5. Laser intensity I versus z coordinate as from Eq.(23) for different values of the initial intensity I_0 compared to the coherence intensity I_c, with $I_s=1000I_c$.

7. Conclusion

The use of the coupled and noncoupled basis has been examined for the analysis of CPT preparation in presence of relaxation processes in the ground state and for the analysis of laser radiation propagation through the CPT medium. The overall behaviour of the atomic system and of the laser propagating through an atomic medium in CPT conditions is analyzed in more simple terms. A comparison with experimental investigations will allow to test the influence of different relaxation processes that may produce a decay of the ground state coherence, or in equivalent terms the conservation of the atoms in the noncoupled state.

----

[1]E. Arimondo and G. Orriols, Nuovo Cimento Lett. 17, 333 (1976); H. R. Gray, R. M. Whitley and C.R. Stroud Jr., Opt. Lett. 3, 218 (1978); G. Orriols, Nuovo Cimento B53, 1 (1979).

[2]G. Alzetta, A. Gozzini, L. Moi and G. Orriols, Nuovo Cimento B36, 5 (1976); R.E. Tench, B.W. Peuse, P.R. Hemmer, J.E. Thomas, S. Ezekiel, C.C. Leiby Jr, R.H. Picard and C.R. Willis J. Phys. Paris 42, 8 (1981); M. Kaivola, P. Thorsen and O. Poulsen, Phys. Rev. A32, 207 (1985).

[3]A.M. Akulshin, A.A. Celikov and V.L. Velichansky, Opt. Commun. 84, 139 (1991).

[4]P. Knight, Nature (London) 297, 16 (1982); P.R. Hemmer, S. Ezekiel and C.C. Leiby Jr., Opt. Lett. 8, 440 (1983); G. Theobald, N. Dimarcq, V. Giordano and P. Cerez, Opt. Commun. 71, 256 (1989).

[5]W.E. Schultz, W.R. MacGillivray, and M.C. Standage, Opt. Commun.45, 67 (1983); J. Mlynek, F. Mitschke, R. Deserno and W. Lange, Phys. Rev. A29, 1297 (1984).

[6]A. Aspect, E. Arimondo, R. Kaiser, N. Vansteenkiste and C. Cohen-Tannnoudji, Phys. Rev. Lett. 61, 826 (1988) and J. Opt. Soc. Am. B6, 2112 (1989).

[7]O. Kocharovskaya and Ya. I. Khanin, Sov. Phys. JETP 63, 945 (1986); M.O. Scully, Shi-Yao Zhu and A. Gavrielides, Phys. Rev. Lett. 62, 2813 (1989).

[8]O. Kocharovskaya and P. Mandel, Phys. Rev. A42, 523 (1990).

[9]O. Kocharovskaya, F. Mauri and E. Arimondo, Opt. Commun. 84, 393 (1991).

[10]F. Papoff, F. Mauri and E. Arimondo, J. Opt. Soc. Amer. B9, 321 (1992).

[11]J. Dalibard, S. Reynaud and C. Cohen-Tannoudji, in *Interaction of Radiation with Matter*, a volume in honour of Adriano Gozzini (Scuola Normale Superiore, Pisa, Italy 1987) pp. 29-48.

[12]B.J. Dalton and P.L. Knight, Opt. Commun. 42, 411 (1982); J. Phys. B: At. Mol. Phys. 15, 3997 (1982).

[13]M.B. Gornyi , B.G. Matisov, and Yu.V. Rozhdestvenskii, Sov. Phys. JEPT 68, 728 (1989); E.A. Korsunskii, B.G. Matisov and Yu.V. Rozhdestvenskii, *ibidem* 73, 797 (1991); B.G. Matisov and I.E. Mazets, Opt. Commun. 92, 247 (1992).

[14]V. Finkelstein, Phys. Rev. A43, 4901 (1991); M.A. Bouchene, A. Débarre, V. Finkelstein, J.-C. Keller, J.-L. Le Gouet, P. Tchenio and P. R. Berman, Europhys. Lett. 18, 409 (1992); M.A. Bouchene, A. Débarre, J.-C. Keller, J.-L. Le Gouet, P. Tchenio, V. Finkelstein and P. R. Berman, J. Phys. II France 2, 621 (1992).

Adiabatic atomic cooling in cavity QED

T. Zaugg, G. Lenz and P. Meystre
Optical Sciences Center, University of Arizona
Tucson, Az 85721, USA

1   Introduction

This lecture discusses an adiabatic atomic cooling technique that does not rely on spontaneous emission as a dissipative mechanism and hence is applicable to cavity QED situations [1]. It exploits the inevitable cavity losses, in combination with the light gradient force, to irreversibly dispose of the atomic center of mass energy.

Experimental evidence for adiabatic cooling in free space has previously been reported. Shevy et al [2] observed adiabatic cooling of atoms in optical molasses as a function of the light turn-off rate, but the interpretation of their results was questioned by Lett et al [3], who attributed a similar effect to sub-Doppler cooling. More recently, Chen et al [4] unambiguously demonstrated adiabatic cooling as atoms moved away from the focus of a laser beam, but the achieved temperature was recoil-limited, as the atoms underwent free-space spontaneous decay. In contrast, the proposed use of adiabatic cooling in microwave cavities is not limited by spontaneous emission or the standard recoil limit.

These notes are organized as follows: Section 2 gives a phenomenological discussion of adiabatic cooling, and Section 3 outlines a formal development of the problem. Selected numerical results are presented in Section 4. A more detailed presentation of this work will be published in a paper in preparation [6].

2   Phenomenology

Adiabatic cooling is a generic cooling mechanism that can be expected to occur whenever particles are subjected to a potential slowly decaying in time, and is a direct consequence of the adiabatic theorem. In its quantum version, this theorem states that if a Hamiltonian with a *discrete spectrum* evolves continuously and infinitely slowly from an initial value H_0 at time t_0 to a final value H_1 at time t_1, then, a system initially in an eigenstate $|\psi_0\rangle$ of H_0 evolves to the eigenstate $|\psi_1\rangle$ of H_1 obtained from $|\psi_0\rangle$ by continuity.

Consider for instance a particle in a one-dimensional harmonic potential

$$V_h(x,t) = \frac{1}{2}M\omega^2(t)x^2, \tag{1}$$

where the frequency ω is assumed to exponentially decay in time,

$$\omega(t) = \omega_0 e^{-\kappa t/2}. \tag{2}$$

If the decay rate κ is sufficiently slow for the adiabatic theorem to be valid, then an oscillator initially in a mixture of states $|n_0\rangle$ of the eigenstates of the system with the potential $V(x,0)$ will remain in exactly the same mixture of states $|n_t\rangle$ of the system with the potential $V(x,t)$. For $t \to \infty$, the separation between these states becomes infinitely small, and the system has effectively been cooled to zero temperature. The classical analog of this system is a pendulum whose length is increased infinitely slowly. A simple experiment verifies that this will indeed lead to the pendulum coming to rest.

An immediate limitation of this cooling scheme is that it requires the oscillations of the particle in the harmonic potential to remain fast compared to the decay rate $\kappa/2$,

$$\omega(t) \gg \kappa/2 \tag{3}$$

With Eq. (2), this indicates that cooling will cease at the latest after a time on the order of

$$t_{max} \simeq \frac{2}{\kappa} \ln\left(\frac{2\omega_0}{\kappa}\right), \tag{4}$$

which occurs when $\omega(t) \simeq \kappa/2$. The ratio of final to initial temperatures is roughly given by the ratio of final to initial oscillator frequencies, that is

$$\frac{T_f}{T_i} \simeq \frac{\kappa}{2\omega_0}, \tag{5}$$

so that substantial cooling can be achieved in principle.

Although intuitively appealing, a harmonic oscillator does not adequately describe the motion of an atom in a decaying standing wave field. A more accurate model is given by a scalar particle in a periodic potential

$$V_s(x,t) = -\hbar\mathcal{R}(t)\cos qx, \tag{6}$$

where

$$\mathcal{R}(t) = \mathcal{R}(0)e^{-\kappa t/2} \tag{7}$$

is the Rabi frequency of the field and q is its wave number. As we see in the next Section, the interaction between a two-level atom and a standing wave field, when expressed in terms of dressed states, is governed to an excellent approximation by two such scalar equations. In contrast to the harmonic potential $V_h(x,t)$, the spectrum of the Hamiltonian consists of a discrete set of energy bands and a quasi-continuum, the successive bandgaps narrowing down as the depth of the potential decreases. Hence, some caution must be used when trying to apply the adiabatic theorem in this case, and the extent to which it remains valid is analyzed numerically in Section 4.

Close to the bottom of the potential wells, however, inter-well tunneling remains negligible and the bands can be approximated by discrete levels. Assuming that the initial atomic wave function is well localized compared to an optical wavelength, one can restrict the discussion to one well only and approximate it as an harmonic oscillator (where the potential has been renormalized to zero at the bottom of the well)

$$V_s \simeq \frac{1}{2}\hbar \mathcal{R}(t)q^2 x^2 \qquad (8)$$

with oscillation frequency

$$\omega(t) = \sqrt{2\mathcal{R}(t)\left(\frac{\hbar q^2}{2M}\right)} = \sqrt{2\mathcal{R}(t)\omega_r}, \qquad (9)$$

where $\omega_r = \hbar q^2/2M$ is the recoil frequency of the particle. In this case, the adiabaticity condition (3) becomes

$$\kappa/2 \ll \sqrt{2\mathcal{R}(t)\omega_r}, \qquad (10)$$

and, with Eq.(5),

$$\frac{T_f}{T_i} \simeq \frac{\kappa}{\sqrt{8\mathcal{R}(0)\omega_r}} \qquad (11)$$

Assuming that the adiabaticity condition remains fulfilled until the equality is reached in Eq. (10), the depth of the potential well at the end of cooling is $\hbar\kappa^2/4\omega_r$, corresponding to a temperature

$$T_{ad} = \frac{\hbar\kappa^2}{4\omega_r k_B} = \frac{M\kappa^2}{2k_B q^2}, \qquad (12)$$

where k_B is Boltzmann's constant.

It is important at this point to note that in addition to the violation of the adiabaticity condition, there is another limit to the achievable cooling: for weak fields, a classical treatment ceases to be valid, and as is well-known, the minimum achievable potential is given by the vacuum Rabi frequency $\mathcal{R}_0$, leading to the temperature

$$T_{vac} = \frac{2\hbar\mathcal{R}_0}{k_B}. \qquad (13)$$

If $T_{vac} > T_{ad}$ then adiabatic cooling is limited only by the vacuum fluctuations of the field.

3 Formal development

3.1 The model

Consider the interaction between an atom of mass M, excited electronic state $|e\rangle$, lower electronic state $|g\rangle$, and Bohr transition frequency ω_0, and a single-mode classical electromagnetic field of frequency ω. In an interaction picture, this system is described by the Hamiltonian

$$H_{sc}(t) = \frac{\mathbf{p}^2}{2M} - \hbar\delta|e\rangle\langle e| - \hbar\mathcal{R}_0(t)\cos q\mathbf{x}\sigma_1, \tag{14}$$

where δ is the detuning between the field and atomic transition frequencies, and $\sigma_i, i = 1, 2, 3$ are the usual pseudo-spin operators. In this Hamiltonian, we introduce the damping of the field by allowing the Rabi frequency $\mathcal{R}(t)$ to be a slowly varying function of time ,

$$\mathcal{R}(t) = \mathcal{R}(0)\exp(-\kappa t/2). \tag{15}$$

In order to connect this Hamiltonian to the discussion of the preceding Section, we note that the *local* eigenvalues of H_{sc} are given by

$$V_-(x,t) = \frac{\hbar}{2}[\Omega(x,t) - \delta] \tag{16}$$

and

$$V_+(x,t) = -\frac{\hbar}{2}[\Omega(x,t) + \delta], \tag{17}$$

where the local, time-dependent generalized Rabi frequency is

$$\Omega(x,t) = \sqrt{\delta^2 + 4\mathcal{R}^2(t)\cos^2 qx}. \tag{18}$$

The corresponding local eigenstates are given by

$$|-\rangle(x,t) = S(x,t)|g\rangle = \cos\frac{\theta(x,t)}{2}|g\rangle - \sin\frac{\theta(x,t)}{2}|e\rangle, \tag{19}$$

and

$$|+\rangle(x,t) = S(x,t)|e\rangle = \sin\frac{\theta(x,t)}{2}|g\rangle + \cos\frac{\theta(x,t)}{2}|e\rangle, \tag{20}$$

respectively, where $S(x,t) = \exp[-i\theta(x,t)\sigma_2/2]$ and the Stückelberg angle θ is defined by $\delta = \Omega(x,t)\cos(\theta(x,t)$ and $2\mathcal{R}(x,t) = \Omega(x,t)\sin\theta(x,t)$.

Assuming that the electronic states adiabatically respond to the dressing provided by the field, *both in space and in time*, we conclude that a particle in the state $|-\rangle$ or $|+\rangle$ behaves effectively as a scalar particle in the potential $V_-(x,t)$ or $V_+(x,t)$, respectively. Assuming for example an atom prepared in the dressed state $|+\rangle$ and initially localized in the vicinity of a potential minimum, we can follow the intuitive argumentation of Section 2 to arrive at the same adiabatic cooling limits T_{ad} and T_{vac}.

Of course, this argumentation is only approximately correct. In particular, in addition to the adiabatic condition (in time) already discussed, it is important to note that the local diagonalization leading to the eigenstates $|+\rangle(x,t)$ and $|-\rangle(x,t)$ neglects a nonadiabatic contribution (in space) described by the Hamiltonian

$$H_{nl} = -\frac{\hbar}{4M}[\mathrm{p}\theta' + \theta'\mathrm{p}]\sigma_2 + \frac{\hbar^2}{8M}\theta'^2, \tag{21}$$

where $\theta' = d\theta/dx$. This Hamiltonian couples the states $|+\rangle$ and $|-\rangle$, if the atom cannot adiabatically follow the local changes in the Rabi frequency $\mathcal{R}(x,t)$. Since a minimum of $V_-(x,t)$ corresponds to a maximum of $V_+(x,t)$, it is quite clear that such nonadiabatic transitions lead to heating rather than cooling, and need therefore to be minimized.

While the locally diagonalized Hamiltonian (14) is useful in gaining an intuitive understanding of the cooling mechanism, our numerical analysis, selected results of which are given in the next Section, uses an exact diagonalization based on the band-theoretical approach of Ref. [5]. The instantaneous eigenstates $|\phi_\nu(k,t)\rangle$, with eigenenergies $E_\nu(k,t)$ of $H_{sc}(t)$

$$H_{sc}|\phi_\nu(k,t)\rangle = E_\nu(k,t)|\phi_\nu(k,t)\rangle, \tag{22}$$

are restricted to the first Brillouin zone, that is, $k \in [-1,1)$, where k is a quasi-momentum, and the band number $\nu \in [0,\infty)$. These eigenstates, which form a complete orthonormal basis, are conveniently expressed in terms of the "bare" electro-translational states

$$|k,n\rangle = \begin{cases} |(k+n)\hbar q, g\rangle \; ; \text{n even} \\ |(k+n)\hbar q, e\rangle \; ; \text{n odd} \end{cases} \tag{23}$$

where $n = 0, \pm1, \pm2, ...$, with

$$|\phi_\nu(k,t)\rangle = \sum_m c_{m\nu}(k,t)|k,m\rangle. \tag{24}$$

Both the expansion coefficients $c_{n\nu}(k,t)$ and the eigenvalues $E_\nu(k,t)$ can readily be obtained numerically by solving a system of tridiagonal recurrence relations [6].

As the field decays to zero, the classical description of the field eventually ceases to be valid, and a full quantum description of the problem becomes necessary. This leads to three major changes. First, an additional quantum number is needed to describe the state of the field. Second, dissipation can not be simply described via a damping constant. Rather, the damping of the field involves its coupling to a thermal bath, and hence a master equation description. Finally, the Hamiltonian describing the complete system (atom + field + reservoir) is now time independent requiring but one diagonalization per excitation level.

The quantum version of the Hamiltonian (14) is

$$H_{qm} = \frac{\mathbf{p}^2}{2M} - \hbar\delta|e\rangle\langle e| - \hbar\mathcal{R}_0\left(a^\dagger\sigma_- + a\sigma_+\right)\cos q\mathbf{x}, \tag{25}$$

where $[a, a^\dagger] = 1$, and σ_-, σ_+ are the usual pseudo-spin lowering and raising operators. The electro-translational states (23) now become

$$|k, n, N\rangle = \begin{cases} |(k+n)\hbar q, g, N\rangle & ;\text{ n even}, \\ |(k+n)\hbar q, e, N-1\rangle & ;\text{ n odd}. \end{cases} \tag{26}$$

The 'N' quantum number on the left-hand side of this equation labels the total number of excitations in the state and differs from the Fock number by one if the atom is excited. The eigenstates $|\phi_{\nu N}(k)\rangle$, with eigenenergies $E_{\nu N}(k)$, which satisfy

$$\mathcal{H}_{qm}|\phi_{\nu N}(k)\rangle = E_{\nu N}(k)|\phi_{\nu N}(k)\rangle, \tag{27}$$

are given in the electro-translational basis as

$$|\phi_{\nu N}(k)\rangle = \sum_n c_{n\nu N}(k)|k, n, N\rangle. \tag{28}$$

The dissipative part of the field evolution is described as usual by coupling it to a markovian bath of harmonic oscillators at zero temperature, and is governed by the master equation

$$\dot{\rho}(t) = -\frac{\kappa}{2}(a^\dagger a\rho + \rho a^\dagger a) + \kappa a\rho a^\dagger, \tag{29}$$

where ρ is the density matrix of the system.

3.2 Initial conditions

It should be clear from the discussion of Section 2 that some care must be given in the description of the initial center-of-mass component ρ_{cm} of the atomic density matrix. We assume that we initially know the mean position $\langle\mathbf{x}\rangle$ and momentum $\langle\mathbf{p}\rangle$ of the atoms as well as the corresponding standard deviations σ_x and σ_p.

The initial center-of-mass density matrix can then be found by maximizing the von Neumann entropy $S = -k_B\text{Tr}\rho_{cm}\ln\rho_{cm}$ subject to the constraints $\text{Tr}\rho_{cm} = 1$,

$$\langle\mathbf{p}\rangle = p_0, \tag{30}$$

$$\langle(\mathbf{p} - p_0)^2\rangle = \sigma_p^2, \tag{31}$$

$$\langle\mathbf{x}\rangle = x_0, \tag{32}$$

$$\langle(\mathbf{x} - x_0)^2\rangle = \sigma_x^2. \tag{33}$$

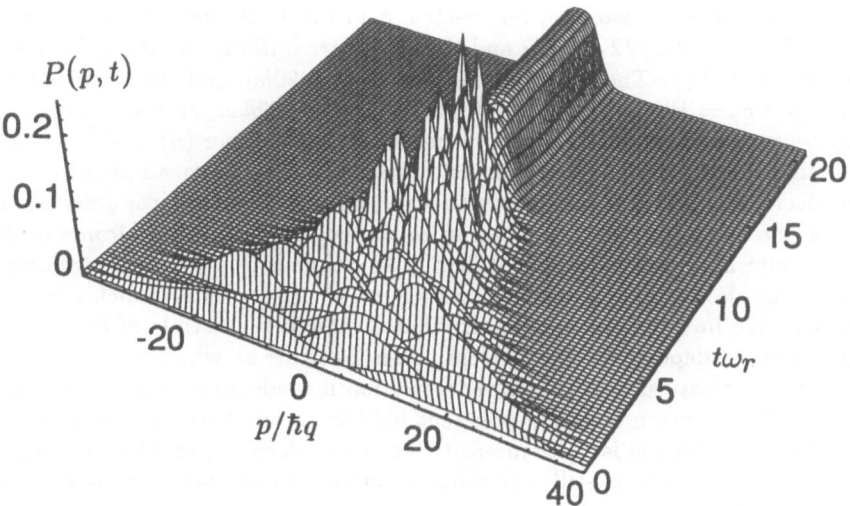

Figure 1: Evolution of the atomic momentum distribution as a function of time, where the initial distribution has $p_0 = 20\hbar q$ and $\sigma_p = 10\hbar q$. There are initially 10^6 photons in the field, $\mathcal{R}_0 = 40\omega_r$ and the field decays at rate $\kappa = \omega_r$. The switch to the full quantum treatment is made when $\omega_r t = 10.75$ at which time the field intensity has decayed to $\langle n \rangle = 20$ photons.

In the momentum representation, this leads to the initial center-of-mass density matrix [6]

$$
\begin{aligned}
\rho_{cm} &= \frac{1}{\sqrt{2\pi\sigma_p^2}} \int_{-\infty}^{\infty} dp \int_{-\infty}^{\infty} dp' \, \exp\left(-\frac{i}{\hbar}(p-p')x_0\right) \\
&\times \exp\left(-\frac{1}{2\sigma_p^2}(\frac{p-p_0}{2} + \frac{p'-p_0}{2})^2\right) \\
&\times \exp\left(-\frac{\sigma_x^2}{2\hbar^2}(p-p')^2\right)|p\rangle\langle p'|.
\end{aligned}
\tag{34}
$$

Note that this density operator is diagonal neither in the momentum, nor in the coordinate representation.

4 Results

Adiabatic atomic cooling is illustrated in Fig. 1. These results were obtained by direct numerical integration of the Schrödinger equation in the semiclassical approximation, and of the master equation in the full quantum-mechanical treatment. In this example, the electric field inside the cavity decays exponentially at a rate $\kappa/2 = \omega_r/2$ and the atoms are initially the their electronic dressed state $|+\rangle$. The mean momentum is $p_0 = 20\hbar q$, and the standard deviation is $\sigma_p = 10\hbar q$. The initial depth of V_+ is $4 \cdot 10^4 \hbar \omega_r$, corresponding to a momentum $p_{max} \simeq 283\hbar q$ and to a mean photon number $\langle n \rangle = 10^6$, so that it is initially justified to treat the field classically. As the mean photon number decreases, however, we switch to a full quantum-mechanical description, using the semi-classical result as an initial condition for the atomic state, and assuming that the quantized field at this time is in a coherent state with the mean photon number corresponding to the appropriate semiclassical intensity. We have found numerically that although the details of the atomic dynamics do depend to some extent upon the time at which the transition from a semiclassical to a quantum description is made, and also on the initial state of the cavity mode (e.g. coherent field vs thermal state), the final momentum distribution is quite insensitive to these details, provided the switch is made while there is still a significant amount of excitation in the mode, say $\langle n \rangle \geq 20$.

In the present example, $\omega_v(t = 0) \simeq 45\omega_r$, so that the adiabaticity condition is initially well fulfilled. As time evolves, the momentum distribution undergoes oscillations corresponding to the motion of the atom inside the potential V_+, as well as a shift toward lower momenta and a narrowing. (These oscillations are not very clear in the figure, due to the large time steps used in the plot.) As is the case for the transition from a semiclassical to a full quantum description, we note that while the details of the oscillations during the cooling stage depend explicitly on the details of the initial condition, the final distribution, which exhibits both a slowing down of the atomic distribution to $\langle p \rangle = 0$ and cooling, is largely independent of the initial momentum distribution, provided it corresponds to a kinetic energy whose mean and standard deviation are small compared to the initial potential depth. These observations are an indication of the robustness of the cooling mechanism.

Equation (5) indicates that for a harmonic oscillator, one would expect the final temperature to be a linear function of the initial temperature. To see to what extent this argument remains valid in our case, we plot in Figure 2 the final variance in atomic momentum as a function of its initial value. For this example we use only the semi-classical approximation, neglecting in effect the quantum cooling limit. For very small initial variances, we observe heating, rather than cooling. This result is easy to understand, since in this case, the initial atomic center-of-mass wave function has an extent larger than a period of the potential. Following the arguments of Section 2, we should indeed expect heating rather than cooling in this case. For larger initial variances, we do observe a linear dependence of the final value on the initial value, indicating that we are in a region of adiabatic cooling. As σ_p^2 is further increased, this linear dependence ceases to be valid. Somewhat surprisingly, however, the cooling seems more effective in this region. Finally,

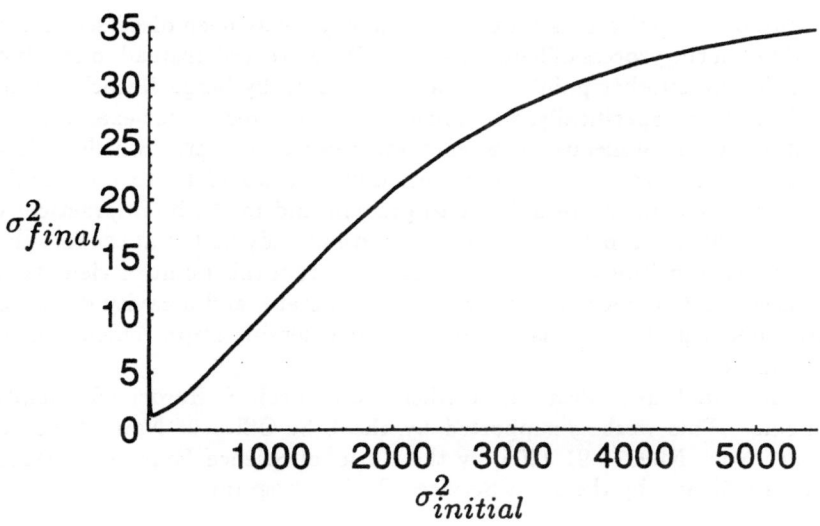

Figure 2: Final momentum variance as a function of the initial momentum variance. In this example, $p_0 = 20$, $\mathcal{R}(0) = 4 \cdot 10^4 \omega_r$, and $\kappa = \omega_r$.

we expect cooling to cease for initial variances corresponding to atomic energy spreads larger than the depth of the initial potential well. However, the numerical simulation of such initial conditions imposes considerable memory requirements on the computer and we have not yet been able to test this regime. In addition, it should be noted that the final momentum variance depends not only on σ_p, but also on p_0, x_0 and σ_x. Detailed results will be given in a paper in preparation. [6]

Typical final temperatures for micromaser atomic transitions have been discussed in Ref. [1]. For the $39S_{1/2} \leftrightarrow 39P_{3/2}$ transition of Rubidium, for example, the vacuum Rabi frequency has been measured [7] to be $\mathcal{R}_0 \simeq 2\pi \cdot 70 KHz$, so that $T_{ad} \simeq 3.5 \cdot 10^{-6}$ K, and for the $63P_{3/2} \leftrightarrow 61D_{3/2}$ transition in ^{85}Rb considered in the Munich micromaser [8], [9] one has $\mathcal{R}_0 \simeq 2\pi \cdot 7$ KHz, corresponding to a theoretical cooling limit of 0.3 microKelvin .

Although the cooling achieved in this example may not seem too impressive, considerable improvement can readily be obtained. The vacuum Rabi frequency which determines the fundamental cooling limit can easily be reduced, e.g. by considering other transitions or by operating off-resonance we have, for $|\delta| \ll \mathcal{R}_0$,

$$T_{vac} \simeq \left(\frac{\mathcal{R}_0}{|\delta|} \right) \frac{2\hbar\mathcal{R}_0}{k_B}, \tag{35}$$

which can easily be one to two orders of magnitude smaller than on resonance.

A number of limitations of our theory have already been discussed in Ref. [1], and are in the process of being removed. Here, we wish instead to conclude by mentioning another point that was brought up by Serge Haroche at the Kühtai meeting. Specifically, the transitions considered in our examples are still subject to spontaneous emission to lower electronic states, which we have neglected. One way to circumvent this difficulty would be to use circular states, but such states are difficult to prepare and might be impractical to use in the proposed scheme. However, we have already hinted at the fact that nonresonant transitions can also be used, and from this point of view, there is no reason not to consider an atomic ground state, and a cavity driven by a slowly decaying field. This would allow considerable improvements in the cooling limit.

We are thankful to Prof. H. Walther and to Prof. S. Haroche for helpful comments. This work is supported by the U.S. Office of Naval Research Contract No. N00014-91-J205, by the National Science Foundation Grant PHY92-13762 and by the Joint Services Optics Program.

References

[1] T. Zaugg, M. Wilkens, P. Meystre, and G. Lenz. *Optics Commun.* **97**, 189 (1993).

[2] Y. Shevy, D. S. Weiss, and S. Chu. In S. Stringari, editor, *Spin polarized quantum systems*, p. 287, World Scientific, Teaneck, N.J., 1989.

[3] P. D. Lett, W. D. Phillips, S. L. Rolston, C. E. Tanner, R. N. Watts, and C. I. Westerbrook. *J. Opt. Soc. Am. B* **6**, 2084 (1989).

[4] J. Chen, J. G. Story, J. J. Tollett, and R. G. Hulet. *Phys. Rev. Lett.* **69**, 1344 (1992).

[5] M. Wilkens, E. Schumacher, and P. Meystre. *Phys. Rev. A* **44**, 3130 (1991).

[6] T. Zaugg, G. Lenz, P. Meystre, and M. Wilkens. *Phys. Rev. A*, submitted.

[7] F. Bernardot, P. Nussenzweig, M. Brune, J. M. Raimond, and S. Haroche. *Europhys. Lett.* **17**, 33 (1992).

[8] G. Rempe, F. Schmidt-Kaler, and H. Walther. *Phys. Rev. Lett.* **64**, 2783 (1990).

[9] H. Walther. *Phys. Reports* **219**, 263 (1992).

PART III: Cavity Quantum Electrodynamics

Quantum Electrodynamics in an Optical Cavity

G. Rempe[1] ,R. J. Thompson and H. J. Kimble

Norman Bridge Laboratory of Physics 12-33, California Institute of Technology, Pasadena, California 91125, USA

[1] Present address: Fakultät Physik, Universität Konstanz, 7750 Konstanz, Germany

Abstract. Observations of a vacuum-field induced normal-mode splitting, photon antibunching, sub-Poissonian photon statistics and optical bistability are reported for a small number of Cesium atoms strongly coupled to a single mode optical cavity.

1. Introduction: Strong coupling in the optical domain

During the last decade insights into the quantum characteristics of the electromagnetic field and into the generation of nonclassical light has increased tremendously. Interest in manifestly quantum effects has been greatly stimulated by the experimental observation of quantum phenomena like photon antibunching [1], sub-Poissonian photon statistics [2] and squeezing [3]. Within this context much of modern research in cavity quantum electrodynamics [4] investigates the nonclassical aspects of the interaction of single atoms with a single mode of the electromagnetic field. In the absence of dissipation the Schrödinger equation for such a system describing the electric-dipole coupling of a two-state atom to the electromagnetic field has analytic solutions [5]. As a result this model has become a cornerstone in modern quantum optics.

Over the years it has been quite difficult to realize such a fundamentally simple system experimentally. For example, most laser experiments reported to date have been performed in a regime of weak coupling for which the interaction energy of an atom with a single light quantum is very small. In this case the rate γ of irreversible decay of an excited atom into the continuum of modes of the electromagnetic field is much larger than the frequency scale g associated with the reversible evolution in the interaction of the atom with any single radiation mode. The first experiments [4] to achieve the condition of strong coupling have been performed in the microwave regime with Rydberg atoms. In this case, g>γ>κ and g≈T^{-1}, with T as the atomic transit time and κ as the cavity field decay rate. In contrast to the usual circumstance of weak coupling in the optical domain, recent technical progress in the fabrication of ultra low loss dielectric mirrors [6] has made possible the experimental realization of systems in the optical domain for which the internal coherent coupling rate g is comparable to or larger than the external dissipative rates which are determined by both the cavity field damping κ and the spontaneous decay of the atomic polarization at rate $\gamma_\perp$ and atomic inversion at rate γ due to the interaction of the atom with all other radiation modes. In such a strongly coupled system a photon emitted by an excited atom into the cavity mode is repeatedly absorbed and reemitted before irreversibly escaping into the environment.

Within this context of cavity quantum electrodynamics in the optical domain we report experiments in which a system made of a small collection of two-state atoms strongly coupled to a single mode of a high finesse optical cavity is investigated in a regime where $g > \gamma_\perp > \kappa > T^{-1}$. In a direct spectroscopic measurement we explore the linear susceptibility of the atom-cavity system and observe a normal-mode splitting induced by the atom-field coupling for N≈1 intracavity atom. To address nonclassical aspects of the dynamics of this open quantum system, we investigate the intensity fluctuations of the transmitted field for weak excitation and observe photon antibunching and sub-Poissonian photon statistics. We also record input-output characteristics for steady state operation with 3<N<65 atoms and observe optical bistability for N>15 atoms.

2. Spectroscopy: Vacuum-Rabi splitting for single atoms

The oscillatory exchange of excitation between an atom and one cavity mode was theoretically described by Jaynes and Cummings [5] for a single intracavity atom (N=1) and by Tavis and Cummings [7] for N>1. The Rabi-oscillation for N=1 atom and a few light quanta has recently been studied experimentally in the time domain using Rydberg-atoms in a superconducting cavity [8], where quantum phenomena sensitive to the photon statistics of the cavity field [9] were observed. For N=1 and a single quantum of excitation (for example an excited state atom and a vacuum-field cavity or a single photon and a ground state atom), the Jaynes-Cummings oscillation is associated with a ubiquitous normal-mode splitting in the eigenvalue spectrum of the interaction Hamiltonian. This splitting persists in the presence of dissipation for atom and cavity only in the strong coupling regime where $g > (\gamma_\perp, \kappa) > T^{-1}$. In this case the spectrum of the system is symmetrically split into two peaks separated by $\pm g$ about the otherwise common resonance frequency of atom and cavity with the width of either peak given by $(\gamma_\perp + \kappa)/2 < g$. This doublet has been dubbed the vacuum-Rabi splitting [10]. In previous experiments, mode splittings were first observed for many atoms (N»1) in the microwave domain [11a]. The first observation of this effect in the optical domain was reported in Ref. [11b], with subsequent experiments described in [11c, 11d]. Quite recently this phenomenon was demonstrated with quantum-well excitons in a semiconductor [11e]. Normal-mode splittings for N≈3 atoms have been reported in the microwave domain [12]. We now describe an optical experiment with N≈1 atom [13].

Since the response of the atom-cavity system for weak excitation is given in terms of the eigenvalues for the normal modes, the basic splitting is independent of the particular excitation scheme and initial conditions. Our choice is to measure the steady state cavity transmission for an external probe field of variable frequency. Since for low number of intracavity atoms (N≈1) excitation of the cavity-field instead of the atom can lead to higher excited states of the combined atom-cavity system with two or more quanta of excitation, our investigation is restricted to a regime of weak external probe fields. In this case the quantum mechanical basis is limited to states with only zero or one quantum of excitation so that the coupling induced splitting of the first excited state of the atom-cavity "molecule" is probed; semiclassically the intracavity intensity is held much smaller than the saturation intensity so that the linear susceptibility of the coupled system is probed. As we [11b, 11c] and others [11d] have emphasized, for weak excitation equivalent descriptions follow either from the quantum master equation or from the semiclassical Maxwell-Bloch equations, since in this domain the system behaves as two coupled linear oscillators.

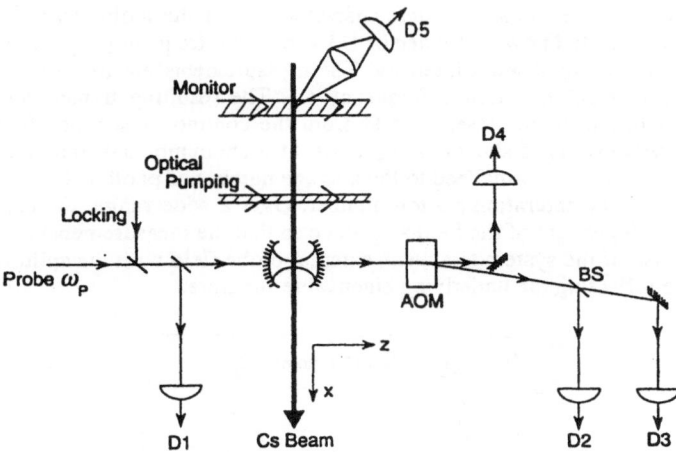

Fig. 1: Diagram of apparatus with atomic beam, monitor detector, optical pumping region, cavity and light detectors as described in the text.

A diagram of the principal elements of the experiment is given in Fig. 1. The cavity formed by two spherical mirrors has length l=1mm, waist ω_0=50 μm and finesse F=80,000. The resonant atomic transition employed is the $6S_{1/2}$, F=4, m_F=4 ↔ $6P_{3/2}$, F=5, m_F=5 transition of the D_2 line of Cesium at 852 nm, with the atoms prepared in the m_F=4 sublevel of the ground state by the optical pumping beam. Of prime importance in the experiment are the single-atom cooperativity parameter $C_1=g^2/2\kappa\gamma_\perp$ and the saturation photon number $n_s=2\gamma\gamma_\perp/3g^2$, where $\kappa/2\pi$=0.9 MHz is the decay rate of the cavity field and $(\gamma_\perp,\gamma)/2\pi$=(2.5,5.0) MHz are the decay rate of the atomic polarization ($\gamma_\perp$) and inversion (γ) into field modes other than the principal cavity mode. $g/2\pi$=3.2 MHz is the optimum coupling rate of an atom to the standing-wave Gaussian cavity mode. Since the solid angle of the Fabry-Perot cavity is small ($\sim 10^{-4}$ sr), the rates for the decay of the atomic polarization and inversion are the same as in free space. While κ and γ can be measured independently, g can be calculated from its definition given below and as well determined from the measured saturation photon number of the system. For our homogeneously broadened system, we have C_1=2.3 and n_s=0.8 so that critical phenomena such as optical bistability are observed with a small number of atoms and only a few photons. Indeed, in a regime of weak excitation, a single intracavity atom which is optimally coupled to the spatially varying cavity mode (at the peak of the standing wave Gaussian mode) reduces the transmission of the cavity by a factor of about 30. The TEM_{00} mode of the cavity is excited and the cavity length is actively stabilized with a chopping technique which uses detector D4. Switching between data collection (with detectors D2 and D3) and stabilization (with detector D4) is accomplished by the acoustooptic modulator AOM shown in Fig. 1 as well as with control of the locking and probe beam at a rate of 2.5 kHz. The signal beam which excites the system is provided by a Titanium-sapphire laser which is locked and tunable relative to a saturated resonance in an auxiliary Cesium cell.

To investigate the steady state characteristics of the atom-cavity system, we record the transmitted power at detector D3 versus the frequency ω_p of the incident probe laser for constant input intensity. The measurements are made with identical cavity frequency ω_c and atomic frequency ω_a. The resulting transmitted intensity versus the detuning of the laser field Ω from the common resonance frequency of atom and cavity $\omega_a=\omega_c$ is shown in Fig. 2 for different numbers of intracavity atoms. The output intensity is normalized to the average number of photons inside the cavity which is below the saturation photon number. Over a wide range, the experimental spectrum is independent of the field amplitude so that the measurements represent the linear response of the system to a weak external probe field with the splitting evident in the figure reflecting the underlying eigenvalue structure.

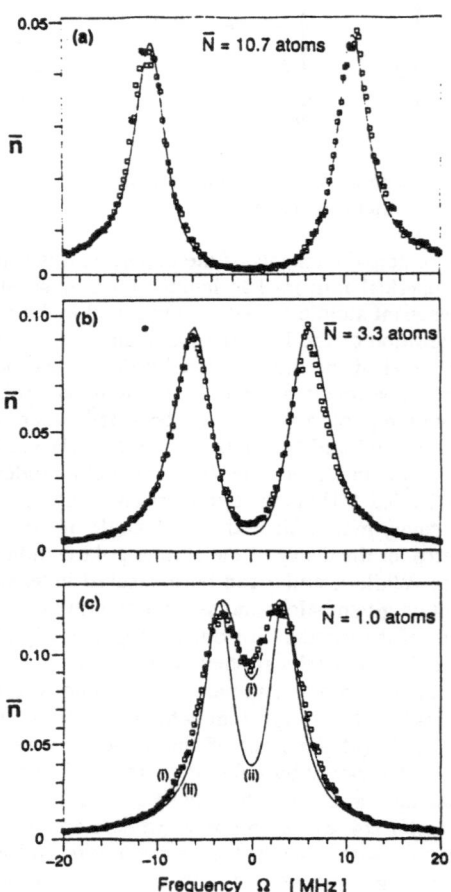

Fig. 2: Normal-mode or vacuum-Rabi splitting for different numbers of intracavity atoms. The transmitted intensity is plotted as a function of the detuning of the probe field Ω from the common atomic and cavity resonance frequency ($\omega_a=\omega_c$).

Also displayed in Fig. 2 as solid curves are fits to the experimental spectra with the measured values of g, γ and κ. Here κ is obtained from a trace as in Fig. 2 with the cavity empty (N=0). Note that the dipole coupling coefficient for an atom of transition moment μ and frequency $\omega_a=\omega_c$ depends on its position r according to $g(r)=g\psi(r)$ with $g=(\pi\mu^2\omega_c/\hbar\varepsilon_0 V)^{1/2}$ as the optimum coupling coefficient. Here $\psi(r)=\sin(kz)\exp[-(x^2+y^2)/\omega_0^2]$ is the Gaussian standing wave cavity mode function with volume $V=\pi\omega_0^2 l/4$. The effective intracavity atom number $\bar{N}=\Sigma_i\psi(r_i)^2$ where the sum is over all possible atomic sites. On resonance ($\omega_a=\omega_c$) the eigenvalues describing the normal modes formed from the "atomic-polarization oscillator" and the "cavity-field oscillator" are given in a rotating frame by [10b,14]

$$\lambda_\pm = \pm \{g^2\bar{N}-[(\gamma_\perp-\kappa)/2]^2\}^{1/2} - i(\gamma_\perp+\kappa)/2. \tag{1}$$

Note that the two eigenvalues characterize a distinctive splitting observable for $g\bar{N}^{1/2}>(\gamma_\perp+\kappa)/2$. The eigenvalue's imaginary part $(\gamma_\perp+\kappa)/2$ determines the width of each symmetrically split peak while $(\gamma_\perp-\kappa)/2$ describes a damping induced shift of the oscillation frequency if the two coupled oscillators have different decay rates. Note that for weak excitation as assumed here the vacuum-Rabi splitting does not depend on the external laser intensity but is uniquely determined by atomic parameters (dipole moment μ and transition frequency $\omega_a=\omega_c$) and the geometry of the cavity (cavity length l and beam waist ω_0). The cavity transmission is [10b,14]

$$T(\Omega) = T_0 \,|\kappa[\gamma_\perp+i\Omega] \,/\, [(\Omega-\lambda_+)(\Omega-\lambda_-)]|^2 \tag{2}$$

where T_0 is the peak transmission of the empty cavity. Note that because the number of atoms is small and an atomic beam is employed, the experimentally observed transmission averages over atomic number fluctuations. In the fits in Fig. 2 these fluctuations are taken into account by randomly distributing atoms over different sites r_i inside a volume much larger than the cavity mode volume and numerically averaging over a large number of realizations. The theoretical fits therefore include fluctuations both in atomic number and position. In fitting the theory to the experiments we point out that the only adjustable parameter is $\bar{N}$ and that the agreement of theory and experiment is quite satisfactory. The value of $\bar{N}$ obtained from the detailed fits agrees well with a much more straightforward determination based upon the frequency separation of the two peaks.

A different approach to the problem of determining $\bar{N}$ is to take advantage of the fluctuations inherent in the atomic beam. The effect of fluctuations in the number and position of intracavity atoms on the recorded spectra is strongly dependent on $\bar{N}$ itself, thus providing a signature in the spectra unique to that value of $\bar{N}$. This signature is particularly distinctive for $\bar{N}\approx1$, as we also illustrate in Fig. 2, where the theoretical spectrum for a single, optimally coupled atom [N=1, g(r)=g] is shown for comparison (line (ii)). It is reasonably clear that a dominant contribution to the measured spectrum arises from the single atom, optimum-coupling result. The principal difference between the result from the single optimally coupled atom (line (ii)) and the average over atom positions (line (i)) in Fig. 2 occurs around $\Omega=0$ and reflects the fact that the spectrum in this region is very sensitive to atomic fluctuations and hence to the precise value of $\bar{N}$. Mathematically, this sensitivity is expressed by the fact that on resonance $\Omega=0$ ($\omega_p=\omega_a=\omega_c$) the transmitted intensity can be calculated from eq. (2) and is given by

$$T(\Omega=0) = T_0/(1+2C_1N)^2 \tag{3}$$

which for $C_1\gg1$ and $N\approx1$ depends strongly on N. Indeed, by keeping the splitting $g\bar{N}^{1/2}$ constant but changing $\bar{N}$ and adjusting g, we are able to determine $\bar{N}$ to within an experimental uncertainty of about ±5%. Note that the definition of $\bar{N}$ as a weighted sum over atoms distributed in a volume larger than the cavity mode volume $V=\pi\omega_o^2 l/4$ (with weighting given by the square modulus of the mode function at the site of the atom) implies that the fluctuations in the effective atomic number $\bar{N}$ are considerably smaller than those of a Poisson distribution with the same mean value.

Additional and independent measurements to determine $\bar{N}$ are obtained by comparing the splitting for different atomic beam densities with the corresponding values of the fluorescence F of an additional laser beam monitor (detector D5) collected upstream at a position where the intersection of the atomic beam and laser light excites a larger number of atoms. The monitor fluorescence serves as a proportional measure of the atomic density, with the constant of proportionality determined from spectra with a large number of intracavity atoms $\bar{N}$ with a negligible role of fluctuations. We found a good correlation between $\bar{N}$ and F over the whole range of atomic densities, and in particular in the region of low $\bar{N}$ where the two peaks of the spectrum begin to merge together. This strongly supports our procedure of finding $\bar{N}$ from fits to the experimental data which relies only upon the independently determined values of g, γ and κ.

Note that the coupling constant g, the cavity field decay rate κ and the natural linewidth of the atom γ are much larger than the inverse transit time of the atom through the cavity mode (400 ns to traverse $2\omega_o$ with average thermal velocity of 250 m/s). Indeed, numerical integration of the Maxwell-Bloch equations for atomic trajectories through the cavity mode indicate that the effects of finite transit time are small and do not need to be included in our model. Thus the features observed are associated with steady states established by individual atoms rather than with states reached incrementally by the successive passage of large numbers of atoms.

3. Nonclassical light: Sub-Poissonian photon statistics and photon antibunching

While thus far we have concentrated on structural aspects, we now change course to discuss measurements of the nonclassical dynamical processes in this coupled system [15]. In the domain of strong coupling in which the critical number of quanta which characterizes the system evolution is small, the usual system-size expansions of quantum statistical physics are not applicable. Dynamical fluctuations at the level of a single photon can have a profound effect on the system's evolution even for large numbers of atoms. In this context, investigations in cavity quantum electrodynamics in a regime of strong coupling provide realizable avenues for the exploration of questions such as the scaling of quantum fluctuations with system size.

Experimentally in cavity quantum electrodynamics, squeezed light was generated by employing the normal-mode structure of the atom-cavity system for $g\bar{N}^{1/2}>(\gamma,\kappa)$ [16]. For the case $g>(\gamma,\kappa)$ nonclassical light has been observed in a strongly coupled Rydberg maser [17a]. To address quantum phenomena of the dynamics in the current optical system the intensity fluctuations of the transmitted field are investigated through measurements of the correlation function $g^{(2)}(\tau)$, where

$$g^{(2)}(\tau) = <:I(t)I(t+\tau):>/<:I(t):>^2 \qquad (4)$$

with the colons denoting normal and time ordering. We observe both photon antibunching and sub-Poissonian photon statistics [18] and interpret these results in terms of quantum state reduction and interference in a dissipative setting [19].

For the measurements of the photon statistics of the transmitted field we employ both detectors D2 and D3 shown in Fig. 1 to accumulate histograms of the number of coincidences $m(\tau)$ versus time delay τ. The detector D2 provides start pulses to a set of two time-to-digital converters while detector D3 provides stop pulses. Time delay $\tau > 0$ ($\tau < 0$) corresponds to a start event followed (preceeded) by a stop event, with $\tau = 0$ referenced to simultaneous events at the beam splitter BS in Fig. 1. For weak excitation with mean intracavity photon number $n \approx 0.04$ as compared to $n_s = 0.8$, results for $g^{(2)}(\tau)$ are shown in Fig. 3 for $18 < \bar{N} < 110$ intracavity atoms. Note that $g^{(2)}(\tau)$ is determined from the ratio $m(\tau)/m_p$, with m_p as the independently measured number of coincidences for a Poisson process. Photon antibunching as expressed by $g^{(2)}(0) < g^{(2)}(\tau)$ and sub-Poissonian photon statistics with $g^{(2)}(0) < 1$ are clearly visible. These characteristics of the data are indicative of the quantum nature of the dynamical processes observed.

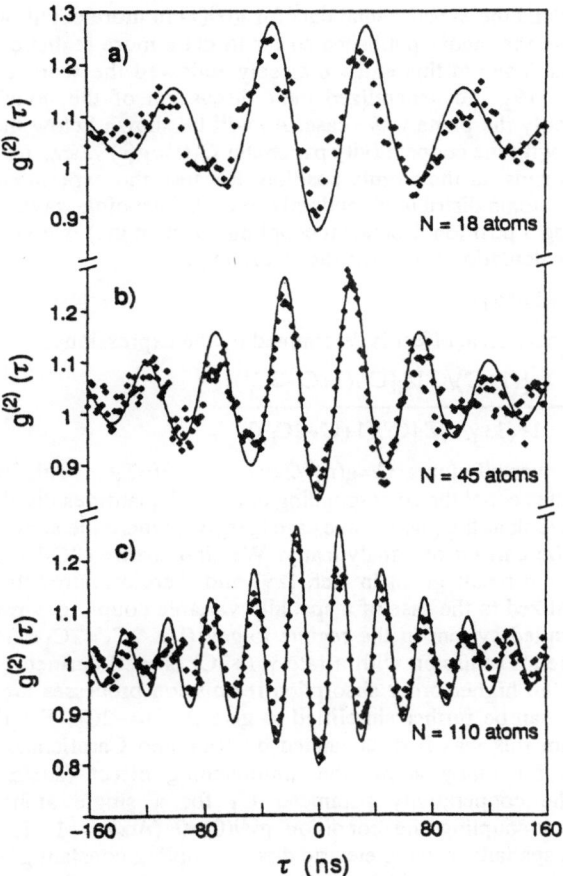

Fig. 3: Intensity correlation function $g^{(2)}(\tau)$ versus time delay τ for $18 < \bar{N} < 110$ atoms. The solid curve is the theoretical result as discussed in the text.

In addition to the behaviour near $\tau=0$, the data also exhibit an oscillatory regression for larger τ as determined by the system eigenvalues. Indeed, Fourier transformations of the experimental data of Fig. 3 show pronounced peaks at frequencies which are in good agreement with estimates based on the knowledge of the coupling constant $g/2\pi=3.2$ MHz and the number of atoms $\bar{N}$. As a result of linewidth averaging [20] the half-widths of the peaks are determined to be 2.1 MHz which is well below the natural half-width of the Cesium atom with $\gamma_\perp/2\pi=2.5$ MHz. This result is not surprising because for the case of strong coupling the excitation is hidden in the weakly damped cavity mode with a decay rate of $\kappa/2\pi=0.9$ MHz on average half of the time. Internal excitation therefore experiences less damping than for an atom in free space. But note also that the observed half-width is not as small as $(1/2)(\gamma_\perp+\kappa)/2\pi=1.7$ MHz expected from the eigenvalue structure (see eq. (1)), an effect which we believe to be mainly due to transit broadening which is not included in the theoretical model.

To understand the experimental data for $g^{(2)}(\tau)$ in more detail, we made some attempt to improve the theory published so far to get a more realistic description of our measurements. Toward this end we closely followed the work of Carmichael, Brecha and Rice [19] and generalized their discussion of the nonclassical effect which considers only the plane wave case. As will be shown below the nonclassical fluctuations scale with the cooperativity parameter $C_1(r)=g(r)^2/2\kappa\gamma_\perp$ for a single atom which in turn depends on the atomic position. Because the experiment employs an atomic beam with atoms distributed randomly over the standing-wave cavity mode, a treatment including a position dependent coupling constant $g(r)$ is highly needed. We omit the detailed calculation and quote the result only:

$$g^{(2)}(0) = \{1+\Delta\alpha/\alpha\}^2. \tag{5}$$

The size of the nonclassical effect is determined by the expression

$$\Delta\alpha/\alpha = 2C\,\frac{1-[(1+C')/C']\Sigma_i[C_i'/(1+C'-2C_i')]}{1+(1+\gamma/\kappa)\Sigma_i[C_i'/(1+C'-2C_i')]} \tag{6}$$

where the sum is over all atoms, $C_i=g(r_i)^2/2\kappa\gamma_\perp$, $C=C_1\bar{N}=\Sigma_iC_i$, with $\bar{N}=\Sigma_i\psi(r_i)^2$ and the primed quantities equal the corresponding unprimed quantities divided by $1+\gamma_\perp/\kappa$. It is instructive to look at the plane-wave case $[g(r_i)=g]$ where the summation over all atomic sites can be carried out analytically. We find $\Delta\alpha/\alpha=-2C_1'2C/(1+2C-2C_1')$. This agrees with the result given in Ref. [19] and therefore strongly supports our calculation generalized to the case of a spatially variable coupling constant. For $N\gg1$ and a weakly coupled system in the regime $\kappa\gg g\gg\gamma$ (i.e. $2C_1'=2C_1\ll1$) where cavity enhanced spontaneous emission with a rate $\gamma(1+2C)$ is well-defined and where the complications of the higher order absorption-reemission processes are avoided, the plane-wave result can be further simplified to give $\Delta\alpha/\alpha=-2C_1$. For the bad-cavity limit and one atom this was first calculated by Rice and Carmichael [21]. Here it shows that even for many atoms the antibunching effect ($\Delta\alpha/\alpha$ negative) is determined by the cooperativity parameter C_1 for a single atom. Lifting the restriction of weak-coupling the condition $g^{(2)}(0)=0$ ($\Delta\alpha/\alpha=-1$) is achieved for $C_1'=1/2$. But for a spatially varying electric dipole coupling constant $g=g(r_i)$, as is the case in the experiment, $C_i=g(r_i)^2/2\kappa\gamma_\perp=C_1\psi(r_i)^2$ depends on the atomic position. Hence it must be expected that the size of any nonclassical effect also depends on the particular distribution of the atoms, which fluctuates in an atomic beam experiment. This clearly justifies the improved calculation which takes into account the cavity geometry and which leads to the general result eq. (6) as shown above. Perfect antibunching can now be observed for $\Sigma_i[C_i'/(1+C'-2C_i')]=1$. But note that the main conclusion is the same as in the plane wave case. That is, the nonclassical character of $g^{(2)}(\tau)$ is effectively determined by single-atom effects even for N»1. This result is in contrast to an intuition based in the standard weak-coupling theory of optical bistability [14], because the quantum fluctuations of the system do not scale as $1/N$ for increasing N. This theoretical prediction was clearly observed in the experiments when the number of intracavity atoms was changed: For $18 < \bar{N} < 110$, the size of the particular nonclassical effect $g^{(2)}(0) < g^{(2)}(\tau)$ (i.e. photon antibunching) is constant.

To gain more insight into this surprising phenomena, we note that in the perturbative limit ($\kappa \gg g \gg \gamma$) there are three indistinguishable paths that can lead to the detection of two photons: (i) the transmission of two photons from the exciting field that pass through the cavity without being absorbed, (ii) the transmission of one unabsorbed photon and one photon emitted by an excited atom and, (iii) the transmission of a pair of photons emitted by two excited atoms. Note that when higher order processes involving absorption and reemission of intracavity photons are important, as they are in our experiment, an infinite number of paths must be considered. However, in the following discussion we restrict ourselves to the perturbative limit to gain a simple qualitative understanding [19]. In this case, the full two-photon probability, given by the sum of the probability amplitudes squared, depends on the quantum interference of these processes, as is reflected by the fact that the expression for $g^{(2)}(\tau)$ is a perfect square. Obviously, the escape of a photon projects the atom-cavity wave function into a reduced state that can deviate significantly from the steady state if C_1 is large. The important point is that the dipole field after state reduction is not simply rescaled by the ratio $1-1/N$ as it would be if we knew that one of the atoms (and indeed which particular one) had returned to its ground state. We do not know if the collapse of the wavefunction was due to the escape of a photon emitted by one of the atoms. The photon may have traversed the cavity without ever being absorbed. From this intrinsic indistinguishability we must conclude that after the photon has left the cavity the polarization field is still radiated by the same N atoms. But a photon has escaped the cavity and the second photon is emitted from a collective atomic oscillator with cavity-enhanced decay rate $\gamma(1+2C-2C_1)$ for $N-1$ atoms (if identical) rather than $\gamma(1+2C)$ for N atoms. Because of the smaller damping rate, the polarization field *increases* in magnitude. Note that since the dipole field is 180° out of phase with the driving field this increase in atomic polarization decreases the cavity field which is the sum of the polarization field radiated by the atoms and the constant external field which would built up inside the cavity if there were no atoms.

To summarize, note that this picture gives an interpretation of the fact that $g^{(2)}(0)$ is largely independent of the atomic number N: Because for constant external driving field both the mean intracavity field (see eq. (3)) and the quantum fluctuations of the macroscopic polarization field (which determine the fluctuations of the intracavity field as discussed above) do scale as $1/N$ for increasing N, the resulting *normalized* intracavity field fluctuations (as measured by the intensity correlation function eq. (4)) do no longer depend on N. They are nonclassical because for the regime of our experiment the intracavity field of the reduced state is smaller than that of the steady state (and hence $g^{(2)}(0) < 1$) because the atomic polarization field near $\tau=0$ increases due to the momentary loss of the reaction field from one of the atoms in the cavity. Note however, that for C_1 sufficiently large it is found possible for the increase in the polarization field to be large enough as to change the sign of the intracavity field and indeed to have $g^{(2)}(0) > 1$ with $g^{(2)}(\tau)=0$ for some later time $\tau > 0$ [19].

In calculating the oscillatory regression to steady state, we neglect the much slower motion of the atoms through the cavity and find

$$g^{(2)}(\tau) = \left\{1+(\Delta\alpha/\alpha)\exp[-(\gamma_\perp+\kappa)\tau/2][\cos(\Lambda\tau)+(\gamma_\perp+\kappa)/(2\Lambda)\sin(\Lambda\tau)]\right\}^2 \qquad (7)$$

where the oscillation frequency Λ is determined by the eigenvalue splitting

$$\Lambda = \left\{g^2\overline{N}-[(\gamma_\perp-\kappa)/2]^2\right\}^{1/2}. \qquad (8)$$

Our theoretical results for $g^{(2)}(\tau)$ are shown graphically in Fig. 3 by the solid curves, which match the experimental results quite well. In plotting these curves, the experimentally determined values g, γ, κ and $\overline{N}$ have been used and a simulation analogous to the one described in the preceeding chapter is employed for evaluating the summations with atoms placed randomly and repeatedly over a volume $V'{\gg}V$, where the average number of atoms in V is given by $\overline{N}$. The caveat in our analysis is that we have scaled the theoretical results for $|g^{(2)}(0)-1|$ down by factor of 4. This discrepancy may be due in part to a small atomic detuning arising from experimental uncertainties in the crossing angle between the cavity axis and the atomic beam direction, or it may arise from a small σ^- left-hand circular polarized field leaking through the cavity when we attempt to excite the σ^+-transition with right-hand circular polarized light. In addition, the transient nature of the atomic motion through the cavity mode (which so far has not been included in the theory) may have a significant effect in decorrelating the otherwise coherent response of the sample to the escape of a photon. Empirically, we also know that $|g^{(2)}(0)-1|$ is reduced somewhat because the weak-field limit is only approximately satisfied in our measurements so that the theory outlined above is not strictly valid. The calculation includes only the first dominant order in the electric field amplitude which allows to truncate the expansion at the level of two-quantum states.

Finally, note that in the experiments with $\overline{N}=110$ atoms the atom-cavity system transmits only about 1 out of 2.5×10^5 incident photons. Interestingly, the system does not work as an ordinary attenuator but it instead transforms the input laser light beam with Poisson photon number fluctuations into an output light beam with sub-Poissonian noise. Even more important for practical applications, the nonclassical light is transmitted as a Gaussian beam and is thus readily available as a source for various experiments in quantum optics.

4. Optical bistability with a few atoms

So far, all experiments that we have discussed were performed in the limit of weak excitation. To investigate a third effect of the steady state operating characteristics, we record the transmitted power at detector D3 versus the incident power monitored with detector D1 [15]. Again, the measurements are made with cavity and atomic detunings approximately zero, that is $\Omega=0$ ($\omega_p=\omega_a=\omega_c$). We then compare the result with the state equation from the semiclassical theory of optical bistability [14]. Because the number of atoms is small, the state equation is averaged over the fluctuations in the number of intracavity atoms N. From a comparison of theory and experiment far from the turning points of the hysteresis cycle of input versus output intensity, we can extract the atomic cooperativity parameter $C=C_1\overline{N}$ and hence an estimate of the effective intracavity atomic number $\overline{N}$. The results of these measurements, which were performed by scanning the laser intensity from very weak to very strong excitation, with the value of C found from the global fit of the state equation to the data, are in reasonable agreement with those obtained in weak field from the observed eigenvalue splitting.

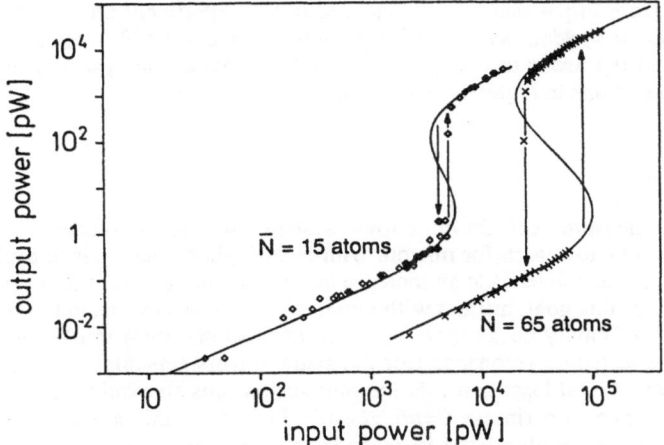

Fig. 4: Transmitted power versus input power for $\bar{N}=15$ and $\bar{N}=65$ atoms. Note that this is a log-log plot with an output power of 1 pW corresponding to an intracavity photon number $n_s=0.8$. The solid curve is from the semiclassical state equation averaged over atomic number and position fluctuations as discussed in the text.

On the other hand, as shown in Fig. 4 we find significant discrepancies near the turning points. This is, for example, evidenced by the fact that the semiclassical theory predicts bistability for $N \geq 5$ atoms ($C \geq 10$ for our Gaussian standing wave geometry) whereas we observe a clearly distinguishable bistability (as indicated by the vertical arrows in Fig. 4) only for $N \geq 15$. Moreover, the theoretical fits are very poor in the neighborhood of the turning points of the hysteresis cycle. In addition, there is dispersion for the value of the switching points in successive sweeps of the input intensity with all other external control parameters held constant. These deviations and the discrepancy with the state equation from optical bistability near the switching points are attributed to the dynamic nature of the fluctuations in the number of intracavity atoms, which occur on a time scale given by the transit time of the atoms through the cavity. For example, when scanning the input intensity at a slow rate close to the turning points, fast atomic number fluctuations will trigger passage from one bistability branch to the other. This means that in the vicinity of the switching point from the low-intensity to the high-intensity branch a decrease in the number of intracavity atoms will cause a transition from the lower to the upper branch. Similarly, close to the other turning point an increase in the number of atoms gives rise to a transition from the upper to the lower branch when slowly decreasing the input intensity along the high-intensity branch. This effect thus leads to an erosion of both sides of the hysteresis loop. As a result of these (and possibly other [14, 22]) microscopic fluctuations, the observed width of the bistability trace in Fig. 4 is considerably narrowed (about 3 times for $\bar{N}=15$ atoms). This may explain that the threshold of optical bistability occurs for approximately 3 times larger atom number than expected from the semiclassical state equation. For the experimental results shown in Fig. 4 we estimate that fluctuations of the number of atoms by two or three standard deviations may explain the discrepancy between the theoretical prediction (solid line) and the experimental observation (vertical arrows).

Note from Fig. 4 that the powers used in the experiment are very low for our strongly coupled system, with switching energies below 10^{-15} J estimated for the transition from the lower to the upper branch. This makes the system a device with potential applications in highly integrated optical electronics.

5. Conclusion

Beyond these measurements for our current system, our effort to increase g relative to $(\kappa, \gamma_\perp)$ has lead us to search for mirrors with even higher finesse since each factor f improvement in finesse leads to an increase in g of roughly $f^{1/2}$ for fixed cavity decay time κ. Towards this goal, mirrors with extremely low loss have been fabricated. Our measurements of cavity decay time (for mirrors separated by 4 mm), linewidth and total power reflected on resonance (for the same mirrors separated by 130 μm) lead to an inference of total loss (transmission plus scatter plus absorption) of $L=1.6\times10^{-6}$, which corresponds to a finesse $F=\pi/L=2\times10^6$! Following this avenue should make possible experiments with a large increase in the coupling constant g. In addition to spectroscopy of higher lying states of the strongly coupled atom-cavity system with more than one quantum of excitation [23], many of the exciting experiments proposed recently seem to be realizable in principle in the optical domain. Some of the possibilities discussed include, for example, the observation of a fundamental change in the radiative properties of the atom-cavity system due to an interaction with squeezed light [24], the trapping of single atoms at an antinode of the cavity mode [25], quantum non-demolition measurements of the photon number inside the cavity and preparation of trapping states of the electromagnetic field [26] or Fock states with a well-defined photon number [27], measurement-induced diffraction and interference of atoms from "virtual slits" [28] and many other interesting applications. To summarize, strongly coupled open quantum systems have many open questions!

This work was supported by the National Science Foundation, the Office of Naval Research, Venture Research International and by a Robert A. Millikan Fellowship (G. R.).

6. References

[1]   R. Loudon, Rep. Prog. Phys. _43_, 913 (1980).
[2]   M. C. Teich, B. E. A. Saleh, in "Progress in Optics", edited by E. Wolf
          (North-Holland, Amsterdam, 1988) Vol. _26_, p. 1.
[3]   See, for example, the special issues on squeezed states of light:
          (a) J. Opt. Soc. Am. B (1987) edited by H. J. Kimble, D. F. Walls;
          (b) J. Mod. Opt. (1987), edited by P. L. Knight, R. Loudon.
[4]   See, for example, the recent review papers:
          (a) E. A. Hinds, in "Advances in Atomic, Molecular and Optical Physics",
              edited by D. Bates and B. Bederson (Academic, 1990), Vol. _28_, 237
          (b) D. Meschede, Phys. Rep. _211_, 201 (1992).
[5]   E. T. Jaynes, F. W. Cummings, Proc. IEEE _51_, 89 (1963).
[6]   G. Rempe, R. J. Thompson, H. J. Kimble, R. Lalezari, Opt. Lett. _17_, 353
          (1992).
[7]   M. Tavis, F. W. Cummings, Phys. Rev. _170_, 379 (1968).

[8]   G. Rempe, N. Klein, H. Walther, Phys. Rev. Lett. 58, 353 (1987).

[9]   J. H. Eberly, N. B. Narozhny, J. J. Sanchez-Mondragon, Phys. Rev. Lett. 44, 1323 (1980).

[10]   (a) J. J. Sanchez-Mondragon, N. B. Narozhny, J. H. Eberly, Phys. Rev. Lett. 51, 550 (1983);
       (b) G. S. Agarwal, Phys. Rev. Lett. 53, 1732 (1984).

[11]   (a) Y. Kaluzny, P. Goy, M. Gross, J. M. Raimond, S. Haroche, Phys. Rev. Lett. 51, 1175 (1983);
       (b) R. J. Brecha, L. A. Orozco, M. G. Raizen, M. Xiao, H. J. Kimble, J. Opt. Soc. Am. B 3, 238 (1986);
       (c) M. G. Raizen, R. J. Thompson, R. J. Brecha, H. J. Kimble, H. J. Carmicheal, Phys. Rev. Lett. 63, 240 (1989);
       (d) Y. Zhu, D. J. Gauthier, S. Morin, Q. Wu, H. J. Carmichael, T. W. Mossberg, Phys. Rev. Lett. 64, 2499 (1990);
       (e) C. Weisbuch, M. Nishioka, A. Ishikawa, Y. Arakawa, Phys. Rev. Lett. 69, 3314 (1992).

[12]   F. Bernardot, P. Nussenzveig, M. Brune, J. M. Raimond, S. Haroche, Europhys. Lett. 17, 33 (1992).

[13]   R. J. Thompson, G. Rempe, H. J. Kimble, Phys. Rev. Lett. 68, 1132 (1992).

[14]   L. A. Lugiato, in "Progress in Optics", edited by E. Wolf (North-Holland, Amsterdam, 1984) Vol. 21, p. 69.

[15]   G. Rempe, R. J. Thompson, R. J. Brecha, W. D. Lee, H. J. Kimble, Phys. Rev. Lett. 67, 1727 (1991).

[16]   M. G. Raizen, L. A. Orozco, M. Xiao, T. L. Boyd, H. J. Kimble, Phys. Rev. Lett. 59, 198 (1987).

[17]   (a) G. Rempe, F. Schmidt-Kaler, H. Walther, Phys. Rev. Lett. 64, 2783 (1990);
       (b) M. Brune, J. M. Raimond, L. Davidovich, S. Haroche, Phys. Rev. Lett. 59, 899 (1987).

[18]   (a) F. Casagrande, L. A. Lugiato, Nuovo Cimento B 55, 173 (1980);
       (b) P. Drummond, D. F. Walls, J. Phys. A 13, 725 (1980).

[19]   H. J. Carmichael, R. J. Brecha, P. R. Rice, Opt. Commun. 82, 73 (1991).

[20]   H. J. Carmichael, R. J. Brecha, M. G. Raizen, H. J. Kimble, P. R. Rice, Phys. Rev. A 40, 5516 (1989).

[21]   P. R. Rice, H. J. Carmichael, IEEE J. Quantum Electron. QE 24, 1351 (1988).

[22]   P. D. Drummond, Phys. Rev. A 33, 4462 (1986).

[23]   (a) J. I. Cirac, H. Ritsch, P. Zoller, Phys. Rev. A 44, 4541 (1991);
       (b) L. Tian, H. J. Carmichael, Quantum Opt. 4, 131 (1992).

[24]   (a) C. W. Gardiner, Phys. Rev. Lett. 56, 1917 (1986);
       (b) A. S. Parkins, P. Zoller, H. J. Carmichael, to be published.

[25]   (a) S. Haroche, M. Brune, J. M. Raimond, Europhys. Lett. 14, 19 (1991);
       (b) B. G. Englert, J. Schwinger, A. O. Barut, M. O. Scully, Europhys. Lett. 14, 25 (1991).

[26]   P. Meystre, G. Rempe, H. Walther, Opt. Lett. 13, 1078 (1988).

[27]   (a) P. Filipowicz, J. Javanainen, P. Meystre, J. Opt. Soc. Am. B3, 906 (1986);
       (b) M Brune, S. Haroche, V. Lefevre, J. M. Raimond, N. Zagury, Phys. Rev. Lett. 65, 976 (1990);
       (c) M. J. Holland, D. F. Walls, P. Zoller, Phys. Rev. Lett. 67, 1716 (1991).

[28]   (a) P. Storey, M. Collett, D. F. Walls, Phys. Rev. Lett. 68, 472 (1992)
       (b) M. A. M. Marte, P. Zoller, Appl. Phys. B 54, 477 (1992).

Fock State Superpositions in Cavity QED with Dark Atoms

A.S. Parkins,[1] P. Marte,[1] P. Zoller[1] and H. J. Kimble[2]

[1] Joint Institute for Laboratory Astrophysics, and Department of Physics, University of Colorado, Boulder, CO 80309-0440

[2] Norman Bridge Laboratory of Physics 12–33, California Institute of Technology, Pasadena, CA 91125

1. Introduction

In this contribution we discuss a novel scheme for the preparation of photon Fock states and *general* superposition states in cavity quantum electrodynamics (CQED). Interest in the preparation of Fock states derives from the fact that these are quantum states of the radiation field with a fixed number of photons, i.e., light with no intensity fluctuations; coherent superposition states, on the other hand, correspond to "Schrödinger cat" states of the photon field. Several proposals for the preparation of Fock states have recently been put forward [1]. These proposals make use of the fascinating and unique possibilities offered by CQED, in which beams of atoms interact strongly with a single quantized field mode of a cavity. Experimental progress in this field over the past few years has been dramatic both in the microwave [2] and optical [3] regimes, with atom–cavity–mode coupling strengths now exceeding dissipative rates (from spontaneous emission and cavity losses) and making the above–mentioned proposals of very direct relevance.

Our scheme is based on the fact that the ground state Zeeman coherence of an atom which is passed through a cavity can be transferred one-to-one to a (coherent) superposition of photon number states of the cavity mode. In particular, initial preparation in a single Zeeman state produces a Fock state. The new scheme is based on adiabatic passage via a superposition of "dark states" of the strongly coupled atom–cavity system involving a resonant coherent pump field, and starting from the cavity mode in the vacuum state. Conditions for adiabatic passage for the generation of an $n = 1, 2, \ldots$–photon Fock state are that the atom–cavity coupling g and the Rabi frequency of the pump field Ω are large compared with the spontaneous decay width Γ of the atom and the cavity damping rate κ,

$$\Omega, 2g \gg \Gamma, n\kappa \quad \text{and} \quad T \ll (n\kappa)^{-1}, \tag{1}$$

and that

$$\Omega T, \quad 2gT \gg 1 \tag{2}$$

with T the interaction time of the atom with the cavity and the laser field. Note, however, that we do *not* require an equality $\Gamma T \ll 1$.

The attractive feature of our adiabatic passage scheme is that the generality of the superpositions that can be produced in the cavity is limited only by the extent to which one can prepare general superpositions of atomic Zeeman sublevels using, for example, optical or radio frequency pumping. Furthermore, since the adiabatic transfer is along a dark state, i.e. the excited atomic states are never populated, there is no spontaneous emission noise. Following the adiabatic transfer, the atomic population is with probability one in a *single* atomic state, thus avoiding the introduction of atomic–state "measurement noise." In this sense, the process of generating a Fock state and superposition state is *deterministic*. We emphasize, however, that Fock states and superposition states will decay due to cavity damping. Finally, we note that the conditions (1) and (2) are *inequalities*, i.e. we do not require a fixed interaction time.

2. Adiabatic Passage in a Three-Level Lambda-System

Adiabatic passage has been studied previously in the context of coherent population transfer [4] and coherent atomic–beam deflection [5] in Λ–systems, in which atomic population is coherently transferred between the atomic ground states via Raman transitions induced by a pair of (overlapping) time delayed laser pulses.

Coherence of the atomic wave function is preserved since the transfer occurs via a dark state: the excited atomic state is never populated and hence spontaneous emission does not play a role. These previous studies of adiabatic passage have dealt with situations in which photons associated with the Raman transitions are absorbed from, and emitted into, light fields that can be regarded as classical fields (i.e. coherent laser fields).

In the present context we consider the field into which photons are emitted to be the field mode of a cavity, and that the requisite time–dependent coupling to this field (i.e. the "pulse") is provided by the atom's motion through the cavity. In other words, we now require that one of the fields be quantized. Aside from this modification, our analysis of the system is completely analogous to previous work on adiabatic passage. To illustrate the basic principle and demonstrate the preparation of Fock states of the cavity mode, we consider first the simplest possible configuration. This consists of a three–level atom with two ground states $|g_1\rangle$ and $|g_2\rangle$ (for simplicity we consider the ground states to have the same energy, but they need not be degenerate) coupled to an excited state $|e\rangle$ via, respectively, a classical laser field $\Omega(t)$ of frequency ω_L and a cavity mode field of frequency ω.

We write the Hamiltonian for this system in the form (for the moment we shall omit dissipation)

$$H(t) = \hbar\omega a^\dagger a + \hbar\omega_{eg}|e\rangle\langle e| - i\hbar g(t)\left(|e\rangle\langle g_2|a - h.c.\right)$$
$$+\frac{1}{2}i\hbar\Omega(t)\left(|e\rangle\langle g_1|e^{-i\omega_L t} - h.c.\right), \qquad (3)$$

where a is the annihilation operator for the cavity mode and $g(t)$ gives the atom–cavity–mode coupling strength. The time dependence of $\Omega(t)$ and $g(t)$ denotes the motion of the atom across the laser–beam and cavity–mode profiles.

The Hamiltonian $H(t)$ has the property that it couples only states within the family, or manifold, $\{|g_1, n\rangle, |e, n\rangle, |g_2, n+1\rangle\}$, where $|g, n\rangle \equiv |g\rangle|n\rangle$, $|e, n\rangle \equiv |e\rangle|n\rangle$, and $|n\rangle$ represents a n–photon Fock state of the cavity mode. Such a family is depicted in Fig. 1(a). The adiabatic energy eigenvalues of the Hamiltonian associated with a particular family of states have the form

$$E_n = n\hbar\omega,$$
$$E_n^\pm = n\hbar\omega + \hbar\left\{\Delta \pm \left[\Delta^2 + g(t)^2(n+1) + \Omega(t)^2/4\right]^{1/2}\right\},$$

where we have assumed that $\omega = \omega_L$, and $\Delta = \omega_{eg} - \omega$ is the detuning. Of particular interest to us is the eigenstate corresponding to $E_n = n\hbar\omega$, which is given by

$$|E_n\rangle = \frac{2g(t)\sqrt{n+1}\,|g_1, n\rangle + \Omega(t)\,|g_2, n+1\rangle}{\sqrt{\Omega(t)^2 + 4g(t)^2(n+1)}}. \qquad (4)$$

This eigenstate does not contain any contribution from the excited state (hence the term "dark state"), and is independent of the detuning

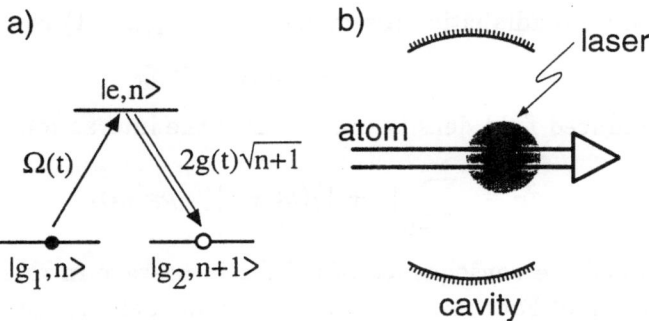

Fig. 1. (a) Λ three-level atom. (b) Proposed configuration for the preparation of Fock states using three–level atoms. The propagation direction of the pump laser is perpendicular to the page.

Δ. The possibility for adiabatic passage arises from the following behavior of $|E_n = n\hbar\omega\rangle$:

$$|E_n\rangle \rightarrow \begin{cases} |g_1, n\rangle, & \text{for } \Omega(t)/g(t) \rightarrow 0 \\ |g_2, n+1\rangle, & \text{for } g(t)/\Omega(t) \rightarrow 0. \end{cases} \tag{5}$$

That is, for the "counter–intuitive" pulse sequence in which the $\Omega(t)$ pulse is time–delayed with respect to $g(t)$, the state $|g_1, n\rangle$ may be adiabatically transformed into the state $|g_2, n+1\rangle$. A possible experimental configuration providing such a pulse sequence is shown schematically in Fig. 1(b). If we assume simple Gaussian pulse profiles for $\Omega(t)$ and $g(t)$ of width T (FWHM) and peak intensities Ω_{max} and g_{max}, the necessary condition for adiabatic following is given by Eqs. (1) and (2). This condition results from the requirement that the probability for transitions from $|E_n\rangle$ to other states be very small. Given that the pulses have a significant overlap in time, it ensures that $|E_n\rangle$ is well separated from $|E_n^{\pm}\rangle$ at all times during the interaction, and that nonadiabatic coupling between these eigenstates is not significant.

The effect on the field statistics of passing a single atom through the cavity is to *shift* the field density matrix by *exactly one photon*. Consider a cavity in a mixed state corresponding to the field density matrix ρ_F. The density operator $\rho(t)$ of the combined atom + field system is thus initially prepared in the state $|g_1\rangle\langle g_1| \otimes \rho_F$, and its time evolution due to adiabatic passage $|g_1, n\rangle \to |g_2, n+1\rangle$ is

$$|g_1\rangle\langle g_1| \otimes \rho_F \to |g_2\rangle\langle g_2| \otimes \rho_F' \qquad (6)$$

with the reduced field density matrix after the interaction

$$\rho_F' = \sum_{n,m=0}^{\infty} |n+1\rangle\langle m+1|\langle n|\rho_F|m\rangle. \qquad (7)$$

An immediate consequence of adiabatic passage in this system is the generation of Fock states of the cavity mode, an intriguing point of note being that single–photon Fock states are produced from adiabatic passage "out of the vacuum." In particular, if the atom enters the interaction region in state $|g_1\rangle$ and the cavity mode is initially in the *vacuum* state $|0\rangle$, then by adiabatic passage the initial state $|g_1, 0\rangle$ is transformed into the state $|g_2, 1\rangle$ and a single–photon Fock state is realized. Sequences of atoms can obviously be used to generate Fock states of higher photon number, each atom providing a "shift" of one photon through the transformation $|g_1, n\rangle \to |g_2, n+1\rangle$. More generally, any initial distribution of Fock states will experience such a "shift" as a result of the passage of an atom through the cavity and laser fields.

Dissipation in the form of spontaneous emission and cavity damping is, of course, an important practical issue. As noted before, the adiabatic passage technique is robust against the effects of spontaneous emission, as, in principle, the excited atomic state is never appreciably populated. Cavity damping is certainly a problem as its effects come into play as soon as the cavity mode is excited, leading potentially to degradation of the adiabatic transfer (a simple picture of the effect of dissipation is that it couples manifolds $\{|g_1, n\rangle, |e, n\rangle, |g_2, n+1\rangle\}$ of different n: ideal adiabatic transfer occurs when the passage is across a single manifold). It follows that the technique will be optimized when condition (1) is satisfied. Hence, we require conditions under which "vacuum–Rabi splitting" would be observable in the coupled atom–cavity system. Results of calculations based on a full quantum master equation treatment to account for cavity damping and spontaneous decay will be shown below.

3. Zeeman Atom and Master Equation Treatment

The adiabatic passage procedure is readily generalized to more complicated atomic–level configurations, as for instance in an atom possessing

Zeeman substructure. The associated increase in the possible number of Raman transitions means that a single atom may be used to generate an n–photon Fock state of the cavity, where n is limited only by the degree of Zeeman degeneracy. As we shall see, this also offers the possibility of creating a very general range of superposition states.

As an example, we consider a general model of an atomic $J_g = N \to J_e = N - 1$ transition in the adiabatic passage configuration, with a π–polarized cavity mode field and a σ^+–polarized laser field. Including dissipation, we turn to a master–equation description which takes the form

$$\frac{\partial \rho}{\partial t} = -i(H_{\text{eff}}\rho - h.c.) + \Gamma \sum_{\sigma=0,\pm 1} A_\sigma \rho A_\sigma^\dagger + 2\kappa a \rho a^\dagger, \qquad (8)$$

where $\rho(t)$ is the reduced density operator of the system, and

$$
\begin{aligned}
H_{\text{eff}} = & (\Delta - i\Gamma/2) \sum_{m_e} |J_e m_e\rangle\langle J_e m_e| - i\kappa a^\dagger a \\
& -i\Omega(t)(A_{+1} - A_{+1}^\dagger) + ig(t)(a^\dagger A_0 - A_0^\dagger a). \qquad (9)
\end{aligned}
$$

The atomic lowering operators A_σ are given by

$$A_\sigma = \sum_{m_e, m_g} |J_g m_g\rangle\langle J_g m_g; 1\sigma|J_e m_e\rangle\langle J_e m_e|, \qquad (10)$$

where $\langle J_g m_g; 1\sigma|J_e m_e\rangle$ is a Clebsch–Gordon coefficient for the dipole transition $|e\rangle \to |g\rangle$ with polarization $\sigma = 0, \pm 1$.

Provided that the conditions for adiabatic transfer are satisfied, i.e., that we have a pulse sequence in which $g(t)$ precedes $\Omega(t)$, and conditions (1) and (2) are satisfied, passage may then occur along a dark state, which for the general system described above, and for the particular transformation $|g_{-N}, n\rangle \to |g_{N-1}, n + 2N - 1\rangle$ (i.e. population initially in the atomic ground state $|g_{-N}\rangle$ is coherently transferred to the state $|g_{N-1}\rangle$, and the initial cavity mode Fock state $|n\rangle$ is transformed into the Fock state $|n + 2N - 1\rangle$) takes the form

$$
\begin{aligned}
|E_n = & \; n\hbar\omega, g_{-N}\rangle = \\
\mathcal{N} & \left\{ |g_{-N}, n\rangle G_{-N+1}^{(n)} G_{-N+2}^{(n)} \cdots G_{N-1}^{(n)} \right. \\
+ & \; |g_{-N+1}, n + 1\rangle \Omega_{-N+1} G_{-N+2}^{(n)} G_{-N+3}^{(n)} \cdots G_{N-1}^{(n)} \\
+ & \; \cdots \\
+ & \left. \; |g_{N-1}, n + 2N - 1\rangle \Omega_{-N+1} \Omega_{-N+2} \cdots \Omega_{N-1} \right\}. \qquad (11)
\end{aligned}
$$

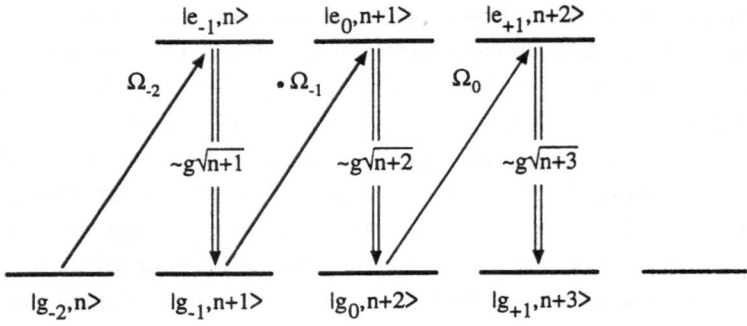

Fig. 2. Adiabatic passage in $J_g = 2 \to J_e = 1$ system. The pump laser is σ^+ polarized, the cavity mode is π-polarized. The family of bare states coupled in the Jaynes Cummings model are $|g_{-2}, n\rangle$, $|e_{-1}, n\rangle$, $|g_{-1}, n + 1\rangle$, ..., $|g_{+1}, n + 3\rangle$. Adiabatic passage corresponds to the transfer $|g_{-2}, n\rangle \to |g_{+1}, n + 3\rangle$.

Here $\mathcal{N}$ is a normalization factor, and

$$
\begin{aligned}
G_k^{(n)} &= g(t)\sqrt{n + N + k}\langle J_g(m_g = k); 10 | J_e(m_e = k)\rangle \ , \quad k < N \quad (12) \\
\Omega_k &= \Omega(t)\langle J_g(m_g = k - 1); 11 | J_e(m_e = k)\rangle \ , \quad k > -N. \quad (13)
\end{aligned}
$$

Hence, beginning with a vacuum state of the cavity mode ($n = 0$), passage of a single atom yields a $(2N - 1)$–photon Fock state in the cavity, and each subsequent atom (entering in the state $|g_{-N}\rangle$) increases the photon number by $(2N - 1)$.

As a specific example we consider a $J_g = 2 \to J_e = 1$ transition ($N = 2$), as depicted in Figs. 2 and 3. Figure 2 shows the bare atom+field states in the Jaynes–Cummings model, $|g_{-2}, n\rangle$, $|e_{-1}, n\rangle$, $|g_{-1}, n + 1\rangle$ ... $|g_{+1}, n + 3\rangle$. Adiabatic passage transfers population $|g_{-2}, n\rangle \to |g_{+1}, n + 3\rangle$. In Fig. 3(a) we have drawn the (seven) states in the manifold of which the initial state $|g_{-2}, 0\rangle$ is a member. Adiabatic passage in this system corresponds to $|g_{-2}, 0\rangle \to |g_{+1}, 3\rangle$ and hence to the preparation of the three–photon Fock state $|3\rangle$.

Results of a numerical solution of the master equation for the preparation of a three–photon Fock state in a $J_g = 2$ to $J_e = 1$ system are shown in Fig. 4. There we display the time variation of (a) the exciting pulses, (b) the (seven) energy eigenvalues associated with the particular manifold involved (compare Fig. 3(a)), (c) the

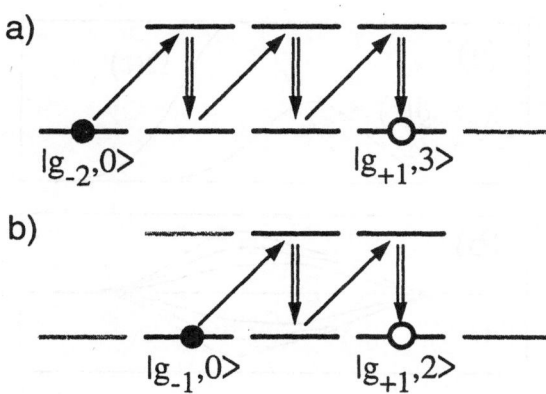

Fig. 3. Manifolds of states for a $J_g = 2 \rightarrow J_e = 1$ transition. Adiabatic passage "from the vacuum" produces the transformations (a) $|g_{-2},0\rangle \rightarrow |g_{+1},3\rangle$, and (b) $|g_{-1},0\rangle \rightarrow |g_{+1},2\rangle$. (The rest of the family members of states $|g_0,0\rangle \rightarrow |g_{+1},1\rangle$ and $|g_{+1},0\rangle \rightarrow |g_{+1},0\rangle$ are not shown.)

populations of the atomic ground–state sublevels, and (d) the mean cavity photon number, $\langle a^\dagger a \rangle$, and the Mandel Q parameter. The Q parameter is a measure of intensity fluctuations in the cavity field ($Q = -1 + [\langle (a^\dagger a)^2 \rangle - \langle a^\dagger a \rangle^2]/\langle a^\dagger a \rangle$) and is equal to -1 for a pure Fock state. The parameters for this figure are $gT = 25$, $\Omega T = 50$, $\Gamma T = 5$. In Fig. 4(d) the mean cavity photon number is the solid curve, while the Mandel Q parameter is plotted as a dashed line. The curves in Fig. 4(d) labeled "1" correspond to no cavity damping. In Fig. 4(d), we have also included results obtained with nonzero cavity damping (curves labeled "2"). This damping clearly limits the maximum obtainable cavity photon number and the maximum reduction in intensity fluctuations.

4.   Preparation of Superposition States

The situation in which the cavity mode is initially in the vacuum state is of special interest because, in this case, the system states $|g_m,0\rangle$, with $m = -N, -N+1, \ldots, N-1$, each belong to different (orthogonal) manifolds of the system Hamiltonian, and hence they evolve (under

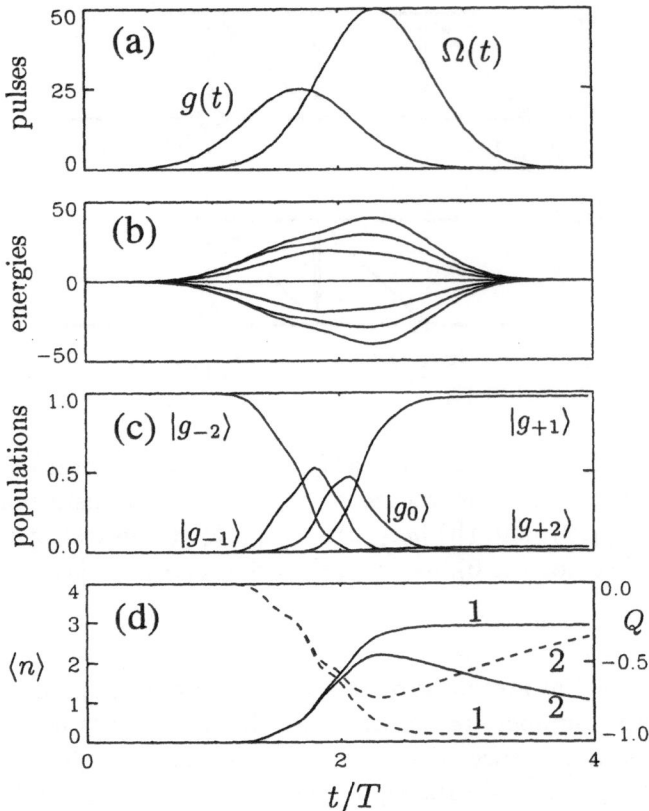

Fig. 4. Preparation of a three–photon Fock state via the adiabatic transformation $|g_{-2}, 0\rangle \to |g_{+1}, 3\rangle$ ($J_g = 2 \to J_e = 1$). Time variation of (a) exciting pulses, (b) energy eigenvalues for the manifold shown in Fig. 3(a), (c) populations of the atomic ground–state sublevels, (d) mean cavity photon number (solid) and Mandel Q parameter (dashed) for $2\kappa T = 0$ (1) and $2\kappa T = 0.5$ (2). Other parameters for this figure are $gT = 25$, $\Omega T = 50$, $\Gamma T = 5$. Plots (c) and (d) were obtained by numerical solution of the master equation of the coupled atom–cavity system.

ideal conditions) independently of each other. Furthermore, each of these manifolds possesses a dark state along which adiabatic passage may proceed, yielding the general transformation $|g_m, 0\rangle \rightarrow |g_{N-1}, N-m-1\rangle$. For the $J_g = 2 \rightarrow J_e = 1$ transition, this means that we can also consider adiabatic passage in, for example, the manifold of states beginning with $|g_{-1}, 0\rangle$, as shown in Fig. 3(b). This corresponds to the preparation of the two-photon Fock state $|2\rangle$.

This possibility of independent adiabatic passage along dark states belonging to different manifolds immediately suggests a means of preparing superposition states of the cavity mode. In particular, given that the atom is prepared in a coherent superposition of ground–state Zeeman sublevels, and that the cavity is initially in the vacuum state, adiabatic passage will coherently "map" this superposition onto the cavity mode Fock states. This is summarized, for a $J_g = N \rightarrow J_e = N - 1$ atomic transition, by the transformation equation

$$\left(\sum_{m=-N}^{N-1} c_m |g_m\rangle \right) |0\rangle \rightarrow |g_{N-1}\rangle \left(\sum_{m=-N}^{N-1} c_m |N-m-1\rangle \right). \qquad (14)$$

This illustrates a means by which we can, in principle, *generate arbitrary superpositions of Fock states*, limited only by the Zeeman degeneracy of the atom, and by the extent to which one can prepare arbitrary superpositions of atomic ground–state Zeeman sublevels (e.g. via optical or radio frequency pumping [6]). A variety of interesting possibilities come to mind, such as (truncated) phase states of the electromagnetic field [7].

As an example we plot in Fig. 5 the time evolution of the reduced field density matrix $\rho_F(t) = Tr_{\text{Atom}}\rho(t)$ as a function of the photon numbers $n = 0, 1, 2, 3$, $m = 0, 1, 2, 3$ for a $J_g = 2 \rightarrow J_g = 1$ atom. At time $t = 0$ the atom is prepared in a Zeeman superposition $(|g_{-2}\rangle + |g_{+1}\rangle)/\sqrt{2}$, and the cavity in the vacuum state. The plot was obtained by integrating the master equation with parameters $gT = \Omega T = 50$, $\Gamma T = 5$, and $\kappa T = 0$ (no cavity damping). After the interaction the field mode is in the superposition state $(|0\rangle + |3\rangle)/\sqrt{2}$.

5. Two-Cavity Superposition States

A variation on the ideas presented above can also be used to prepare superposition states of *macroscopically separated cavity fields* [8]. Consider, for example, two separated cavity modes, each of which overlaps

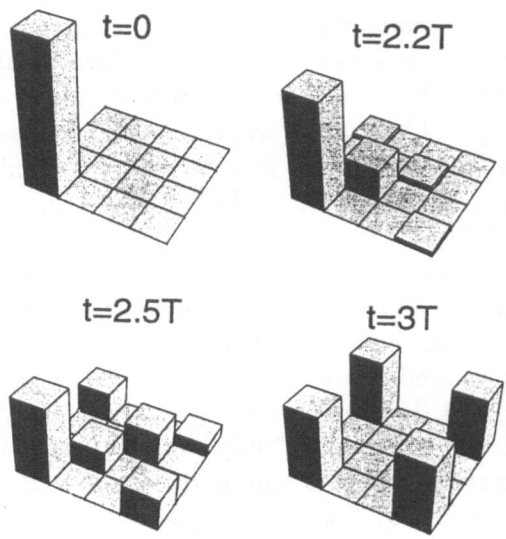

Fig. 5. Preparation of a $(|0\rangle + |3\rangle)/\sqrt{2}$ superposition state in a $J_g = 2$ to $J_e = 1$ system. We plot the time evolution of the reduced field density matrix $\rho_F = Tr_A\rho(t)$. The density matrix in the Fock basis $\langle n|\rho_F(t)|m\rangle$ is plotted as a 3D–histogram with x–axis $n = 0, 1, 2, 3$ and y–axis $m = 0, 1, 2, 3$ (the origin $n = m = 0$ is in the far left corner of each histogram). The parameters are $gT = \Omega T = 50$, $\Gamma T = 5$, and $\kappa T = 0$ (no cavity damping). At time $t = 0$ the system is in the vacuum.

a laser field in the adiabatic passage configuration.

The two cavity fields are of identical linear polarization and are coupled to the $J_g = 1 \rightarrow J_e = 0$ transition of a single atom passing through the cavities. The laser fields in the first and second cavities are σ^+- and σ^--polarized respectively. The atom is assumed to be initially prepared in a coherent superposition of the ground states $|g_{-1}\rangle$ and $|g_{+1}\rangle$. It is straightforward to show that passage of the atom through the two (consecutive) interaction zones produces the transformation

$$\frac{1}{\sqrt{2}}(|g_{-1}\rangle + |g_{+1}\rangle)|0,0\rangle$$

$$\rightarrow \frac{1}{\sqrt{2}}(|g_0\rangle|1,0\rangle + |g_{+1}\rangle|0,0\rangle) \quad \text{(1st cavity)}$$

$$\rightarrow \quad \frac{1}{\sqrt{2}}|g_0\rangle(|1,0\rangle + |0,1\rangle) \quad \text{(2nd cavity)}, \tag{15}$$

where $|a,b\rangle \equiv |a\rangle|b\rangle$ denotes the state with a photons in the first cavity and b photons in the second cavity. More general superposition states of two, or even more, cavity fields can be constructed by considering atomic transitions of higher Zeeman degeneracy.

6. Conclusions

We have proposed a new scheme for generating Fock states in a cavity by "adiabatic passage from the vacuum." In addition we have shown that an atomic Zeeman coherence can be mapped one–to–one to a photon coherence in a cavity. To the extent that arbitrary Zeeman coherences can be prepared in an atom by radio frequency fields and optical pumping, a "general superposition state" can be prepared in the cavity. The maximum number of photons in these states is limited by the Zeeman degeneracy. We emphasize that the scheme is robust against atomic spontaneous emission, and that the method of preparing the photon states is deterministic, since the atoms leave the cavity in a single final (pure) state.

Finally, we note that all of the adiabatic passage schemes presented above are reversible, i.e., by reversing the order of $g(t)$ and $\Omega(t)$, states of the cavity field(s) can be mapped onto atomic ground-state sublevels. Probing these sublevels following the passage of the atom through a cavity would thus provide detailed information on the field that was present in the cavity. Hence, we would also like to propose adiabatic transfer as a powerful tool for the *measurement* of cavity fields.

We thank C.M. Savage for discussions. The work at JILA is supported in part by the National Science Foundation.

References

1. M. Brune *et al.*, Phys. Rev. Lett. **65**, 976 (1990); M.J. Holland, D.F. Walls, and P. Zoller, *ibid.* **67**, 1716 (1991); J.I. Cirac *et al.*, *ibid.* **70**, 762 (1993); J.J. Slosser, P. Meystre, and E.M. Wright, Opt. Lett. **15**, 233 (1990).

2. G. Rempe, F. Schmidtkaler, and H. Walther, Phys. Rev. Lett. **64**, 2783 (1990).

3. G. Rempe *et al.*, Phys. Rev. Lett. **67**, 1727 (1991).

4. J. Oreg, F.T. Hioe, and J.H. Eberly, Phys. Rev. A **29**, 690 (1984); F.T. Hioe and C.E. Carroll, *ibid.* **37**, 3000 (1988); J.R. Kuklinski *et al.*, *ibid.* **40**, 6741 (1989); U. Gaubatz *et al.*, J. Chem. Phys. **92**, 5363 (1990).

5. P. Marte, P. Zoller, and J.L. Hall, Phys. Rev. A **44**, R4118 (1991); the coherent beam–deflection scheme proposed by these authors has recently been implemented experimentally: L. Goldner, private communication; J. Lawall and M. Prentiss, private communication.

6. One means of preparing atomic ground–state superpositions is to pump the atom into a single Zeeman ground state and introduce a rotation of the quantization axis between the preparatory and interaction regions.

7. D.T. Pegg and S.M. Barnett, Phys. Rev. A **39**, 1665 (1989).

8. P. Meystre, in *Progress in Optics*, Vol. XXX, ed. E. Wolf (North–Holland, Amsterdam, 1992).

MESOSCOPIC QUANTUM COHERENCES IN CAVITY QED

S. Haroche[1], M. Brune[1], J.-M. Raimond[1] and L. Davidovich[2]

[1] Laboratoire de Spectroscopie Hertzienne de l'Ecole Normale Supérieure [3]
24, rue Lhomond, 75231 Paris Cedex 05.

[2] Departamento de Fisica, Pontificia Universidade Catolica, 22453 Rio de Janeiro, Brazil.

1 Introduction

The quantum theory of measurement is still vigorously debated among physicists, in spite of the unquestionable success of quantum mechanics at explaining all phenomena in the physical world. These debates are fueled by the apparent contradiction between a theory which is based on the existence of probability amplitude interferences, and our daily experience of the classical world, where probabilities merely add up. A very clear exposition of this problem, with its historical background, along with a convincing discussion of how probability amplitude coherence is lost in classical physics, can be found in a recent Physics Today article by Zurek [1].

The basic point exposed by Zurek and others [2] is that decoherence follows from the irreversible coupling of the observed system to the outside world "reservoir". This coupling results in a very fast decay of non classical coherences, on a time scale much shorter than the relaxation time of the physical observables of the system under study. Within this short decoherence time, the system evolves into a statistical mixture of states, whose basis is selected by the apparatus used for the experiment. In this mixture, all the information in the system can be described in classical terms and our usual perception of the world is recovered.

Systems exhibiting quantum coherence between parts differing by some macroscopic classical physical parameter have been given the generic name of "Schrödinger cats", after the famous discussion carried out by this physicist in the thirtees [3]. The general argument exposed above implies that, due to unavoidable friction forces, the coherence between the two parts decays to zero within a time which is of the order of the system classical relaxation time divided by a dimensionless number measuring the "separation" between

---

[3]Unité de Recherche de l'Ecole Normale Supérieure et de l'Université Pierre et Marie Curie, associée au CNRS

the two parts. To state that the system is "macroscopic" amounts to saying that this separation is an astronomically large number, as it undoubtely is when considering for example biological systems such as "cats" made of huge number of molecules. In the simple case of a particle split into two spatially separated wave packets by a distance d, the dimensionless measure of the separation is $(d/\lambda_{dB})^2$, where λ_{dB} is the particle de Broglie wavelength. For a massive particle at not too low a temperature, this number is huge and the decoherence is for all purposes instantaneous.

It is possible however to envision "mesoscopic" systems, at the frontier between the classical and the quantum worlds, in which decoherence could occur within a physically observable time scale. The experimental study of such systems is extremely interesting, since it provides a testing ground for fundamental ideas which are at the heart of the measurement theory in quantum mechanics. One of these systems, whose study has been pioneered by the works of Calderia and Leggett [4], involves electric currents flowing in Josephson junctions and tunneling across an effective potential barrier. An even conceptually simpler system involves an electromagnetic field prepared into a superposition of states with different classical amplitudes or phases [5]. Various situations have been theoretically discussed in the past ten years or so, which generally involve some kind of non linear coupling between different field modes, bringing the system in an entangled state. A measurement on one of the fields then projects the other into a superposition state. The difficulty of most of the schemes presented so far is that they involve traveling fields, which escape at speed of light from the experimental area, so that tests of the mesoscopic coherence are hard to achieve, even conceptually.

We have recently proposed a variant of these non classical field experiments [6], which we are now trying to perform at ENS on a microwave field stored in a very high Q superconducting cavity. The field should thus remain there to be observed for an extended period of time. The cavity field is prepared into a Schrödinger cat state and detected by a beam of Rydberg atoms excited in radiatively long lived "circular states" [7]. Both the atom and the field parts of the system have very long relaxation times, typically in the 10^{-2}s to 10^{-3}s range, which should allow quantum coherences to be observed for an appreciable time, even on systems made of a relatively large number of photons. The aim of this chapter is to describe this experiment in its simplest possible form and especially to show how the decoherence of the Schrödinger cat state could be monitored so to speak in real time, by means of a two atom correlation measurement. The general idea of this experiment has been presented elsewhere [6], but the decoherence detection scheme presented here is much simpler than the one discussed earlier and should be easier to implement.

2 Quantum superpositions of classical fields with different phases

Assume that by some means to be specified later, one has prepared in a cavity a field in the quantum state:

$$|\Psi_{field}\rangle = \frac{|\alpha_0\rangle + e^{i\psi}|-\alpha_0\rangle}{\sqrt{2(1 + \cos\psi\exp(-2|\alpha_0|^2))}} \tag{1}$$

linear superposition of two classical fields of opposite phases. In this expression, α_0 is a c-number representing the complex amplitude of the classical field and ψ is a quantum phase taking any value between 0 and 2π. The denominator is a normalization constant, very close to $1/\sqrt{2}$ when the modulus of α_0 exceeds a few units ($\langle\alpha_0| - \alpha_0\rangle \simeq 0$). Each component in Eq.(1) is the field equivalent of a minimum particle wave packet oscillating in an harmonic potential (α_0 being the oscillator position in phase space). The superposition (1) is thus the exact counterpart of a particle split into two parts, oscillating symmetrically in the potential. Such a state, alternatively non local in position and momentum spaces, is highly non classical. The quantum phase ψ must not be confused with the classical phase of the field components, which differ by π from each other.

Of special interest are the even and odd cat state superpositions ($\psi = 0$ and π respectively). The photon number probability distribution $P(n)$ is non zero only for even n values in the $\psi = 0$ cat state, only for odd n values in the $\psi = \pi$ cat [6]. The photon number distributions thus exhibit fringes versus n, which are due to interference between photon number amplitudes associated to the two parts of the field wavefunction.

Figure 1a shows the Wigner distribution representing the density matrix of the $\psi = 0$ cat state in phase space [6]. This distribution appears as a real function of a complex amplitude α, which can take negative values in limited regions of phase space (quasi-probability distribution function). The two gaussian peaks are associated to the $|\alpha_0\rangle$ and $|-\alpha_0\rangle$ classical fields and the oscillatory part in between is directly related to the coherence between them. The existence of the oscillations, associated with negative quasi-probability values, is a signature of a non-classical state. Such a state is very different from a statistical mixture involving the same classical field parts, for which the Wigner distribution merely exhibits two separated peaks without superimposed oscillations (see Fig.1b).

This system is simple enough so that one can derive an exact analytical expression for its evolution, from a quantum superposition to a statistical mixture. Assuming that the field is weakly coupled to a reservoir of oscillators

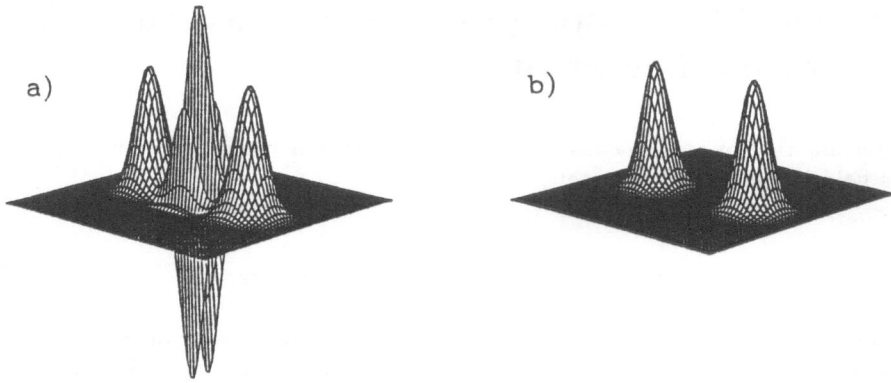

Figure 1: Representations of the Wigner distribution of a coherent superposition of two opposite phase classical fields (a) and of a statistical mixture of the same fields (b). The average number of photons is 5.

in the cavity walls, one finds the following exact expression for the system Wigner function, as a function of time (here again we assume $\psi = 0$) [6]:

$$
\begin{aligned}
W(\alpha, t) = \frac{2}{\pi(1+e^{-2|\alpha_0|^2})} \Big\{ &\exp(-2|\alpha - \alpha_0 \ e^{-t/2T_{cav}}|^2) \\
&+ \exp(-2|\alpha + \alpha_0 \ e^{-t/2T_{cav}}|^2) \\
+ 2e^{-2|\alpha|^2} \cos\left[4\Im(\alpha\alpha_0^*)e^{-t/2T_{cav}}\right] &\exp\left[-2|\alpha_0|^2(1 - e^{-t/2T_{cav}})\right]
\end{aligned}
\tag{2}
$$

The two gaussian functions in Eq.(2) are associated with the two classical field components. Their center evolves towards the origin of coordinates (vacuum state) within a time constant $2T_{cav}$, where T_{cav} is the damping time of the electromagnetic energy in the cavity. On the other hand, the oscillatory part of the Wigner function (cosine term in Eq.(2)) decays, when $t \ll T_{cav}$, with the time constant $T_{cav}/|\alpha_0|^2$, i.e. $T_{cav}/\bar{n}$ where $\bar{n}$ is the initial mean number of photons in each field component. This results could have been guessed from the general ideas outlined above, since $2|\alpha_0|$ measures in dimensionless units the "distance" in phase space between the two classical field components. After a decoherence time of the order of $T_{cav}/\bar{n}$, the system is described by a double peaked Wigner distribution of the kind shown in Fig.(1b) and has practically become an incoherent statistical mixture.

The above exact results can be qualitatively summarized in a simpler form, in terms of the quantum phase ψ. Initially ψ is well defined and the field is in a pure quantum state. Relaxation makes it evolve however into a statistical mixture, for which ψ must be described as a random quantity. At the end of the decoherence process, at time t longer than $T_{cav}/\bar{n}$, ψ has been completely scrambled so that :

$$\langle \exp(i\psi) \rangle = 0 \qquad (t \gg T_{cav}/n) \qquad (3)$$

The information left in the system is then the same one would get from an ensemble of wavefunctions of the kind described by Eq.(1), with an averaging over a random ψ phase. This is of course equivalent to the results obtained with an incoherent normalized density matrix:

$$\rho = (|\alpha_0\rangle\langle\alpha_0| + |-\alpha_0\rangle\langle-\alpha_0|)/[2(1 + \cos\psi \exp(-2|\alpha_0|^2)], \qquad (4)$$

described by the Wigner distribution represented in Fig.1b.

The time scale of the quantum phase randomization can be retrieved by a simple physical argument, allowing us to understand —qualitatively at least— why the system decoheres much faster than it decays. The argument goes as follows. As the quantum coherence is lost, the interference fringes in the photon number distribution disappear, and $P(n)$ evolves into a simple Poisson distribution, characteristic of a coherent field, or of an incoherent mixture of such fields. The decoherence time can thus be assimilated with the time it takes for the relaxation processes to "fill in" the odd or even n value "holes" in the $P(n)$ distribution. This "hole filling" occurs because relaxation admixes photon number states with adjacing near values, an n state being soon contaminated by relaxation from the $n + 1$ state and so on. Obviously, the field energy relaxation occurs as a result of the cumulative effect of such elementary one photon processes. If it takes a time of the order of T_{cav} for the photon number to go from $\bar{n}$ to zero, it will take only a time of the order $T_{cav}/\bar{n}$ for the field to loose one quantum of energy, and thus a time of the order of $T_{cav}/\bar{n}$ also for the fringes in the $P(n)$ distribution to vanish.

3 Experimental scheme for Schrödinger cat state preparation

Microwave superconducting cavities coupled to long lived circular Rydberg atoms are ideal systems to prepare these Schrödinger cat states and study the decoherence process as it unfolds. The scheme of our experiment is shown in Fig.2. A high Q niobium superconducting cavity, whose field energy damping time is typically 10^{-2} seconds, is excited by a classical source S coupled to the

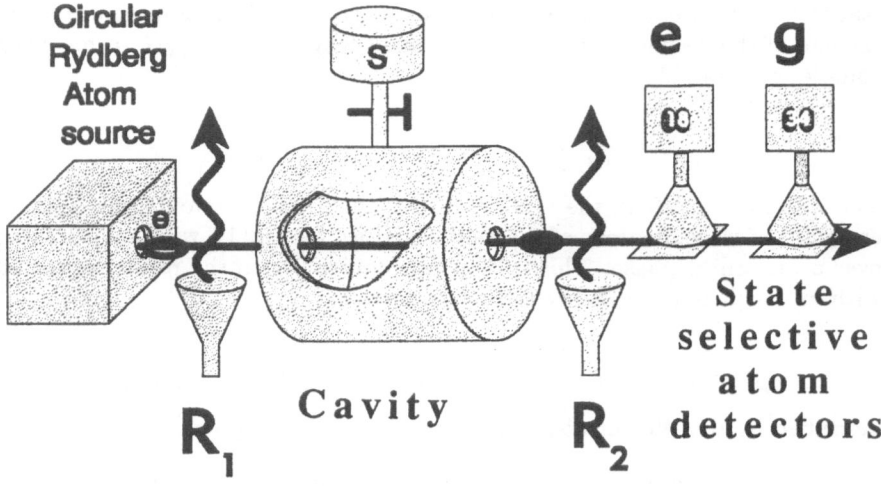

Figure 2: General scheme of the "Schrödinger cat field" preparation and detection experiment.

cavity by a wave guide. One thus prepares a coherent field whose amplitude α_0 is determined by the time t during which the connection between the source and the cavity is open. The source is then suddenly switched off and the field allowed to relax freely in the cavity. Before this occurs, though, a single atom is sent across the cavity, in a linear superposition of two circular Rydberg states, which writes as :

$$|\psi_{atom}\rangle = \frac{1}{\sqrt{2}} [|e\rangle + |g\rangle] \tag{5}$$

This superposition is obtained by having the atom cross, before the cavity, a classical microwave field zone R_1 essentially performing a $\pi/2$ pulse on the atom, initially prepared in the circular state e. The microwave applied in R_1 is resonant with the $e \rightarrow g$ transition towards an adjacing circular state g and, by adjusting the microwave phase, a superposition of the form described by Eq.(5) can be generated. The frequency of all the allowed transitions originating from levels e and g are somewhat detuned from the cavity resonant frequency, so that the atom in the linear superposition (5) cannot absorb or emit resonantly energy in the field. The detuning of one of these transitions is very small though so that the dispersive effects taking place during the atomic passage may induce an appreciable phase shift on the field in the cavity. An essential ingredient of the experiment is that the shift has to be different for states e and g. Various schemes are possible. Conceptually the simplest one amounts to assuming that the atom produces a large shift if it is in level e, and no shift at all if the level g (all transitions originating from g being for example largely detuned from the cavity mode).

By adjusting the atom velocity, the e state phase shift can be tailored to any value between zero and a few radians. Typically, the atom principal quantum number is 50 in level e, with a cavity tuned close to the 50→51 circular to circular state transition around 50GHz. An atom crossing the centimeter size cavity at a velocity of about 100m/s in state e typically produces a π phase shift on the field [6]. Another important feature of the design is that the coupling of the atom to the cavity field is adiabatic, with a slow rise and decrease of the cavity field amplitude along the atom trajectory, thus preventing any unwanted atomic transitions.

After the atom has crossed the cavity, in a time of the order of 10^{-4} seconds, short compared to the field relaxation time, the combined system has simply evolved into the entangled state :

$$|\psi_{atom+field}\rangle = \frac{1}{\sqrt{2}}\left[|e; -\alpha_0\rangle + |g; \alpha_0\rangle\right] \tag{6}$$

in which the strong atom-field correlations are the result of purely dispersive effects. We should note at this point the analogy with the pair of correlated spin particles in the Einstein-Podolski-Rosen (EPR) paradox experiment [8]. The atom and field states are indeed similar to spin 1/2 states particles (e, g and $+\alpha_0$, $-\alpha_0$ respectively) presenting non local correlations long after the two systems have separated. It is essential for this correlation to survive that the field and atomic relaxations could be neglected during the time of flight of an atom across the apparatus. We have seen that this condition is largely fulfilled for the field stored in the superconducting cavity. Circular Rydberg atoms, on the other hand, have radiative damping times in the range of milliseconds and can thus survive in the $e - g$ excited state superposition during the 10^{-4}s flight time. This is one reason why these states are required for this experiment, which would be impossible if e and g were mere low angular momentum Rydberg states with typically 10^{-5}s life times.

As in the EPR experiment, we next perform a measurement on one of the separated and correlated subsystems (the atom) to reduce the state of the other (the field). It is important to notice that one can freely choose, again as in the EPR situation, the atomic observable to detect. If one would decide to detect the atomic energy, with a state selective field ionization detector located directly behing the cavity, one would reduce the state described by Eq.(6) into one of its two components and the field would turn out to be either in state $|\alpha_0\rangle$ or $|-\alpha_0\rangle$ immediately after the measurement. More interestingly, one can also choose to detect linear superpositions of the form $|\pm\rangle = (1/\sqrt{2})(|e\rangle \pm |g\rangle)$ instead. This can be readily achieved by sending before detection the emerging atom through a second microwave zone R_2 identical to the first one, which produces another $\pi/2$ rotation of the spin one half like atomic system (see Fig.2). Detecting the atomic state (e or g) behind R_2 is obviously equivalent to detecting the superpositions $|\pm\rangle$ directly behind the cavity. The atom-field system behind R_2 has become :

$$|\psi_{atom+field}\rangle = \frac{1}{\sqrt{2}} [|e;-\alpha_0\rangle + |g;-\alpha_0\rangle + |g;\alpha_0\rangle - |e;\alpha_0\rangle] \qquad (7)$$

and the detection of the atom in level $|e\rangle$ or $|g\rangle$ behind R_2 (i.e. in level $|\pm\rangle$ behind the cavity) reduces the field into a state described by Eq.(1) with $\psi=0$ for the atom detected in g and $\psi=\pi$ for the atom detected in e. Schrödinger cats of the field are thus obtained as a result of the interaction of the initial coherent field with a single atom, conveniently prepared and detected in a superposition of its energy eigenstates.

4 Observing the decoherence by a two–atom correlation measurement

It is of course essential for a convincing demonstration of a "Schrödinger cat" to be able to experimentally distinguish between such a state and a mundane statistical mixture of two classical fields with opposite phases. Even more interesting would be the direct observation of the evolution of the former state into the latter, which would amount to witnessing the decoherence process "in real time".

We have proposed in [6] one possible method which would consist in adding to the cat in the cavity a second "reference" field either in phase or in quadrature with one of its components, and measuring the resulting photon number distributions by having the cavity interact subsequently with a sequence of atoms, performing a quantum nondemolition measurement of its energy [6]. Although it should work in principle, such a method could prove difficult to implement, because it requires a minimal time to mix the "cat state" with the reference and to carry out the subsequent QND measurement, which would involve the detection of at least five or six atoms.

We propose now a much simpler experiment (whose idea was indeed implicit in reference [6], although it was not noticed clearly at the time). We merely have to send a second atom after the first one and measure the probability to detect this atom in e or g behind R_2. Atom #2 has the same velocity as atom #1, so that it produces exactly the same phase shift effects on the field in the cavity. Measuring the probability of detecting it in e or g of course implies a repetitive experiment, resuming again and again the same cat state preparation and sending a second atom at a given time after the first one. What we propose to do here is in fact detecting correlated pairs of atoms, the first one preparing the "cat state" in the cavity and the second probing it. We will thus measure the conditional probability that a first measurement in level e or g will be followed by a second measurement in one level or the other. The outcome of the second measurement must be correlated with the first one because the nature (odd or even) of the "cat state" stored in the cavity between the passages of the two atoms depends on it.

What would be the result of such an experiment if there were a statistical mixture of $|\alpha_0\rangle$ and $|-\alpha_0\rangle$ fields in the cavity ? If the two fields are non overlapping ($\langle\alpha_0|-\alpha_0\rangle = 0$), classical probability laws apply in this case and predict that the probabilities of detecting the second atom in e or g must be $1/2$. The second atom is indeed prepared with equal probabilities in both states by the first microwave pulse R_1. After cavity crossing, all interferences between the amplitudes of finding the atom in level e and g are washed out since the correlated field states are orthogonal. Furthermore, the cavity field, whether it has one phase or the other, cannot change the e/g probabilities since it is not resonant and does not make the atomic pseudospin "flip" from $|e\rangle$ to $|g\rangle$). The second microwave zone, which again admises $|e\rangle$ and $|g\rangle$ in equal amounts, cannot change this state of affairs. Since the probabilities are $1/2$ for each of the two possible fields in the cavity ($|\alpha_0\rangle$ or $|-\alpha_0\rangle$), they must keep the same values for any incoherent admixture of such states (in particular the equal weight admixture considered here).

The situation is however completely different if there is a coherent "cat state" in the cavity between the two atoms. The initial state of the "second atom + cavity field", representing the system as atom #2 leaves the first microwave zone R_1, is now :

$$|\psi_{atom2+field}\rangle = \frac{1}{\sqrt{2}} \left[|e_2\rangle + |g_2\rangle\right] \otimes \frac{1}{\sqrt{2}} \left[|\alpha_0\rangle + e^{i\psi_1}|-\alpha_0\rangle\right] \qquad (8)$$

The index 2 refers to the probe atom and the index 1 in the phase ψ of the Schrödinger cat recalls that this phase depends on the result of the first atom measurement ($\psi = 0$ or π depending whether the first atom has been detected in g or e). We have also assumed for sake of simplicity in Eq.(8) that the distance between $|\alpha_0\rangle$ and $|-\alpha_0\rangle$ is large enough so that one can neglect the overlap integral between the two coherent fields in the normalization of the field state. This is practically the case when $\bar{n}$ exceeds a few units.

After the second atom has left the cavity, and just before atom #2 interacts with R_2, the system state has obviously evolved into :

$$|\psi_{atom2+field}\rangle = \frac{1}{2} \left[|e_2; -\alpha_0\rangle + |g_2; \alpha_0\rangle + e^{i\psi_1} (|e_2; \alpha_0\rangle + |g_2; -\alpha_0\rangle)\right] \quad (9)$$

since atom #2 in $|e\rangle$ dephases the field by an angle π, whereas the same atom in $|g\rangle$ does not dephase the field at all. The mixing effect of R_2 is readily accounted for and when atom #2 is behind R_2, the system has evolved into the final state :

$$|\psi_{atom2+field}\rangle = \frac{1}{2\sqrt{2}}\Big[|e_2; -\alpha_0\rangle + |g_2; -\alpha_0\rangle + |g_2; \alpha_0\rangle - |e_2; \alpha_0\rangle$$
$$e^{i\psi_1} (|e_2; \alpha_0\rangle + |g_2; \alpha_0\rangle + |g_2; -\alpha_0\rangle - |e_2; -\alpha_0\rangle))\Big] \quad (10)$$

Assuming again that the $|\alpha_0\rangle$ and $|-\alpha_0\rangle$ states are practically orthogonal, we readily get from Eq.(10) the probability of detecting the second atom in level g or e :

$$P(g_2) = \frac{1}{2} [1 + cos\psi_1]; \quad P(e_2) = \frac{1}{2} [1 - cos\psi_1] \quad (11)$$

We immediately conclude from Eq.(11) that the second atom is always detected in the same level as the first one (complete correlation between the two atoms).

This result was already obtained in [6], but we have shown explicitly in the above derivation that it is an effect of quantum interference. The probability of detecting atom #2 in e or g oscillates between 0 and 1 with the quantum phase of the Schrödinger cat, which is a clear evidence of an interference process. Each double atom detection event can indeed be analyzed as follows. A pair of atoms initially excited in level e "collides" with the "R_1 + cavity + R_2 system and gets detected in a combined state $e - e$, $e - g$, $g - e$ or $g - g$ with the (unread) field in the cavity being left in either state $|\alpha_0\rangle$ or $|-\alpha_0\rangle$. Each possible outcome may occur via two possible paths. For example the outcome $e - e$ with the field left in state $|\alpha_0\rangle$ may be associated with the process where the first atom has crossed the cavity in level $|e\rangle$ (thus dephasing the field to $|-\alpha_0\rangle$) and has remained in level $|e\rangle$ after the second microwave zone, while the second atom has also crossed the cavity in level $|e\rangle$ (thus bringing back the field to $|\alpha_0\rangle$) and has also remained in level $|e\rangle$ after the second microwave zone. Alternatively, it may be that both atoms have crossed the cavity in level $|g\rangle$ (thus leaving the field unchanged) and that both have flipped back to $|e\rangle$ in the second microwave zone. Of course, these two pathes are totally undistinguishable and their amplitudes must thus interfere in the expression of the probability for this outcome. Since the two possible histories differ by the state of the field in the cavity between the two atoms, which are related to each other by the quantum phase factor $e^{(i\psi)}$, it is clear that this phase factor must show up in the interference signal. The incoherent mixture result discussed above is also contained in Eq.(11). We merely have to average over a random ψ variable to find that both probabilities reduce indeed to 1/2. The quantum interferences are of course lost in this case.

The above simplified argument can easily be extended to treat the general situation where the field in the cavity has relaxed between the two atoms into an intermediate state, retaining partial coherence. From the exact expression for the field density matrix at all times (Eq.(2)), we can get an analytical expression for the joint atomic probability as a function of the time delay between the two atoms. We merely give here the result of the calculation, whose details will be presented elsewhere. We call $P(g, a; t)$ the conditional probability that, the first atom being detected in state g, the second one is found in state a a time t later (a stands for e or g). We obtain :

$$P(g,a,t) = \frac{1}{2}\left[1 \pm \frac{e^{-2|\alpha_0|^2 e^{-\gamma t}} + e^{-2|\alpha_0|^2(1-e^{-\gamma t})}}{1 + e^{-2|\alpha_0|^2}}\right] \qquad (12)$$

with the + and - sign in the right hand side expression corresponding to the cases $a = g$ and e respectively. In this equation, we have defined $\gamma = 1/T_{cav}$. Obviously the general expression (12) reduces for $t = 0$ (pure cat state) and for $T_{cav}/\bar{n} < t < T_{cav}$ (statistical mixture) to the limiting cases discussed above. We have represented in Fig.(3) the probability $P(g, e; t)$ as a function of t up to time $t = T_{cav}/2$. In this calculation, the value $\bar{n} = 10$ has been choosen. The probability evolves, on a time scale of the order of $T_{cav}/10$ from 0 (quantum result) to 1/2 (classical result). Observing such a behavior of the two atom photon count would amount to a direct detection of the quantum decoherence on this simple field system. Note that for large delays t (not represented on Fig.3), the probability $P(g, e; t)$ departs from the value 1/2 and tends towards zero (as can be checked directly on expression 12). This occurs when the two parts of the field, which are relaxing towards the vacuum, start to "overlap" with each other. As soon as this happens the classical argument presented above (for the statistical mixture case) breaks down and again, interference effects, associated this time to the evolution of the atomic coherence between the two microwave zones, become important. When the field has relaxed to the vacuum, the cavity has no longer any effect on the atoms and the combined effects of the two $\pi/2$ pulses amounts to a π pulse. The second atom is then found in level g with unit probability and $P(g, e)$ is equal to zero.

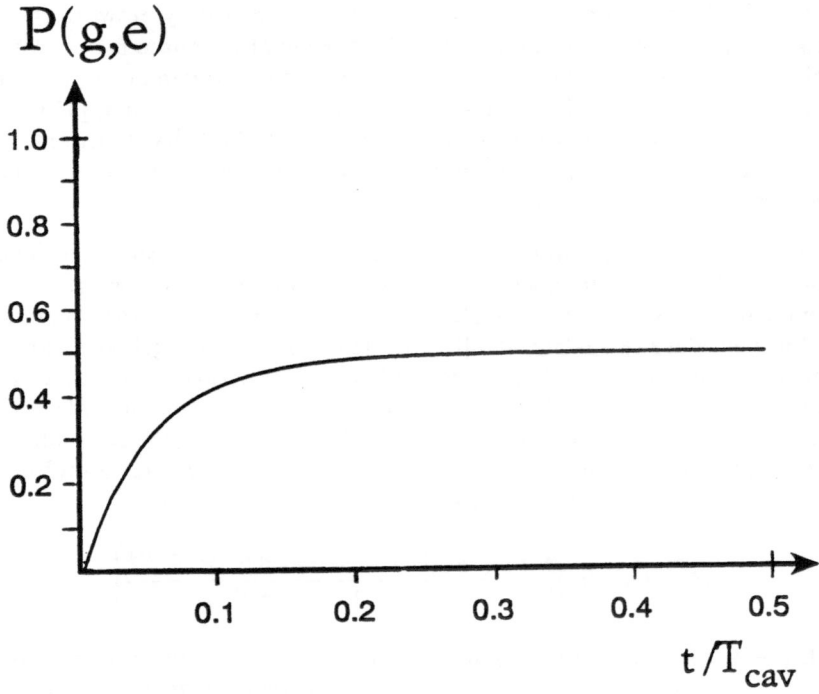

Figure 3: Conditional probability $P(g, e)$ to detect the second atom in level e after having detected the first one in level g, as a function of the delay t between the two atoms (10 photons on average in the cavity). The variation of $P(g, e)$ from 0 to 0.5 over a time of the order of $T_{cav}/10$ reveals the decay of the mesoscopic quantum coherence in the cavity. $P(g, e)$ evolves towards zero for larger t values.

5 Conclusion

We have shown in this chapter that the preparation of a quantum coherence between two classical subsystems differing by an adjustable parameter going continuously from a microscopic to a macroscopic value is a realistically achievable goal in cavity QED. The two subsystems are classical fields of opposite phase and the square of the "distance" between these fields is measured by the photon number, which can be varied from zero to a large absolute value. For small enough systems ($\bar{n}$ in the range of 10 or so) the decoherence time should be within experimental observation. By increasing $\bar{n}$ beyond this value, we should get into the classical regime where the decoherence will occur before any quantum coherence could be observed. The experiment would thus be a striking illustration of the transition between the quantum and the classical worlds.

The example discussed here is only one of many possible schemes for preparing and detecting Schrödinger cats in cavity QED. Other particularly interesting experiments could involve superpositions of field in two or more cavities, crossed successively by circular Rydberg atoms ensuring the preparation and detection of the quantum coherence. There again a single atom in a proper state superposition should be enough to prepare the cat state. The detection of interferences in the probability to detect a second probe atom in one of its quantum states could similarly demonstrate the existence of quantum coherences between the delocalized field states prepared by the first atom. These schemes will be discussed elsewhere.

It is remarkable to notice that the nonclassical character of the field system appears in these experiments as a correlation signal in a two atom detection event. This can be related to a well known feature in quantum optics, where photon count signals are used to observe the radiation of atomic sources [9]. When these sources exhibit a nonclassical behavior (such as quantum jumps), this behavior can be recorded in photon counting correlation experiments. On the other hand, single photon count signals do not permit a clear distinction between classical and quantum effects. In cavity QED experiments, the roles of field and matter are reversed. It is the field which is investigated by detecting the atoms interacting with it. Here again, however, only correlation signals involving at least two detection events seem to be required to distinguish a classical from a quantum behavior of the system.

Aknowledgments: Work supported in part by Direction des Recherches et Etudes Techniques. grant 90/186

References

1 W.H. Zurek, Physics Today, October 1991, p.36.

2 Quantum Theory and Measurements, J.A. Wheeler and W.H. Zurek editors, Princeton Univ. Press, Princeton, N.J. (1983); H.D. Zeh, Found. Phys. 1, 69 (1970).

3 E. Schrödinger, Naturwissenschaften 23, 807, 823 and 844 (1935); (English Translation by Trimmer J.D., Proc. Am. Phys. Soc. 124, 3235 (1980)).

4 A.J. Leggett et al, Rev. Mod. Phys. 51, 1 (1987); A.D. Calderia and A.J. Leggett, Ann. Phys. (N.Y.) 149, 374 (1983).

5 B. Yurke and D. Stoler, Phys. Rev. Lett. 57, 13 (1986); B. Yurke, W. Schleich and D.F. Walls, Phys. Rev. A42, 1703 (1990); G. Milburn. Phys. Rev. A33, 674 (1986).

6 M. Brune et al., Phys. Rev. A45, 5193 (1992).

7 R.G. Hulet and D. Kleppner, Phys. Rev. Lett. 51, 1430 (1983); A. Nussenzveig et al., Euro. Phys. Lett. 14, 755 (1991).

8 A. Einstein, B.Podolski and N. Rosen, Phys. Rev 47, 777 (1935)

9 See for example, R. Loudon, Th. Quantum Theory of Light, Clarendon Press Oxford, 1983.

10 S. Haroche in Fundamental Systems in Quantum Optics, Les Houches Summer School Session LIII; J. Dalibard, J.-M. Raimond and J. Zinn-Justin editors, North Holland (1992), and references in.

PĂRT IV: Quantum Statistics of Light

Using the Positive P-Representation.

C W Gardiner[1], A Gilchrist[1] and P D Drummond[2]

[1]Physics Department, University of Waikato, Hamilton, New Zealand.
[2]Physics Department, University of Queensland, Queensland 4072, Australia.

1   Introduction: what is the positive P-representation?

The development of quantum mechanics was strongly influenced by the recognition that there is a clear formal similarity between classical and quantum theories. This, in fact, led Schrödinger to his discovery of the coherent state of the harmonic oscillator, as a nearly classical state. In this case, the coherent state is labelled by a complex parameter $\alpha = x + ip$. The state is a minimum uncertainty state in the sense that it minimises the Heisenberg uncertainty product. In addition to this desirable property, the coherent state remains a coherent state under time-evolution, with the classical paths for the variables x and p. More technically, a coherent state is usually defined today as an eigenstate of the annihilation operator a:

$$a|\alpha\rangle = \alpha|\alpha\rangle \tag{1}$$

Despite the formal similarity of quantum and classical mechanics, there is no practical way of representing quantum uncertainties using a positive distribution on a classical phase space. This fact was discovered by von Neumann, who showed that no classical, positive phase-space distribution could exist for all quantum states, with correct marginal distributions in both x and p simultaneously. This is perhaps the earliest "no hidden variables theorem" in quantum mechanics, and can be regarded as the precursor of more powerful results of Einstein, Podolsky and Rosen, and the famous Bell inequalities. In a fundamental sense, this theorem shows that there is an intrinsic difference between classical and quantum uncertainties in measurement.

There are many reasons why classical-like phase space distribution techniques are popular in theoretical quantum physics. These techniques take advantage of the well-developed theory of Fokker-Planck equations and stochastic equations, and are often able to give results where other methods fail. It is clear, however, that all quantum phase space distributions must have properties differing from those of classical phase-space distributions.

Probably the most widely known phase-space method is that of the Wigner distribution function. This has correct marginal distributions in x and in p, at the expense of developing negative values for certain quantum states. An alternative that is sometimes used is the Q-distribution, defined by: $Q(\alpha) = \langle\alpha|\rho|\alpha\rangle$ This is always positive, and has the physical interpretation that it corresponds to the probability of measuring x and p simultaneously, giving $\alpha = x + ip$. Simultaneous measurement of non-commuting operators like x and p *is* possible, but entails a larger variance than any single measurement of x or p by itself. Accordingly, the Q-distribution pays for its positivity by having larger marginal variances than the Wigner function, for isolated measurements on the canonical variables. This property is to be expected in view of von Neumann's fundamental result, and can be corrected for when necessary.

Another approach to this problem was that of Glauber and Sudarshan [1], who independently achieved a significant breakthrough in the early 1960s. These workers recognised that the classical evolution of coherent states made them the natural basis for the study of radiation fields with nearly classical coherence properties. A single mode radiation field is characterized by just one set of harmonic oscillator operators. Accordingly, it was proposed that an arbitrary density operator ρ be represented by:

$$\rho = \int d^2\alpha |\alpha\rangle\langle\alpha| P(\alpha) \tag{2}$$

Unfortunately, the function $P(\alpha)$ often does not exist except as a highly singular generalised function. The reason for this is not hard to find, since it is possible to show that $P(\alpha)$ can be convolved with a two-dimensional Gaussian distribution, to give the Wigner function. Thus, the Wigner function always has larger marginal variances than the P-function. This leads to substantial difficulties if the quantum state is a position eigenstate, having a Wigner function of zero marginal variance on the x-axis: and hence a P-function of negative variance! Despite this, the technique proved to have many computational advantages in laser physics, and was widely used in this field.

That certain types of nonlinear interactions would cause problems was not widely understood until the late 1970s, when techniques were developed to obtain the corresponding nonclassical states of the radiation field in the laboratory. In these nonclassical states, the Glauber-Sudarshan P-representation is negative or singular. The more general nonlinear Hamiltonians used in these cases have Fokker-Planck equations which have a non-positive-semidefinite diffusion, and hence no corresponding stochastic process.

It might seem that this could be overcome by returning to the use of the Q-function, which has a classical phase space, and is always positive. In fact, the Q-function—and the Wigner function—can always be calculated from the P-function. However, these distributions on a classical phase space generally do *not* have positive-semidefinite diffusion in their Fokker-Planck equations, and frequently have time-evolution equations with higher than second order derivatives. In such cases, there is no stochastic interpretation of the time evolution. This leads to substantial difficulties in actual calculations, even when the distribution is well defined.

An often more useful resolution of the problem was provided by the observation that, in cases of nonpositive-definite diffusion, a naive application of standard techniques would result in the following generic stochastic equation:

$$\dot{\alpha} = -i\omega\alpha + \xi_1(t), \qquad \dot{\alpha}^* = -i\omega\alpha^* + \xi_2(t) \tag{3}$$

Here the two noise terms $\xi_1(t)$ and $\xi_2(t)$ are uncorrelated random noise terms, describing a departure from the classical path of Schrödinger. They are present because the

Hamiltonian of these nonlinear systems has extra terms beyond the normal harmonic oscillator terms, meaning that the quantum coherent state neither remains a coherent state, nor follows a classical path.

Since the extra stochastic terms are uncorrelated, they are not conjugate, which leads to an inconsistency in (3). However it was realised that the resulting equations could be re-interpreted as applying to two *independent* complex variables, which were only complex conjugate in the mean. This required the introduction of a normalised representation, having off-diagonal terms in the coherent state expansion of the density operator. The new representation, although on an enlarged phase space, can be shown to have at least one positive distribution function for all quantum states or density operators. This new representation—one of a class of generalised P-representations—is called the positive P-representation. The detailed properties and development of the new representation are given in the following sections.

2 The Glauber-Sudarshan P-Representation

The Glauber-Sudarshan P-representation is defined by

$$\rho = \int d^2\alpha \, P(\alpha, \alpha^*)|\alpha\rangle\langle\alpha|. \tag{4}$$

Glauber and *Sudarshan* [1] both showed that such a representation of the density operator did exist for a large class of ρ, and indeed *Klauder* et al. [2] showed that, provided $P(\alpha, \alpha^*)$ is permitted to be a sufficiently singular generalised function, such a representation exists for any density operator ρ. It is easily deduced that for any normal product of creation and destruction operators we can write

$$\langle (a^\dagger)^r a^s \rangle = \int d^2\alpha \, (\alpha^*)^r \alpha^s P(\alpha, \alpha^*). \tag{5}$$

Thus the quantity $P(\alpha, \alpha^*)$ plays the role of a probability density for the variables α and α^*, in that the means of *normally ordered* products of quantum operators are simple moments of $P(\alpha, \alpha^*)$.

The usefulness of this representation can be shown by the example of the driven damped harmonic oscillator, for which the master equation may be written

$$\begin{aligned} \dot{\rho} &= -i\omega[a^\dagger a, \rho] + \tfrac{1}{2}K(\bar{N}+1)(2a\rho a^\dagger - a^\dagger a\rho - \rho a^\dagger a) \\ &+ \tfrac{1}{2}K\bar{N}(2a^\dagger\rho a - aa^\dagger\rho - \rho aa^\dagger). \end{aligned} \tag{6}$$

The procedure for using the P-representation to solve this equation is as follows. We first note that the action of the creation and destruction operators on $|\alpha\rangle$ is

$$a|\alpha\rangle = \alpha|\alpha\rangle, \qquad a^\dagger|\alpha\rangle = (\alpha^* + \partial/\partial\alpha)|\alpha\rangle \tag{7}$$

with the corresponding Hermitian conjugate. We then insert the expansion (4) into the master equation, and use the correspondences (7). We then integrate by parts, and *discard any surface terms.* This leads to a correspondence of the form

$$
\begin{aligned}
a\rho &\leftrightarrow \alpha P(\boldsymbol{\alpha}), & a^\dagger \rho &\leftrightarrow (\alpha^* - \partial/\partial\alpha)P(\boldsymbol{\alpha}) \\
\rho a^\dagger &\leftrightarrow \alpha^* P(\boldsymbol{\alpha}), & \rho a &\leftrightarrow (\alpha - \partial/\partial\alpha^*)P(\boldsymbol{\alpha}).
\end{aligned}
\tag{8}
$$

Thus, wherever a creation or destruction operator occurs in the master equation, one can translate this into the corresponding operation on $P(\alpha)$. The equation which results is a Fokker-Planck equation :

$$
\frac{\partial P}{\partial t} = \left\{ \tfrac{1}{2}K\left(\frac{\partial}{\partial\alpha}\alpha + \frac{\partial}{\partial\alpha^*}\alpha^*\right) - i\omega\left(\frac{\partial}{\partial\alpha}\alpha - \frac{\partial}{\partial\alpha^*}\alpha^*\right) + K\bar{N}\frac{\partial^2}{\partial\alpha\partial\alpha^*} \right\} P.
\tag{9}
$$

Classical stochastic theory [3] shows that this Fokker-Planck equation is equivalent to the Stochastic differential equation in the form

$$
\frac{d\alpha}{dt} = -\left(\tfrac{1}{2}K - i\omega\right)\alpha + \sqrt{K\bar{N}}\,\xi(t),
\tag{10}
$$

and the corresponding complex conjugate equation. The quantity $\xi(t)$ is a white noise fluctuating force with the correlation properties

$$
\langle\xi(t)\rangle = 0, \qquad \langle\xi(t)\xi^*(t')\rangle = \delta(t - t')
$$
$$
\langle\xi(t)\xi(t')\rangle = \langle\xi^*(t)\xi^*(t')\rangle = 0
\tag{11}
$$

3 Generalised P-Representations

The equivalence to a classical SDE is not always so straightforward. It is not unusual to find an equation which appears to be of the Fokker-Planck form, but which has a diffusion matrix with some negative eigenvalues. It was this fact that motivated *Drummond* and *Gardiner* [4] to introduce a class of generalised P-representations, by expanding in non-diagonal coherent state projection operators. The generalised P-representations are defined as follows. We set

$$
\rho = \int_D \Lambda(\alpha, \alpha^+)P(\alpha, \alpha^+)d\mu(\alpha, \alpha^+),
\tag{12}
$$

where

$$
\Lambda(\alpha, \alpha^+) = \frac{|\alpha\rangle\langle\alpha^{+*}|}{\langle\alpha^{+*}|\alpha\rangle},
\tag{13}
$$

$d\mu(\alpha, \alpha^+)$ is the integration measure which may be chosen to define different classes of possible representations and $\mathcal{D}$ is the domain of integration. The projection operator $\Lambda(\alpha, \alpha^+)$ is analytic in (α, α^+). In the case of the Positive P-Representation we choose $d\mu(\alpha, \alpha^+) = d^2\alpha d^2\alpha^+$. This representation allows (α, α^+) to vary independently over the whole complex plane. In this case the normally ordered moments are given by

$$
\langle(a^\dagger)^m a^n\rangle = \int d^2\alpha d^2\alpha^+ \alpha^{+m}\alpha^n P(\alpha, \alpha^+).
\tag{14}
$$

The positive P-representation has quite strong existence properties—in fact it can be shown [4] that a positive P-representation exists for any quantum density operator ρ, with

$$P(\alpha) = (1/4\pi^2)\exp\left(-\frac{|\alpha - \alpha^{+*}|^2}{4}\right)\left\langle\frac{\alpha + \alpha^{+*}}{2}\left|\rho\right|\frac{\alpha + \alpha^{+*}}{2}\right\rangle. \tag{15}$$

NB: from now on for brevity, we use the notation $\alpha = (\alpha, \alpha^+)$.

3.1 Definition of the Positive P-Representation by Means of the Quantum Characteristic Function

The direct definition of the Positive P-representation as an expansion in operators makes the examination of convergence and the neglect of boundary terms very difficult. A better definition is given by saying that $P(\alpha)$ is a positive P-function corresponding to a density operator ρ provided

$$\chi_P(\lambda, \lambda^*) \equiv \int\int P(\alpha)\exp(\lambda\alpha^+ - \lambda^*\alpha)d^2\alpha d^2\alpha^+ \tag{16}$$

is identical with the quantum characteristic function of the density operator,

$$\chi(\lambda, \lambda^*) \equiv \mathrm{Tr}\left\{\rho e^{\lambda a^\dagger}e^{-\lambda^* a}\right\}. \tag{17}$$

It can be shown [5] that the integral (16) always converges provided $P(\alpha)$ is normalisable. To do this requires a careful definition of the four dimensional integral, which must be understood as the limit of an integral over disks of radii R, R' in the α, α^+ planes as $R, R' \to \infty$.

This is much milder than one might expect from (16), which would lead one to expect that $P(\alpha)$ would need to drop off exponentially as $|\alpha|, |\alpha^+| \to \infty$ in the direction of positive $\lambda\alpha^+$, $-\lambda^*\alpha$.

3.2 Operator Identities

From the definitions (13) of the nondiagonal coherent state projection operators, the following identities can be obtained. Again, α is used to denote (α, α^+):

$$\begin{aligned}
a\Lambda(\alpha) &= \alpha\Lambda(\alpha) \\
a^\dagger\Lambda(\alpha) &= (\alpha^+ + \partial/\partial\alpha)\Lambda(\alpha) \\
\Lambda(\alpha)a^\dagger &= \alpha^+\Lambda(\alpha) \\
\Lambda(\alpha)a &= (\partial/\partial\alpha^+ + \alpha)\Lambda(\alpha).
\end{aligned} \tag{18}$$

By substituting the above identities into (12), which defines the generalised P-representation and using partial integration (provided the boundary terms vanish), these identities can be used to generate operations on the P-function depending on the representation.

In order to derive time development equations for the Positive P-Representation we use the analyticity of $\Lambda(\alpha)$ and note that if $\alpha = \alpha_x + i\alpha_y$, $\alpha^+ = \alpha^+{}_x + i\alpha^+{}_y$ then

$$
\begin{aligned}
(\partial/\partial\alpha)\Lambda(\alpha) &= (\partial/\partial\alpha_x)\Lambda(\alpha) = (-i\partial/\partial\alpha_y)\Lambda(\alpha) \\
(\partial/\partial\alpha^+)\Lambda(\alpha) &= (\partial/\partial\alpha_x^+)\Lambda(\alpha) = (-i\partial/\partial\alpha_y^+)\Lambda(\alpha).
\end{aligned}
\tag{19}
$$

3.3 Time Development Equations

A time development equation for the density operator in the form of a master equation can be reduced to a c-number Fokker-Planck equation by using the mappings (18). The basic procedure is as follows

i) By use of the operator identities (18) write the master equation in the form [where $(\alpha, \alpha^+) = \boldsymbol{\alpha} \equiv (\alpha^{(1)}, \alpha^{(2)}); \mu = 1, 2]$:

$$
\begin{aligned}
\frac{\partial\rho}{\partial t} &= \int \Lambda(\boldsymbol{\alpha})\frac{\partial P(\boldsymbol{\alpha})}{\partial t} d^2\alpha\, d^2\alpha^+ \\
&= \int \left\{ \left[A^\mu(\boldsymbol{\alpha})\frac{\partial}{\partial\alpha^\mu} + \tfrac{1}{2}D^{\mu\nu}(\boldsymbol{\alpha})\frac{\partial}{\partial\alpha^\mu}\frac{\partial}{\partial\alpha^\nu} \right] \Lambda(\boldsymbol{\alpha}) \right\} P(\boldsymbol{\alpha})d^2\alpha\, d^2\alpha^+.
\end{aligned}
\tag{20}
$$

ii) The symmetric matrix $\mathbf{D}(\boldsymbol{\alpha})$ can always be factorised into the form

$$
\mathbf{D}(\boldsymbol{\alpha}) = \mathbf{B}(\boldsymbol{\alpha})\mathbf{B}^T(\boldsymbol{\alpha}).
\tag{21}
$$

and we split A and B into real and imaginary parts: $\mathbf{A}(\boldsymbol{\alpha}) = \mathbf{A}_x(\boldsymbol{\alpha}) + i\mathbf{A}_y(\boldsymbol{\alpha})$, $\mathbf{B}(\boldsymbol{\alpha}) = \mathbf{B}_x(\boldsymbol{\alpha}) + i\mathbf{B}_y(\boldsymbol{\alpha})$

iii) Using the identities (19), as well as *partial integration, with neglect of surface terms*, the Master equation then yields a Fokker-Planck equation which is equivalent to the Ito stochastic differential equations

$$
\frac{d}{dt}\begin{pmatrix} \alpha_x \\ \alpha_y \end{pmatrix} = \begin{pmatrix} A_x(\boldsymbol{\alpha}) \\ A_y(\boldsymbol{\alpha}) \end{pmatrix} + \begin{pmatrix} \mathbf{B}_x(\boldsymbol{\alpha})\xi(t) \\ \mathbf{B}_y(\boldsymbol{\alpha})\xi(t) \end{pmatrix},
\tag{22}
$$

or recombining real and imaginary parts

$$
d\boldsymbol{\alpha}/dt = A(\boldsymbol{\alpha}) + \mathbf{B}(\boldsymbol{\alpha})\xi(t).
\tag{23}
$$

Apart from the substitution $\alpha^* \to \alpha^+$, (23) is just the stochastic differential equation which would be obtained by using the Glauber-Sudarshan representation and naively converting the Fokker-Planck equation with a non-positive-definite diffusion matrix into an Ito stochastic differential equation.

In our derivation, the two formal variables (α, α^*) have been replaced by variables in the complex plane (α, α^+) that are allowed to fluctuate independently. The positive P-representation as defined here thus appears as a mathematical justification of this procedure.

4 The Effect of the Boundary Terms

4.1 Neglect of Boundary Terms

In the derivation of the positive P-representation Fokker-Planck equation in the previous section, partial integration is used, and it was assumed that boundary terms at infinity could be neglected. The direct derivation in terms of projection operators $\Lambda(\alpha)$ makes it very difficult to assess the magnitude of these. However, defining the positive P-representation through the characteristic function as in (16) makes the problem merely one of calculus. The equation of motion for the characteristic function is obtained by the rules

$$\rho a^\dagger \quad \leftrightarrow \quad \frac{\partial \chi}{\partial \lambda}, \qquad a^\dagger \rho \quad \leftrightarrow \quad \left(-\lambda^* + \frac{\partial}{\partial \lambda}\right)\chi$$

$$a\rho \quad \leftrightarrow \quad -\frac{\partial \chi}{\partial \lambda^*}, \qquad \rho a \quad \leftrightarrow \quad \left(\lambda - \frac{\partial}{\partial \lambda^*}\right)\chi \qquad (24)$$

which easily gives the correspondences required *provided the boundary terms can be neglected.* We can estimate these boundary terms for any particular case, provided we know the behaviour of $\mathbf{A}(\alpha)$, $\mathbf{B}(\alpha)$ as $\alpha, \alpha^+ \to \infty$. In most cases of interest, these coefficients are low order polynomials in α, α^+, and we will find we require a slightly faster drop off as $|\alpha|, |\alpha^+| \to \infty$ than $|\alpha\alpha^+|^{-2}$. [5] The neglect of surface terms which is necessary to derive all the Fokker-Planck equations equivalent to (22,23) is a deceptively simple step. Yet in the derivation of the Fokker-Planck equations, it is the only step which can be subject to any kind of doubt—if this neglect of surface terms is valid in any particular case, there is no doubt that the resulting Fokker-Planck equation and stochastic differential equations truly represent the original master equation.

In all cases of bilinear Hamiltonians, the requirement is trivially satisfied. Thus, for example, the linear harmonic oscillator has no boundary terms, so that the treatment of damping by the Glauber-Sudarshan P-representation can easily be extended to cover any non-classical quantum state. Similarly, the linear or linearised equations often used in treating squeezed or correlated states in a cavity have no boundary corrections.

However from the very beginning of the use of the positive P-function for numerical simulations, there were found to be problems associated with some trajectories escaping to infinity. And more recently a few examples have been found which actually give the wrong results. On the other hand there are numerous applications which give no problems [6]. Indeed Wolinsky and Carmichael [7] were able to show a specific example in which the domain of integration was finite, where boundary terms could be explicitly shown to vanish, and where the results were unqustionably correct.

In the following we give the results of some empirical investigations which, though not rigorous mathematically, do clearly solve the problem. We consider the two models which have shown incorrect results, and investigate the nature of the trajectories which "escape". In both cases the problem is the result of the underlying deterministic equations of motion, and yields a simple procedure to determine where and whether the positive representation is useful for the particular case.

We also consider the case of the anharmonic oscillator, which gives no escapes to infinity, and for which the positive P-representation always gives correct results.

4.2 Two photon absorption model

Smith and Gardiner [8] investigated a model of a harmonic oscillator damped by both one photon and two photon absorption. In an interaction picture the master equation is

$$\frac{\partial \rho}{\partial t} = \frac{1}{2}\kappa(2a^2\rho a^{\dagger 2} - a^{\dagger 2}a^2\rho - \rho a^2 a^{\dagger 2}) + \frac{1}{2}\gamma\kappa(2a\rho a^\dagger - a^\dagger a\rho - \rho aa^\dagger). \tag{25}$$

Carrying out the positive-P procedures, we arrive at the Ito stochastic differential equations

$$\frac{d\alpha}{dt} = -\kappa(\tfrac{1}{2}\gamma\alpha + \alpha^+\alpha^2) + i\sqrt{\kappa}\alpha\xi_1(t) \tag{26}$$

$$\frac{d\alpha^+}{dt} = -\kappa(\tfrac{1}{2}\gamma\alpha^+ + \alpha\alpha^{+2}) - i\sqrt{\kappa}\alpha^+\xi_2(t). \tag{27}$$

By defining $N = \alpha^+\alpha$ and using Ito rules [3], we obtain a single equation

$$\frac{dN}{d\tau} = -(\tfrac{1}{2}\gamma N + N^2) + iN\xi(\tau) \tag{28}$$

where

$$\tau = 2\kappa t, \qquad \xi(\tau) = \frac{\xi_1(t) + \xi_2(t)}{\sqrt{2\kappa}} \tag{29}$$

Simulations of the stochastic differential equation (28) were found to give correct results only for $\gamma > 2$—indeed for $\gamma < 1$, that is, sufficiently large 2-photon damping, (28) predicts that the stationary value of $\langle N \rangle$ is not even zero! The kind of behaviour obtained is shown in fig.1(a). (Notice that $\gamma < 2$ corresponds in practice to a totally unreachable magnitude of nonlnear absorption, so that the problem is one of principle only)

Smith and Gardiner, and indeed others who have used the positive P-representation, always found some trajectories in which the photon number approached very large (usually negative) values. Initially most authors ([9] [10]) thought that these "spikes" were the result of numerical problems, but in fact (28) can be solved exactly, and it can be conclusively demonstrated that these spikes are actually present in some solutions. We have carried out an investigation into this model to obtain quantitative statistics. We solved the equation (28) for a range of values of γ, with an initial condition of $N = 5$. We decided to investigate the statistics of the "spikes", and defined a spike to have occurred if $\mathrm{Re}(N)$ became less that -2. In fig.1(b) we plot the probability of escape before time T as a function of γ. It can be seen that for $\gamma < 1$, escape appears to be eventually certain, but not for $\gamma > 1$, where there is a small probability of escape, which is only in the initial stages. This behaviour is not difficult to understand. The simulation algorithm actually solves the equivalent Stratonovich equation

$$(S) \qquad \frac{dN}{dt} = \tfrac{1}{2}(1 - \gamma)N - N^2 + iN\xi(t) \tag{30}$$

The deterministic part of this equation has two stationary solutions, $N = 0$, and $N = \frac{1}{2}(1 - \gamma)$, and the greater of these is stable. When $\gamma > 1$, this is $N = 0$. A trajectory with initial condition of N real and positive will, with high probability, approach the point $N = 0$, and since the noise vanishes as $N \to 0$, escape is thereafter unlikely. If escape occurs at all, it will be the result of the improbable occurrence of a trajectory reaching a point with $\mathrm{Re}(N) < 0$ without having come near to $N = 0$, and from then on it is clear that escape to large negative $\mathrm{Re}(N)$ is almost inevitable.

However, if $\gamma < 1$, the stable attractor is $N = -\frac{1}{2}(1 - \gamma)$, at which point the noise is not zero; hence escape is always possible to $\mathrm{Re}(N) < 0$ by the process of activated escape over a barrier. We thus see that the existence of these escapes correlates well with the accuracy of the solutions.

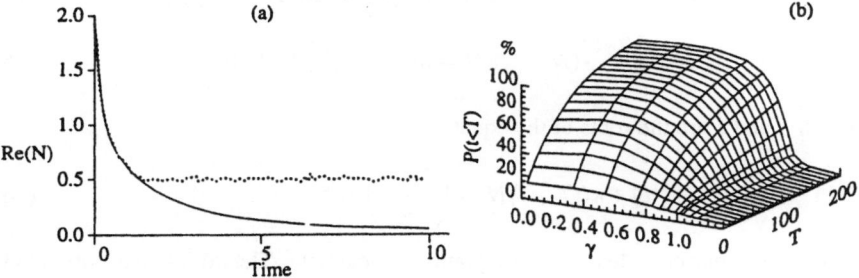

Fig.1
(a) Ensemble average of N against time: Solid line—exact solution; Dotted line—simulation of positive-P equations.
(b) Probability of an escape to $\mathrm{Re}(N) < -2$ before time T as a function of γ.

4.3 The laser

Schenzle and Schack [11] considered the application of the positive P-representation to the Laser, for which they take the rotating wave van der Pol equation

$$d\alpha = \left[\kappa\alpha(C - 1) - \left(\frac{\kappa C}{n_0} \right) \alpha^2 \alpha^+ \right] dt + \sqrt{2Q}\,dW(t) \tag{31}$$

They found that the vacuum initial condition, when represented by

$$P(\alpha, \alpha^+, 0) = \delta^4(\alpha) \tag{32}$$

gives (of course) the usual results, but that the choice corresponding to (15) of

$$P(\alpha, \alpha^+, 0) = \frac{1}{4\pi^2} \exp\left(-\frac{|\alpha - \alpha^+|^2}{4} \right) \left\langle \frac{\alpha + \alpha^+}{2} \Big| 0 \right\rangle \left\langle 0 \Big| \frac{\alpha + \alpha^+}{2} \right\rangle \tag{33}$$

gave different results.

Objections may be raised that the laser equation in this form is the result of many approximations, and that any problems which arise could well be the result of these approximations, which may not be valid in the regions of phase space explored by the positive-P simulations, and that the values of parameters chosen were totally unrealistic. These are valid reasons, but nevertheless we will see that a study of the mathematical processes underlying this behaviour is very illuminating.

The problem considered was the evolution of the vacuum to the lasing states. We will show why this happens, and that it is really a deterministic problem. If we define

$$N = \alpha^+ \alpha \tag{34}$$

then we can derive the stochastic differential equation for N

$$dN = -\frac{2\kappa C}{n_0}(N - a)(N - b)dt + 2\sqrt{Q}N\,dW(t) \tag{35}$$

in which a and b are the roots of the equation

$$Q + \kappa(C - 1)N - \left(\frac{2\kappa C}{n_0}\right)N^2 = 0 \tag{36}$$

We now have a one complex variable equation to deal with. The initial distribution (33) is equivalent to

$$P(\alpha, \alpha^+, 0) = \frac{1}{4\pi^2} \exp\left\{-\frac{|\alpha|^2}{2} - \frac{|\alpha^+|^2}{2}\right\} \tag{37}$$

By writing

$$N = Re^{i\phi} \tag{38}$$

we can eliminate the extra degrees of freedom, to show that the initial distribution in R, and ϕ is

$$P(R, \phi, 0) = \frac{R}{4\pi} \int_0^\infty dz \exp(-\frac{1}{2}R^2/z - \frac{1}{2}z) \equiv K(R)/2\pi \tag{39}$$

It is not necessary to evaluate this distribution explicitly in order to understand the behaviour of solutions—the only significant point at the moment is the fact that the distribution is independent of ϕ, and can thus be considered as a superposition of ring shaped distributions.

All physical quantities can be expressed as the averages of analytic functions of α, α^+—thus if we are interested in functions of N, we would like to consider the averages of analytic functions of N, i.e., of $f(Re^{i\phi})$. For an initial ring shaped distribution of radius R, we therefore want to consider

$$\langle f \rangle = \int d\phi f(Re^{i\phi}) = \oint \frac{dN}{N} f(N) \tag{40}$$

where the contour integral is over the circle $|N| = R$.

Now let us consider what happens if we decide to consider the deterministic laser, obtained by dropping the noise term in (35). The solution of this equation can be written

$$N(t, n) = b + \frac{(a - b)[n - b]}{n(1 - e^{\lambda t}) + (ae^{\lambda t} - b)} \tag{41}$$

$$\text{where} \quad N(0, n) = n, \quad \text{and} \quad \lambda = \frac{2\kappa C}{n_0}(a - b) \tag{42}$$

and we can choose $a > 0, b < 0$.

The formula (40) can be used to obtain

$$\langle f(t) \rangle = \oint \frac{dn}{n} f\left(N(t, n)\right). \tag{43}$$

As long as $N(t, n)$ is analytic in n inside the circle $|n| = R$ and $f\left(N(t, n)\right)$ is also analytic inside that circle, we can use contour integration to get

$$\langle f(t) \rangle = N(t, 0) \tag{44}$$

Thus, the mean value will be the solution for the deterministic equation with the vacuum initial condition. But $N(t, n)$ does have a singularity at $n = (b - ae^{-\lambda t})/(1 - e^{-\lambda t})$ and this starts initially at $-\infty$, and moves along the negative real axis until it hits the circle of radius R at the time

$$t = t_e \equiv \frac{1}{\lambda} \log\left(\frac{R + a}{R + b}\right) \tag{45}$$

which is positive, since $a > b$, provided $R + b > 0$. (If $R + b < 0$, the singularity does not hit the circle). At the same time, we find that the solution for the initial condition $N(0) = -R$, namely $N(t, -R)$ escapes to infinity.

To evaluate the averages after t_e is a problem which depends on the nature of f. However, we can evaluate $\langle N(t) \rangle$, and find that we have the discontinuous solution

$$\langle N(t) \rangle = N(t, 0) + \frac{(a - b)^2 e^{-\lambda t}}{(ae^{-\lambda t} - b)(1 - e^{\lambda t})} u(t - t_0) \tag{46}$$

where $u(t)$ is a step-function,

$$u(t) = \begin{cases} 0, & t < 0 \\ 1, & t \geq 0. \end{cases} \tag{47}$$

We can now return to the full distribution (39), and simply integrate over the various R, to find that in this case

$$\langle N(t) \rangle = N(t, 0) + \frac{(a - b)^2 e^{-\lambda t}}{(ae^{-\lambda t} - b)(1 - e^{-\lambda t})} \int_{R(t)}^{\infty} \frac{1}{2} RK(R)dR \tag{48}$$

(The function $K(R)$ can in fact be evaluated in terms of the Bessel functions K_0, K_1 and $\check{K}_2$.)

These results can be checked by simulations. The discontinuous solution (46) is simulated by starting with initial conditions $Re^{i\phi}$, where ϕ is uniformly distributed on $(0, 2\pi)$. Each solution is smooth and continuous unless $\phi = \pi$, which never happened in the cases tried. The results are seen in fig.2(a), where the formula (46) (dotted line) is compared with the mean of 1000 trials (solid line, real part, dashed line imaginary part). The agreement is remarkable, especially the rather accurate reproduction of the discontinuity by the sum of continuous functions.

When we choose an initial ensemble corresponding to the Gaussian distribution (37) we get the results of fig.2(b), which are very similar as those of Schenzle and Schack, who simulated the full stochastic equations. The dotted line here is the correct solution.

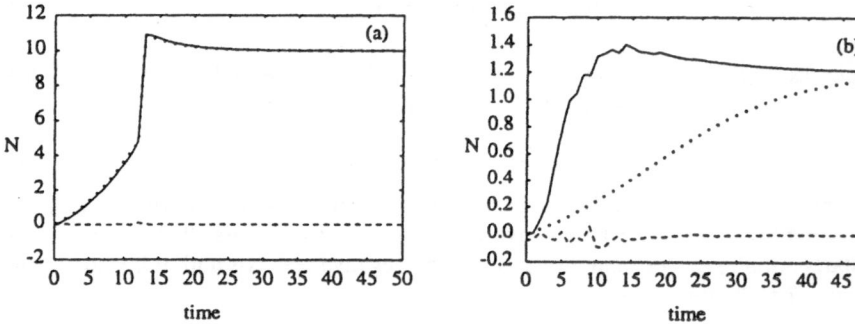

Fig.2
(a) Solid line—Simulation of the deterministic equations for the laser with an initial condition uniformly distributed on a circle of radius 2 centred on the origin; Dotted line—theoretical curve.
(b) Solid line—Simulation of the deterministic equations for the laser with the Gaussian initial condition in α, α^+ centred on the origin (real part); Dashed line (imaginary part); Dotted line—Exact deterministic solution.

The source of the problem is the fact that there is a trajectory which can escape to infinity in a finite time. If the ensemble of initial conditions includes points which lie on this trajectory, then the solution will be incorrect. As a check, we can consider the simulation of the laser equations with an initial condition which is such that there is no significant probability of any trajectory ever escaping during the full time evolution. In the case of the parameters chosen by Schenzle and Schack, such trajectories are not obviously available, since the mean photon number in the steady state is only one photon, and even from this state it does not seem unreasonable to expect a fluctuation which might induce an escape. Against this is the dissipative nature of the equations of motion, and the fact that the noises in this case are actually conjugate. This means that any initial condition in which the variables α, α^+ are non-conjugate yields a trajectory in which eventually $\alpha^* = \alpha^+$ to a high degree of accuracy, and such trajectories cannot yield any spikes.

Bearing this in mind, we chose situations with a higher mean photon number in the steady state, and simulated with initial conditions given by the formula (15). As expected, problems are only found for very low initial photon numbers, in which there is a significant probability of an N with a negative real part. These small photon numbers are of course incompatible with the semiclassical equations we have used here. In practise spontaneous emission noise is relatively large when these equations are are valid, so no significant problems occur for realistic parameter values.

All of this is obviously related to the dropping of boundary terms. If we consider the ring shaped initial distribution it can be seen that there is no current at infinity until the critical trajectory reaches infinity. At that stage there is a finite contribution to the boundary terms, and the equation ceases to represent the physics of the problem.

4.4 The anharmonic oscillator

This is a Hamiltonian system, with Hamilton

$$H = \hbar \omega a^\dagger a + \hbar \frac{\epsilon}{2} a^{\dagger 2} a^2 \tag{49}$$

to which is added *linear* damping; the positive-P Ito stochastic equations are (at zero temperature)

$$\dot{\alpha} = (-i\omega - \kappa)\alpha - i\epsilon\alpha^+\alpha^2 + \alpha\sqrt{i\epsilon}\,\xi_1(t) \tag{50}$$

$$\dot{\alpha} = (i\omega - \kappa)\alpha + i\epsilon\alpha^{+2}\alpha + \alpha^+\sqrt{-i\epsilon}\,\xi_2(t) \tag{51}$$

and these yield a stochastic equation for the $N = \alpha^+\alpha$ variable

$$\frac{dN}{dt} = -2\kappa N + \frac{1}{2}\sqrt{\epsilon}N(\xi_a(t) + i\xi_b(t)) \tag{52}$$

$$\xi_{a,b} = -\sqrt{\frac{1}{2}}(\xi_1 \pm \xi_2) \tag{53}$$

which is linear, and exactly solvable. No spikes can therefore occur in the N variable— it is not difficult to see that no escapes can thereafter occur in the α, α^+ variables.

Notice that these equations are of essentially the same kind as those for nonlinear damping, apart from the fact that the nonlinear deterministic terms are pure imaginary, rather than real.

Simulations of this kind of equation yield no escapes and results which are in agreement with other methods of solving the problem. The reason for this seems to be the fact that the deterministic equations have a constant of the motion, $N \equiv \alpha^+\alpha$, so that the unbounded trajectories of the previous two problems do not occur. It should be noted that this is not a generic property—there are pronlems in which the nonlinearity occurs only in the Hamiltonian part of the equations of motion which nevertheless give spikes. These are usually multimode problems. It is useful, again, to compare this situation with the more traditional representations. None of these have stochastic equations. The P- and Q- representations have non-positive-definite Fokker-Planck equations, while the Wigner equation is of third order—which automatically prevents any possible interpretation of the propagator.

4.5 Subharmonic generation in a cavity

A nonlinear example[7] where the boundary terms do vanish is the case of subharmonic generation in a cavity, with the driving field adiabatically eliminated. This has a finite phase-space manifold at zero temperature, and it is straightforward to show that the boundary terms vanish. It is known in this case that the steady-state distribution has an exact potential solution. All quantum moments are calculable in the steady-state. In addition, the dynamics can be readily simulated numerically, and even the above-threshold tunnelling [12] can be analytically calculated. This problem can also be treated, with some difficulty, using a numerical solution of the master equation. Exact agreement is obtained between these number-state results, and the positive-P results. This, of course, is expected, since there are no boundary terms, and no other approximations like system-size expansions are used.

It should be borne in mind that not all physical possible situations can be represented on the bounded manifold—for example, consider a situation in which the mean photon number is many times the maximum of $\alpha\alpha^+$ on the bounded manifold.

It is useful to compare this situation of subharmonic generation in a cavity, with that for the more traditional distributions, like the Wigner, Q-, or Glauber-Sudarshan representations. In none of these is it possible to even define an exact stochastic process, let alone find the steady-state quantum distribution. This appears to be generic when there is a nonlinearity in these classical-like representations, due to the occurrence of higher-order or non-positive definite diffusion terms. In the case of the Wigner distribution, it is often argued that the higher-order derivatives can be neglected at large photon number, leading to a semi-classical theory. However, this procedure does not give the correct tunnelling rates for this problem.[12] Depending on exactly which stage in the calculation the higher derivatives are dropped, the tunnelling rate may be many orders of magnitude too high or too low. It is also impossible to obtain the steady state quantum distribution, without approximations—nor is there any proof that the boundary terms vanish in the Wigner case.

5 Conclusion

The source of all problems (apart from purely numerical or computational problems) in using the positive P-representation is the validity of dropping boundary terms at infinity. If these are not negligible, there will be a finite probability current at infinity, which will manifest itself in simulations as the escaping of some trajectories to infinity. The mechanism for this is always in the deterministic motion, and the same behaviour will be manifested even without any noise. The presence of such spikes can appear dependent on the choice of initial conditions (as in the laser) or it can be independent of the initial conditions (as in the nonlinear absorber). If such spikes can occur for some initial conditions, however, there is always a probability of a spike from *any* initial condition, though the probability may be extremely small.

In this sense, we can view the positive P-function as giving quite accurate, but only asymptotically correct results, in situations in which spikes can occur. We can give, as a criterion for the use of the positive P-representation, the following:

- All problems (apart from computational problems) arise from the fact that neglected boundary terms may in fact not be negligible.

- This only occurs in practice if there is at least one deterministic trajectory in the ensemble being considered which can escape to infinity in a finite time. A numerical calculation will of course always give finite results, but usually spikes will be seen in this case.

- If spikes are seen, then we know that the solutions will be inaccurate for all times after the time of earliest appearance of such spikes. It is not correct to simply discard such trajectories.

- In practical calculations there is often an unstable and a stable manifold, which are usually connected purely by noise induced tunnelling. This tends to reduce the spiking effects at large photon numbers to very small levels.

- As a general rule, one can say that it is possible to use the positive P-representation numerically provided no "spikes" are found in the simulations.

References

[1] R.J. Glauber, Phys. Rev. **131**, 2766 (1963); E.C.G. Sudarshan, Phys. Rev. Lett **10**, 277 (1963)

[2] J.R. Klauder, J. McKenna, D.G. Currie, J. Math. Phys **6**, 734 (1965)

[3] C. W. Gardiner, *Handbook of Stochastic Methods*, (2nd ed., Springer, Heidelberg, 1990)

[4] P.D. Drummond, C.W. Gardiner, J. Phys. A **13**, 2353 (1980)

[5] C.W. Gardiner, *Quantum Noise*, (Springer, Heidelberg, 1991

[6] e.g., S.J. Carter, P.D. Drummond, M.D. Reid, R.M. Shelby, Phys. Rev. Lett. **58**, 1841, (1987)

[7] M. Wolinsky, H.J. Carmichael, Phys. Rev. Lett **60**, 1836 (1988)

[8] A.M. Smith, C.W. Gardiner, Phys. Rev **39**, 3511 (1989)

[9] M.L. Steyn, M.Sc. Thesis, University of Waikato, (1979)

[10] H.J. Carmichael, in *Quantum Optics IV*, ed. D.F. Walls, J.D. Harvey, Springer Proceedings in Physics 12, (Springer, Heidelberg, 1986

[11] R. Schack, A. Schenzle, Phys. Rev. A. **44**, 682 (1991)

[12] P. Kinsler, P.D. Drummond, Phys. Rev. Lett **57**, 2520 (1986)

Generating Number-Phase Squeezed States

M.J.Collett

Physics Department, University of Auckland, Auckland, New Zealand

Abstract. I discuss a number-phase interaction analogous to the usual quadrature squeezing Hamiltonian for a mode of the electromagnetic field. Applied to an initial coherent state, it produces a number-phase squeezed state. For moderate squeezing in either direction the result is close to a minimum uncertainty state. Two different $\chi^{(3)}$ approximations to the ideal interaction are found, appropriate for the different directions of squeezing. Even with the approximate interaction, fluctuations in photon number can be reduced to less than half a photon for an arbitrarily intense field.

1 Introduction

Number states of field modes (Fock states) are the simplest, most basic theoretical representation of the electromagnetic field. Such states are also in principle the most efficient way of conveying information optically. Each individual photon can be significant, regardless of the total intensity; this contrasts with the case of the usual coherent state, in which only intensity changes of the order of the square root of the total are significant. A general method of generating such states, at least to a good approximation, is clearly highly desirable.

A quadrature squeezed state can of course have reduced intensity fluctuations provided that the squeezing is not too large. However, the variance of the photon number n in such a state cannot be reduced below $(\overline{n})^{2/3}$. A scheme suggested by Kitagawa and Yamamoto [1] does somewhat better: by giving coherent light an intensity-dependent phase shift using a Kerr medium and then mixing it with a strong coherent beam, the number fluctuations can be made as small as $(\overline{n})^{1/3}$. This is of course a significant improvement on the value found in a coherent state, $\mathrm{Var}(\widehat{n}) = \overline{n}$, especially for large intensities; but it is still a very long way from being a number state. Other methods suggested for reducing number fluctuations below the squeezed state limit depend on measurement of the number, either in a quantum nondemolition device or as part of a feedback loop [2, 3, 4, 5]. Although each run of such a system can produce a well-defined intensity, the indeterminacy of quantum measurement prevents the specification in advance of what that intensity will be.

If we turn our attention from number states to eigenstates of the photon number's conjugate variable, the optical phase, the situation is less clear. Although the discovery of a hermitian phase operator [6] has given rise to considerable recent interest in phase properties, this has mainly been descriptive or formal in nature, and methods for the generation of phase states have received little or no attention. In fact, an ordinary quadrature squeezed state is a very good approximation to a phase state. The minimum possible phase variance is of order $1/\overline{n}^2$ [7], while a high-intensity quadrature-phase squeezed state has a minimum phase variance of $\ln \overline{n}/4\overline{n}^2$ [8]. Low-intensity quadrature squeezed states are also numerically close to the minimum possible phase variance [9].

In between the extremes of number state and phase state are the so-called *number-phase squeezed states*. These include as an ideal case the family of minimum uncertainty states in $\widehat{n}$ and $\widehat{\sin}\phi$ [10, 11], satisfying

$$\Delta\widehat{n}\Delta\widehat{\sin}\phi = |\langle\widehat{\cos}\phi\rangle| . \tag{1}$$

This family does not include coherent states, and is restricted to discrete values of intensity, so we cannot expect to find a simple method of generating them exactly. However, a high-intensity coherent state is a good approximation to an $\widehat{n}$-$\widehat{\sin}\phi$ mimimum uncertainty state, and in this paper I describe a nonlinear interaction which generates a number-phase squeezed state from an initial coherent state, while remaining close to the minimum uncertainty condition eqn(1).

In Section 2 I introduce the interaction and the resulting equations of motion. Section 3 and Section 4 discuss the noise reduction that can be obtained in the number and phase respectively. Section 5 shows how the theoretically simple but experimentally unrealistic interaction of Section 2 can be approximately realised by a $\chi^{(3)}$ nonlinearity.

2 A generator for number-phase squeezing

Ideal quadrature phase squeezing, satisfying $\Delta X\Delta Y = 1$ with either ΔX or ΔY less than 1, can be generated by the interaction Hamiltonian

$$H_{I,0} = \tfrac{1}{4}\hbar\kappa[X - x_0, Y - y_0]_+ , \tag{2}$$

where the symmetrisation guarantees hermiticity. For an initial coherent state with amplitude α (taken to be real and positive), the two interesting special cases are that of squeezing about the origin ($x_0 = 0$) and squeezing about the initial amplitude ($x_0 = 2\alpha$), with $y_0 = 0$ in both cases.

This suggests by analogy investigating the possiblity of

$$H_I = \tfrac{1}{2}\hbar\kappa[\widehat{n} - n_0, \widehat{\sin}\phi]_+ \tag{3}$$

being a generator of number-phase squeezing, where we may wish to choose either $n_0 = 0$ or $n_0 \simeq \alpha^2$. (The initial coherent amplitude is again, and throughout the rest of this paper, real and positive). Note that since we are now dealing with a nonlinear interaction, including the displacement n_0 does make a real difference to the dynamics and to the states that are generated. This is in contrast to the case of quadrature squeezing, in which the same states may be generated either by squeezing and then displacing or by displacing and then squeezing. On the other hand, including a nonzero phase displacement by using $\phi - \phi_0$ instead of ϕ is just equivalent to a change of reference phase, so this need not be considered (although the most general possible treatment would of course include the case in which the initial amplitude α was not real). The use of $\widehat{\sin}\phi$ instead of $\hat{\phi}$ makes the Hamiltonian continuous in phase space, and also allows use of the Susskind-Glogower phase operator instead of the Pegg-Barnett one. Using $\widehat{\sin}m\phi$ would select m stable phases instead of just one, while if instead of symmetrising, one guaranteed hermiticity of the Hamiltonian by some other ordering rule (e.g. normal ordering), this would merely be equivalent to a small change (of the order of unity) in the value of n_0.

Provided that we may neglect occupation of the vacuum state, we have

$$[\hat{n}, \widehat{\sin}\phi] \simeq i\widehat{\cos}\phi , \tag{4a}$$

$$[\hat{n}, \widehat{\cos}\phi] \simeq -i\widehat{\sin}\phi , \tag{4b}$$

$$[\widehat{\sin}\phi, \widehat{\cos}\phi] \simeq 0 . \tag{4c}$$

Note that eqn(4a) and eqn(4b) are exact for the Susskind-Glogower phase operator [12], while eqn(4c) is exact for the Pegg-Barnett operator [6]. I shall in the rest of this paper consider only the former, since the latter would cause direct transitions between the vacuum and an arbitrarily intense number state. These come from the extra term $|N\rangle\langle 0|$ in the Pegg-Barnett $\exp(i\hat{\phi})$, where N is the cutoff occupation number, eventually taken to infinity. This term is needed to ensure unitarity and gives no problem in *describing* normalised states, since for large enough N the amplitude of $|N\rangle$ must be vanishingly small, but clearly gives unphysical dynamics.

Now let us introduce

$$I_x = \hat{n} , \tag{5a}$$

$$I_y = \tfrac{i}{2}[\hat{n}, \widehat{\cos}\phi]_+ , \tag{5b}$$

$$I_y = -\tfrac{i}{2}[\hat{n}, \widehat{\sin}\phi]_+ . \tag{5c}$$

Then, neglecting the vacuum, these obey angular momentum commutation relations:

$$[I_x, I_y] \simeq I_z \text{ and cyclic permutations.} \tag{6}$$

The combinations

$$I_\pm = I_x \pm iI_y = \tfrac{1}{2}[\hat{n}, (1 \mp \widehat{\cos}\phi)]_+ \tag{7}$$

therefore obey

$$[I_+, I_-] \simeq 2I_z , \quad [I_\pm, I_z] \simeq \mp I_\pm . \tag{8}$$

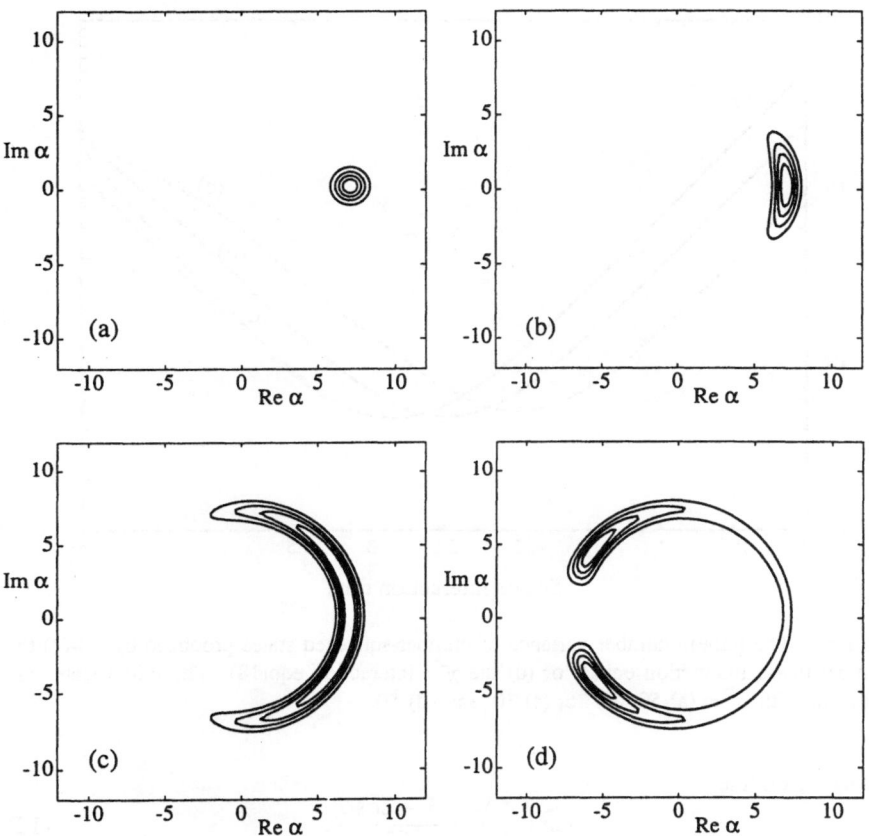

Figure 1: The evolution due to the number-phase interaction eqn(3) of an initial coherent state with mean photon number 50, as depicted by contours of the Q-function. The scaled interaction times are $\kappa t =$ (a) 0.0, (b) 1.4, (c) 2.8 (near the minimum photon noise), and (d) 4.2.

The Hamiltonian eqn(3) is just proportional to I_z, so that

$$I_\pm(t) = \exp\left(\kappa I_z t\right) I_\pm \exp\left(-\kappa I_z t\right) \simeq e^{\pm \kappa t} I_\pm , \qquad (9)$$

and the solution for $\hat{n}$ is

$$
\begin{aligned}
\hat{n} - n_0 &= \tfrac{1}{2}(I_+ + I_-) \\
&\simeq \tfrac{1}{4}[\hat{n}_0 - n_0, 1 + \widehat{\cos}\phi_0]_+ e^{\kappa t} + \tfrac{1}{4}[\hat{n}_0 - n_0, 1 - \widehat{\cos}\phi_0]_+ e^{-\kappa t} . \quad (10)
\end{aligned}
$$

Similarly, the solution for the phase is

$$\widehat{\tan}\left(\frac{\phi}{2}\right) \simeq e^{-\kappa t} \widehat{\tan}\left(\frac{\phi_0}{2}\right) , \qquad (11)$$

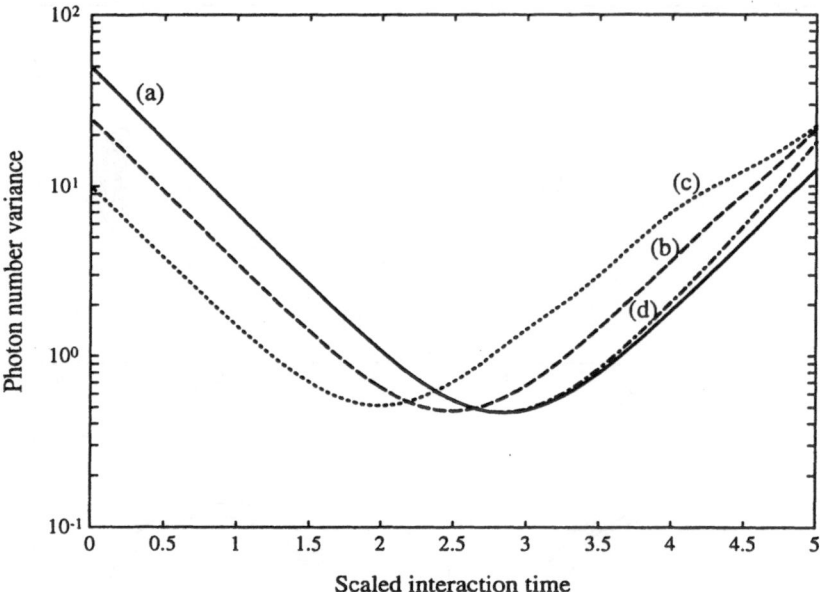

Figure 2: The photon number variance of number-squeezed states produced by (a)-(c) the number-phase interaction eqn(3) or (d) the $\chi^{(3)}$ interaction eqn(18). The initial states are coherent with $\alpha^2 =$ (a) 50, (b) 25, (c) 10, and (d) 50.

where of course

$$\widehat{\tan}\left(\frac{\phi}{2}\right) \simeq \frac{1 - \widehat{\cos}\phi}{\widehat{\sin}\phi}. \tag{12}$$

3 Number squeezing

For $\kappa < 0$, the interaction introduced in Section 2 generates a number-squeezed state (Fig. 1). From eqn(11) we see that the phase variance initially increases as $e^{2|\kappa|t}$, but then levels off as $\tan(\phi/2)$ becomes very different from $\phi/2$ (or $\sin(\phi/2)$). Similarly, the number variance initially decreases, but reaches a minimum when the noise in the small but growing second term of eqn(10) is comparable to that in the initially large first term. For an initial coherent state $|\alpha\rangle$, to leading order in $1/\alpha^2$,

$$\mathrm{Var}\left(\tfrac{1}{2}[\widehat{n}_0 - n_0, 1 + \widehat{\cos}\phi_0]_+\right) \approx 4\alpha^2, \tag{13a}$$

$$\mathrm{Var}\left(\tfrac{1}{2}[\widehat{n}_0 - n_0, 1 - \widehat{\cos}\phi_0]_+\right) \approx \frac{7}{64\alpha^2} + \frac{(\alpha^2 - n_0)^2}{32\alpha^2}, \tag{13b}$$

$$\left\langle \tfrac{1}{2}[\widehat{n}_0 - n_0, 1 + \widehat{\cos}\phi_0]_+, \tfrac{1}{2}[\widehat{n}_0 - n_0, 1 - \widehat{\cos}\phi_0]_+\right\rangle \approx \frac{n_0}{4\alpha^2}. \tag{13c}$$

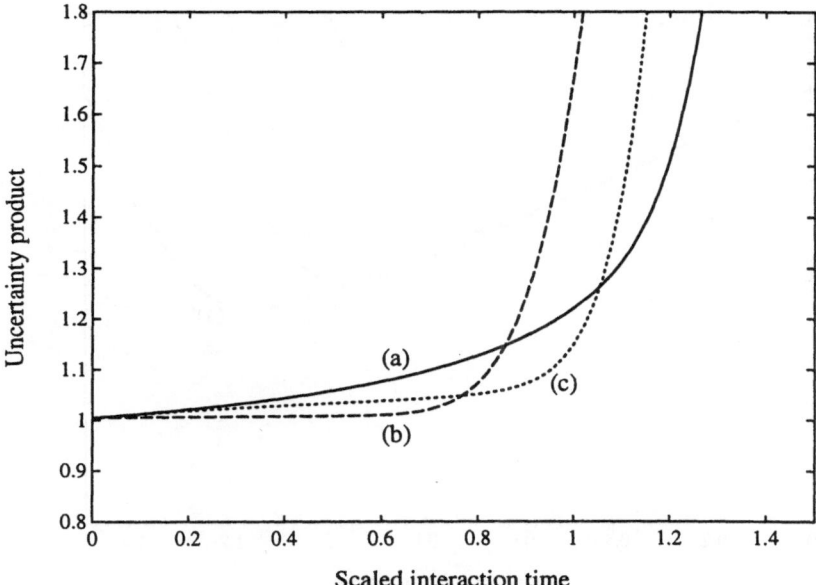

Figure 3: The $\widehat{n}$–$\widehat{\sin\phi}$ uncertainty product of number-squeezed states produced by (a) the number-phase interaction eqn(3) or (b) the $\chi^{(3)}$ interaction eqn(18), compared to (c) its minimum value according to the uncertainty principle (calculated for (a), but not significantly different for (b)). The initial states are coherent with $\alpha^2 = 50$.

The choice $n_0 = \alpha^2$ makes the noise in the second term, and hence the minimum total noise, as small as possible. The minimum is reached when

$$e^{-4|\kappa|t} = \frac{7}{256\alpha^4} , \qquad (14)$$

giving

$$\mathrm{Var}(\widehat{n})_{\mathrm{min}} = 2\sqrt{\tfrac{7}{256}} + \tfrac{1}{8} \simeq 0.4557 . \qquad (15)$$

Thus for an initial coherent state of sufficiently large intensity, the interaction eqn(3) can reduce the variance in the intensity to less than half a photon. For smaller intensities, neglecting the vacuum state occupation no longer gives accurate results, but numerical simulations confirm that even for $\alpha^2 \approx 10$ the intensity variance can still be reduced to approximately half a photon (Fig. 2(c)). As long as the first term in eqn(10) is dominant, the state remains close to minimum uncertainty in $\widehat{n}$ and $\widehat{\sin\phi}$ (Fig. 3).

If the interaction continues longer than the optimum time given by eqn(14), the number variance increases again and the state becomes localised in phase about

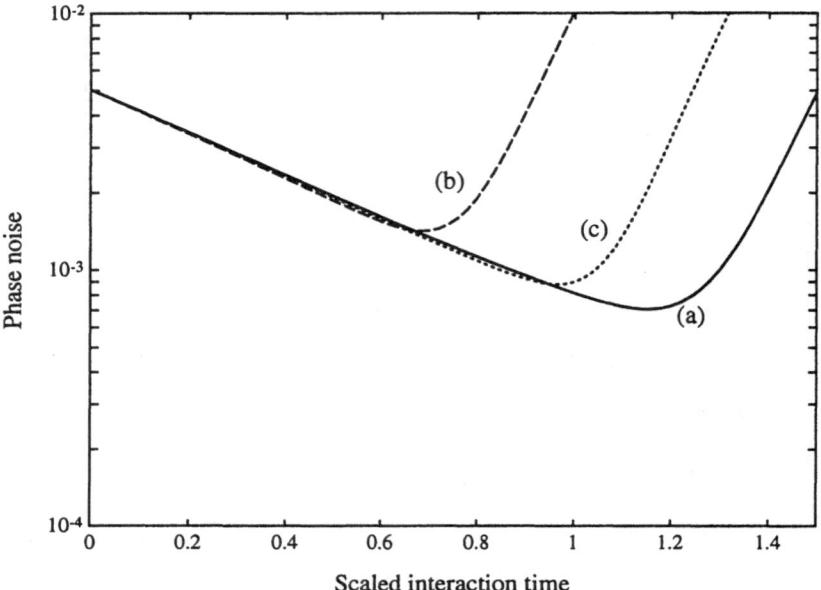

Figure 4: The phase noise of phase-squeezed states produced by (a) quadrature squeezing eqn(2), (b) the number-phase interaction eqn(3) or (c) the $\chi^{(3)}$ interaction eqn(19). The initial states are coherent with $\alpha^2 = 50$.

the opposite point from where it started. To see this, we can rewrite eqn(11) as

$$\widehat{\tan}\left(\pi - \frac{\phi}{2}\right) \simeq e^{-|\kappa|t}\widehat{\tan}\left(\pi - \frac{\phi_0}{2}\right). \tag{16}$$

Since the initial coherent amplitude is an unstable stationary point of the evolution, this localisation about the opposite phase is slow: in the long time limit, the phase variance goes down only as $e^{-|\kappa|t}$, not (as one might expect) as $e^{-2|\kappa|t}$.

4 Phase squeezing

For $\kappa > 0$, the interaction introduced in Section 2 generates a phase-squeezed state. From eqn(11) and eqn(10) we see that the phase variance decreases as $e^{-2\kappa t}$, while the number variance grows as $e^{2\kappa t}$. In this case the second term in eqn(10) is not important: it starts small and becomes smaller.

Since the intensity noise is increasing, whatever the initial intensity there eventually comes a time when the population in the region of the vacuum is no longer negligible. The approximation eqn(11) then no longer holds, and and the phase variance in fact begins to increase again. This increase can be avoided by choosing not $n_0 = \alpha^2$, as for number squeezing, but $n_0 = 0$; however, this also causes the

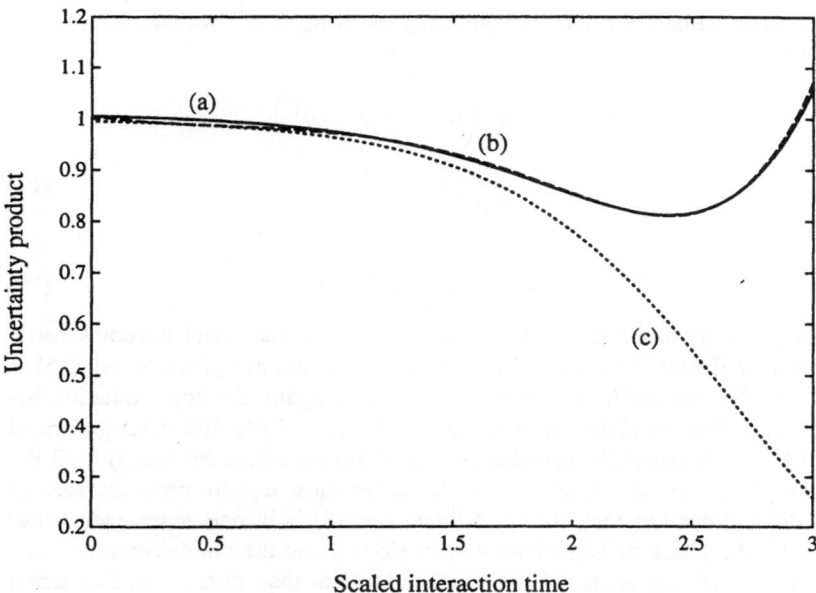

Figure 5: The $\widehat{n}$–$\widehat{\sin\phi}$ uncertainty product of phase-squeezed states produced by (a) quadrature squeezing eqn(2), (b) the number-phase interaction eqn(3) or (c) the $\chi^{(3)}$ interaction eqn (19). For squeezing in this direction, the minimum uncertainty product is not significantly different from 1. The initial states are coherent with $\alpha^2 = 50$.

mean number to increase. Numerical simulation suggests that the minimum phase noise achievable by this technique is approximately $\ln \overline{n}/2\overline{n}^2$ with either choice of n_0 (see Fig. 4). An ordinary quadrature squeezed state can get to half this [8], so if the aim is just to reduce phase noise for a given *mean* intensity, the interaction discussed here has no advantage over quadrature squeezing. However, for small to moderate phase squeezing, the interaction eqn(3) does result in smaller intensity *fluctuations*; that is, it is closer to being a number-phase minimum uncertainty state (Fig. 5).

5 Realisations using $\chi^{(3)}$ nonlinearities

The phase operator, or even a trigonometric function of it, involves infinitely high orders of the field operator, through which all physical interactions occur. Thus eqn(3) cannot be implemented as an exact interaction experimentally. However, we can find approximations to it that involve at most fourth order terms in the field operators (that is, $\chi^{(3)}$ nonlinearities) but still preserve the essential features of the full interaction.

The basic idea is to expand the operator $\widehat{\sin\phi}$ in powers of the field operator about its mean value. For number squeezing (with $n_0 \simeq \alpha^2$) we can use just the leading term,

$$
\begin{aligned}
\widehat{\sin}\phi &= \frac{1}{2i}\left(a\widehat{n}^{-\frac{1}{2}} - \widehat{n}^{-\frac{1}{2}}a^\dagger\right) \\
&\simeq \frac{1}{2i\alpha}\left(a - a^\dagger\right),
\end{aligned}
\tag{17}
$$

giving

$$
H_I \simeq \frac{\hbar\kappa}{4\alpha}[\widehat{n} - n_0, Y]_+ .
\tag{18}
$$

This is a good approximation for the initial coherent state, and becomes better as the number fluctuations are reduced. Once the minimum given by eqn(15) is reached and the number fluctuations begin to grow again, the approximation becomes poorer. Numerical simulations confirm this (Fig. 2(d)): the states generated by eqn(18) are almost indistinguishable from those generated by eqn(3) until the mimimum photon noise is reached, but thereafter show a more rapid increase of number fluctuations. In fact, for short times eqn(18) is in one sense even better than eqn(3): the phase noise increases more slowly, and the number-phase uncertainty product actually comes closer to the minimum than in the initial coherent state (Fig. 3(b)). Note that although eqn(18) is apparently only third-order in the field, a classical driving field must be present to pick out the required phase quadrature; its amplitude is included in the coupling constant.

Phase squeezing increases the number fluctuations, so we cannot expect the leading term in the expansion of $\widehat{\sin}\phi$ to be sufficient for this. But if we just expand $\widehat{n}^{-1/2}$ in powers of $\widehat{n} - \alpha^2$, even the second term already requires a $\chi^{(5)}$ nonlinearity. However, the phase quadrature Y will not contribute significantly to the mean number; we can take $\widehat{n} \simeq \frac{1}{4}X^2$, and then expand to second order in X. This gives (again for $n_0 \simeq \alpha^2$)

$$
\begin{aligned}
H_I &\simeq \tfrac{1}{4}\hbar\kappa[X - 2\alpha, Y]_+ - \frac{\hbar\kappa}{16\alpha}[(X - 2\alpha)^2, Y]_+ \\
&\simeq \tfrac{1}{2}\hbar\kappa[X - 2\alpha, Y]_+ - \frac{\hbar\kappa}{4\alpha}[\widehat{n} - n_0, Y]_+ .
\end{aligned}
\tag{19}
$$

Not surprisingly, the leading term in the first line of eqn(19) is just the usual quadrature squeezing interaction; we already know that this is good for phase squeezing. The second term increases the rate of squeezing closer to the origin and decreases it further away, giving the resulting state an approximation to the "wedge" or "sector" shape expected for a true number-phase minimum uncertainty state. The $\chi^{(3)}$ realisation of it is *twice* the usual quadrature interaction *minus* the approximation eqn(19) useful for number squeezing. Even with the second term included, this is still a poorer approximation to the $\widehat{n}$-$\widehat{\sin}\phi$ interaction eqn(3) than eqn(18) is for squeezing the other way: numerical results for both the phase noise and the number-phase uncertainty product of states generated by eqn(19) lie roughly midway between those for quadrature squeezing and those for eqn(3) (Fig. 4(c) and Fig. 5(c)).

I have not considered in this paper the effects of losses or other nonidealities. Usual linear damping tends to drive a mode towards a classical (coherent) state, and would obviously reduce the effectiveness of the interactions considered here. On the other hand, with a driving field to maintain the same mean intensity, and perhaps a different choice of the parameter n_0, such a damped system should allow the generation of a stable number-phase squeezed state, rather than the transient effects described here.

This research has been supported by the University of Auckland Research Committee and the New Zealand Lottery Grants Board.

[1] M.Kitagawa and Y.Yamamoto, Phys Rev.A **34**, 3974 (1986).
[2] M.Brune, S.Haroche, V.Lefevre, J.M.Raimond and N.Zagury, Phys Rev.Lett. **65**, 976 (1990).
[3] M.J.Holland, D.F.Walls and P.Zoller, Phys Rev.Lett. **67**, 1716 (1991).
[4] Y.Yamamoto, N.Imoto and S.Machida, Phys Rev.A **33**, 3243 (1986).
[5] Y.Yamamoto, S.Machida and O.Nilsson, Phys Rev.A **34**, 4025 (1986).
[6] D.T. Pegg and S.M. Barnett, Phys.Rev.A **39**, 1665 (1989).
[7] G.S.Summy and D.T. Pegg, Optics Comm. **77**, 75 (1990).
[8] M.J.Collett, to appear in Physica Scripta (1993).
[9] A.Bandilla, Optics Comm. **94**, 237 (1992).
[10] P.Carruthers and M.M.Nieto, Phys.Rev.Lett **14**, 387 (1965).
[11] R.Jackiw, J.Math.Phys. **9**, 339 (1968).
[12] L.Susskind and J.Glogower, Physics **1**, 49 (1964).

Suppression of Photon Fluctuations in the Light Flow under Non-linear Amplification in a Resonant Medium Layer

Yuri. M. Golubev

St. Petersburg State University, Physics Institute
198904, St. Petersburg, Petershof, Russia

ABSTRACT   Saturation of an amplifying medium in combination with regular pumping of active atoms leads to a suppression of surplus noise as well as shot-noise in the inital super-Poissonian light.

1   Introduction

The non-linear amplifier has unique properties compared to a linear device. A sufficiently thick layer of an amplifying medium is able to suppress surplus noise within the initial light distribution completely. This is impossible for a linear amplifier. There, surplus-noise increases only at the expense of inevitable spontaneous emission [1]. Here, spontaneous emission turns out to be of minor importance compared to stimulated emission and a ceiling effect occurs in the saturated regime i.e. when stimulated emission excels spontaneous processes by a large extent. This effect can be exploited for (sub-Poissonian) quantum light production. Recent proposals predict the conversion of Poissonian light into light with sub-Poissonian statistics (for example [2]). The preliminary suppression of the surplus-noise can be realized by an amplifier in the saturation regime.

In addition to that, we expect that a combination of saturation with regular pumping of the amplifying atoms (the stationary population exhibits no fluctuations) can give rise to an even more radical effect: surplus- and shot-noise will be suppressed simultaneously. Consequently, there is now a possibility for converting super-Poissonian light directly into light with sub-Poissonian statistics without any supplementary steps.

Of course there are serious difficulties here because regular pumping has to be realized in a thick layer of the medium. But it is very clear that there are no contradictions with any physical principles [3].

The simple consideration, based on the following equation, provides us with the essential knowledge of the physical processes taking place here

$$\dot{n} = \frac{An}{1 + \beta n} + F(t). \tag{1}$$

This is the Langevin-equation for the stochastic photon-number of a single mode of an ideal cavity with a non-linear medium. The stochastic source $F(t)$ has been discussed in detail [4]. It is determined by spontaneous emission into the cavity mode. It is known that in the case of random excitation of the active atoms $F(t) \to 0$, if $\beta n \gg 1$. However, $F(t)$ must be taken into account in the absence of absorbing atoms.

We are facing the following situation: photon-number fluctuations are very small i.e.

$$n = \bar{n} + \epsilon, \tag{2}$$

with $\epsilon \ll \bar{n}$. Then we can set up an equation for ϵ:

$$\dot{\epsilon} = \frac{A\bar{n}}{(1 + \beta\bar{n})^2}\epsilon + F(t), \tag{3}$$

provided that there is an ordinary semi-classical equation for $\bar{n}$

$$\bar{n} = \frac{A\bar{n}}{(1 + \beta\bar{n})}. \tag{4}$$

In the case of saturation when $\beta\bar{n} \gg 1$ and $F(t) = 0$, we have

$$\dot{\epsilon} = 0. \tag{5}$$

Because ϵ describes surplus-fluctuations, it is justified to assume that no surplus-noise develops with time. At the same time, $\bar{n}$ (and shot-noise consequently) is increasing because

$$\bar{n} = \frac{A}{\beta}. \tag{6}$$

This means that the relative role of surplus-noise has diminished compared to shot-noise. What can be expected when there is regular pumping without fluctuations? We assume that shot-noise is suppressed in this situation, too, since this was the case for laser-systems [3].

Our main attention will be focused on theoretical problems of the spatial field evolution and not so much on statistical aspects. It is well known that there are significant difficulties in this range. But they can be circumvented in several ways.

266

For example, it was stated [5] that laser-theory could be converted into a spatial theory by means of the substitution $t \to z$. On this basis, we can convert laser-theory according to Lamb-Scully [6] into a statistical theory of an amplifier (in the absence of a cavity). I think, this can be very usefully for some problems. But under description of light statistical pattern we cannot be sure at least that this is all right. Moreover, it is impossible to calculate multiple time averages because the time variance has fallen outside the theory completely.

In this paper, we will exploit the quantum radiation transfer theory in the form of the kinetic equation for the time-spatial electromagnetic field density matrix. It was formulated for the first time in 1974 [7].

2 The physical setup and the observed signal

We will consider the transmission of a laser beam emerging from source (I) through a resonant medium layer (II) and a filter with bandwidth $\Delta\omega$ (III). The current from the photo-detector (IV) will be analyzed.

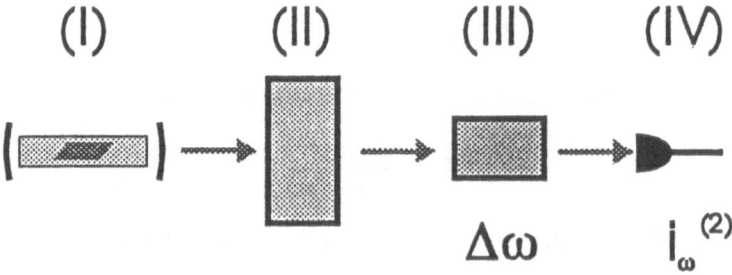

FIGURE 1.

In accordance with well known results [8], [9], we use the following formula for the photocurrent-spectrum

$$i_\omega^{(2)} = \frac{q}{\omega} \int_{(S)} d^{(2)}r \left\langle E^+(\mathbf{r}, t) E(\mathbf{r}, t) \right\rangle + \left(\frac{q}{\omega}\right)^2 \int\int_{(S)} d^{(2)}r_1 d^{(2)}r_2 2\Re \int_0^\infty d\tau \exp(i\omega\tau)$$
$$\left\langle E^+(\mathbf{r}_1, t) E^{(+)}(\mathbf{r}_2, t+\tau) E(\mathbf{r}_2, t+\tau) E(\mathbf{r}_1, t) \right\rangle, \quad (7)$$

with S denoting the surface of the photocathode, q the quantum efficiency of the detector and $E^{(+)}$ (E) the negative (positive) frequency part of the electric field operator in the Heisenberg representation.

We can write the ordinary superposition

$$E(\mathbf{r}, t) = \sum_{\mathbf{k}} i \left(\frac{\omega_{\mathbf{k}}}{2V}\right)^{\frac{1}{2}} a_{\mathbf{k}}(t) \exp{(i\mathbf{kr})}, \qquad (8)$$

where $k_i = \frac{2\pi}{L_i} n_i$, $n_i = 0, \pm 1, \pm 2, \ldots$, $i = x, y, z$ and $V = L_x L_y L_z$ is an auxiliary volume (in final results V is infinite). The operators $a_{\mathbf{k}}, a_{\mathbf{k}}^{\dagger}$ are annihilation and creation operators because

$$[a_{\mathbf{k}}, a_{\mathbf{k'}}^{\dagger}] = \delta_{\mathbf{kk'}} \qquad (9)$$

According to [7] we will produce the following operator packets:

$$a_{\mathbf{m}}(\mathbf{r}, t) = \sum_{\mathbf{k} \sim \mathbf{m}} \left(\frac{v}{V}\right)^{\frac{1}{2}} a_{\mathbf{k}}(t) \exp{(i(\mathbf{k} - \mathbf{m})\mathbf{r})}, \qquad (10)$$

$$a_{\mathbf{m}}^{\dagger}(\mathbf{r}, t) = \sum_{\mathbf{k} \sim \mathbf{m}} \left(\frac{v}{V}\right)^{\frac{1}{2}} a_{\mathbf{k}}^{\dagger}(t) \exp{(-i(\mathbf{k} - \mathbf{m})\mathbf{r})},$$

Here the set m is a sub-set of k. We have divided the space of all k-vectors into cells, m are the centers of the cells. In Eq.(10), we sum up in the limit of a single m-cell. We can fix m-sets in the following way:
$m_i = \frac{2\pi}{l_i} n_i$, $n_i = 0, \pm 1, \pm 2, \ldots$, $i = x, y, z$ and $v = l_x l_y l_z$ is a new auxiliary volume, moreover $l_i \ll L_i$. The commutation relation for the new operators defined in Eq.(10) are

$$[a_{\mathbf{m}}(\mathbf{r}, t), a_{\mathbf{m'}}^{\dagger}(\mathbf{r'}, t)] = \delta_{\mathbf{mm'}} \, v \delta_v(\mathbf{r} - \mathbf{r'}), \qquad (11)$$

where

$$v\delta_v(\mathbf{r}) = l_x \delta_{l_x}(x) l_y \delta_{l_y}(y) l_z \delta_{l_z}(z), \qquad (12)$$

and

$$l_i \, \delta_{l_i}(z) = \sum_{k_i \sim m_i} \left(\frac{l_i}{L_i}\right) \exp{[(k_z - m_z)z]} = \frac{\sin(\pi z/l_z)}{(\pi z/l_z)}. \qquad (13)$$

If we interpret this as the step function in the form:

$$l\delta_l(z) = \begin{cases} 1, & z < 1 \\ 0, & z > 1 \end{cases}, \qquad (14)$$

then

$$[a_{\mathbf{m}}(\mathbf{r}), a_{\mathbf{m'}}(\mathbf{r'})] = \begin{cases} \delta_{\mathbf{mm'}}, & \text{when } |\mathbf{r} - \mathbf{r'}| < l_i \\ 0, & \text{when } |\mathbf{r} - \mathbf{r'}| > l_i \end{cases}. \qquad (15)$$

On a coarse grained spatial scale which is characterized by l_i, it is possible to use the concept of a local field oscillator (LFO) since we can interpret the new operators $a_\mathbf{m}(\mathbf{r})$ and $a_\mathbf{m}^\dagger(\mathbf{r})$ as annihilation and creation operators of photons inside a spatial v-cell. Now we can define properties for a spectral filter (see Fig. 1). By assuming that $\Delta\omega = l^{(-1)}$ we have chosen a model in which the photocathode can be reached by a single m-mode. Within this approximation the photocurrent spectrum simplifies to

$$i_\omega^{(2)} = i_{\text{shot}}^{(2)} \left[1 + \frac{2q}{l\,n(z)} \Re \int_0^\infty g(z,t) \exp{(i\omega t)} dt \right], \qquad (16)$$

with $\bar{n} = \langle a^\dagger(z) a(z) \rangle$, $a(z) = a(z, t = 0)$, $g(z,t) = \langle a^\dagger(z) a^\dagger(z,t) a(z,t) a(z) \rangle$ and $i_{\text{shot}}^{(2)} = \frac{q}{2l} \bar{n}(z)$.

We have assumed that there is a single plane m-wave running along the z-axis therefore only a single spatial parameter $l = l_z$ is kept in the theory. For notational convenience we will suppress the operator indexes m. Eq.(16) will be the basis for our further calculations.

The quantities $\bar{n}(z)$ are auxiliary. It is the LFO photon number, i.e. the photon number which can be found in the auxiliary spatial v-cell. In the final results it will be replaced by the more physical light power $W(z)$ (see Eq.(47)).

3 The transfer equation for the field density matrix

It is not difficult to set up the equation for the $a_\mathbf{m}$ operators.

$$\frac{\partial}{\partial t} a_\mathbf{m} + \frac{\mathbf{m}}{m} \frac{\partial}{\partial \mathbf{r}} a_\mathbf{m} = -i\omega_\mathbf{m} a_\mathbf{m}, \qquad (17)$$

which follows from an initial well known equation

$$\dot{a} = -i\omega_\mathbf{k} a. \qquad (18)$$

Equation(17) in not exact generally speaking because we have neglected higher spatial derivatives. It is accurate as long as the equality

$$k = m + \frac{\mathbf{m}}{m} (\mathbf{k} - \mathbf{m}), \qquad (19)$$

holds true. In case of a one-dimensional problem Eq.(19) is correct and consequently Eq.(17) too. However, the reason for ignoring diffusion has to be discussed.

As stated above, we assume that the operators $a_\mathbf{m}, a_\mathbf{m}^\dagger$ are LFO. In the Heisenberg representation these operators depend on space and time while the density matrix of the LFO $\rho_\mathbf{m}$ is independent of t and r.

Consequently it is possible to move a slow time-space dependence form the operators to the density-matrix by means of a suitable unitary transformation. Therefore we have

$$\left[\frac{\partial}{\partial t} + \frac{\mathbf{m}}{m}\frac{\partial}{\partial \mathbf{r}}\right]\rho\mathbf{m} = 0, \tag{20}$$

$$\dot{a}_{\mathbf{k}} = -i\omega_{\mathbf{k}} a_{\mathbf{k}},$$

which describes a free electromagnetic field. But this can be generalized to a situation when also a medium is present. According to kinetic theory, the density-matrix of the sub-system also changes by the interaction with other sub-systems. In our situation the LFO is this selected sub-system and we are interested in its state. It interacts with the atoms in the same spatial cell. On the other hand we have to consider the radiation transfer i.e. the interaction between different LFOs. Therefore we have

$$\dot{\rho}\mathbf{m} = [\dot{\rho}\mathbf{m}]_{\text{transfer}} + [\dot{\rho}\mathbf{m}]_{\text{atoms}}, \tag{21}$$

and in accordance with Eq.(20)

$$[\dot{\rho}\mathbf{m}]_{\text{transfer}} = -\frac{\mathbf{m}}{m}\frac{\partial}{\partial \mathbf{r}}\rho\mathbf{m},$$

and $[\dot{\rho}\mathbf{m}]_{\text{atoms}}$ is calculated in each physical situation separately.

In our situation we can exploit the same expressions as in the laser theory according to Lamb-Scully [6]. Let us recall how this is done. At first the change of the field oscillator state (a single eigenstate of an optical cavity)

$$\delta\rho = \rho(t+T) - \rho(t),$$

is calculated for a single active atom. If T is long enough $\delta\rho$ will become independent of t and T. In case of a large sample of atoms the total change will be given by

$$\Delta\rho = rT\delta\rho,$$

with r denoting the mean rate of atomic excitation. The quantity $\delta\rho$ was calculated on the basis of the equations for the physical system: 'single atom plus field oscillator' with fixed initial conditions for the atom.

In principle our situation is the same when we have to calculate the change of the density matrix in the presence of the medium. There is only one difference: there the field oscillator was in an eigenstate of the optical cavity, here it is the LFO. The initial basic equation will be the same. Consequently the final equation for $\Delta\rho$ will be the same, too. We will use the form[10] of

$$[\dot{\rho}]_{\text{atoms}} = \left[r_a\hat{L}_a + r_b\hat{L}_b\right]\rho. \tag{22}$$

Here

$$\widehat{L_a} = \left(\underset{\rightarrow}{a^\dagger}\,\underset{\leftarrow}{a} - \underset{\rightarrow}{aa^\dagger}\right)\widehat{R_a} + h.c,$$

$$\widehat{L_b} = \left(\underset{\rightarrow}{a}\,\underset{\leftarrow}{a^\dagger} - \underset{\rightarrow}{a^\dagger a}\right)\widehat{R_b} + h.c,$$

$$\widehat{R_a} = \tfrac{1}{2}\beta_a\left[1 + \tfrac{1}{2}\beta_b\left(\underset{\rightarrow}{aa^\dagger} - \underset{\leftarrow}{aa^\dagger}\right)\right]$$

$$\left[1 + \tfrac{1}{2}\beta\left(\underset{\rightarrow}{aa^\dagger} + \underset{\leftarrow}{aa^\dagger}\right) + \tfrac{1}{4}\beta_a\beta_b\left(\underset{\rightarrow}{aa^\dagger} - \underset{\leftarrow}{aa^\dagger}\right)^2\right]^{(-1)} \quad (23)$$

and $\hat{R}_a \leftrightarrow \hat{R}_b$ by mutual replacements $a \rightarrow b, b \rightarrow a$ and $aa^\dagger \rightarrow a^\dagger a$. The arrows under the operators indicate the direction of action.

$$\beta_a = \tfrac{2|g|^2}{\gamma_a\gamma_{ab}}, \quad \beta_b = \tfrac{2|g|^2}{\gamma_b\gamma_{ab}}, \quad \beta = \beta_a + \beta_b, \quad (24)$$

are saturation parameters, r_a, r_b are the mean pump rates of the active atoms, γ_a, γ_b are decay rates from the active states to other non-resonant atomic states (see Fig. 2)

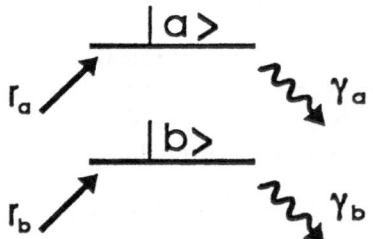

FIGURE 2.

and γ_{ab} is the transverse relaxation constant.

$$g = i\left[\frac{\omega}{2V}\right]^{\frac{1}{2}} d_{ab}\exp{(imz)} \quad (25)$$

is the interaction constant between the atom and the m-wave.

As a result we have an equation for the time-spatial field density matrix $\rho(z, t)$. To simplify the notation we drop the index m of the operators ρ, a and $a^\dagger$.

For the one-dimensional theory when the wave is running along the z-axis our equation is

$$\left(\frac{\partial}{\partial t} + \frac{\partial}{\partial z}\right)\rho = \left[r_a\hat{L}_a + r_b\hat{L}_b\right]\rho. \quad (26)$$

4　The Diagonal Representation

The operator equation(26) is not very useful for further application. In this connection we prefer to exploit the Glauber (normal) diagonal representation (NDR) which is introduced by means of the following integral relation [8]

$$\rho(z,t) = \int P(\alpha, z, t) \, |\alpha\rangle \, \langle\alpha| \, d^{(2)}\alpha, \qquad (27)$$

with $a|\alpha\rangle = \alpha|\alpha\rangle$ and $d^{(2)} = d(\Re\alpha)d(\Im\alpha)$. Let me remind that the a-operator concerns LFO in our theory in contrast to other approaches. However this does not introduce any difficulties.

In order to obtain Eq.(26) in diagonal presentation i.e the equation for $P(\alpha, z, t)$ we take into account the following rules

$$\underset{\rightarrow}{a} \to \alpha, \quad \underset{\rightarrow}{a^\dagger} \to \alpha^* + \frac{\partial}{\partial\alpha}, \quad \underset{\leftarrow}{a} \to \alpha + \frac{\partial}{\partial\alpha^*}, \quad \underset{\leftarrow}{a^\dagger} \to \alpha^*, \qquad (28)$$

It is not difficult to see that the operators $\hat{L}$ have a form which is very convenient for similar transformations.

The NDR equation following straight from Eq.(26) has still a very complicated form and therefore we will not write it down here. But it has to be mentioned that it contains an infinite number of derivatives for the complex field amplitudes α, α^*. For ordinary classical fields $P(\alpha)$ is a sufficiently smooth function of α and that is why we have the right to ignore all higher derivatives except the first and second (the diffusion approximation). But quantum fields are not so well behaved and in principle we have to keep all derivatives. At the same time we will demonstrate that the high derivatives do not contribute to the photocurrent spectrum Eq.(8).

Under our conditions the photon fluctuations are relatively small

$$|\alpha|^2 = \bar{n} + \epsilon, \quad \epsilon \ll \bar{n}, \qquad (29)$$

Generally speaking $\bar{n} = \bar{n}(z,t)$. By the approximation Eq.(29) the NDR equation takes on a much simpler form. We can write then for the photon density matrix

$$R(\epsilon, z, t) = \int_0^\infty d\phi P\left(\alpha = \sqrt{\bar{n} + \epsilon}\exp\left(i\phi\right); z, t\right), \qquad (30)$$

the following equation

$$\left(\frac{\partial}{\partial t} + \frac{\partial}{\partial z}\right) R = -\Gamma \frac{\partial}{\partial\epsilon}(\epsilon R) + D\frac{\partial^2 R}{\partial\epsilon^2} + \sum_{i=3}^{\infty} D_i \frac{\partial^i R}{\partial\epsilon^i}, \qquad (31)$$

where

$$\Gamma = \frac{A(z,t)}{1 + I(z,t)}, \quad A = r_a\beta_a - r_b\beta_b, \quad I = \beta\bar{n}(z,t), \qquad (32)$$

$$N_a = \frac{r_a}{\gamma_a}, \ N_b = \frac{r_b}{\gamma_b}, \ N = N_a - N_b,$$

$$D = \frac{A(z,t)\bar{n}}{(1+I)^2}\left(\frac{N_a}{N} + \frac{N_b}{N}I\right). \tag{33}$$

We could write down all other coefficients D just as Γ and D. But as stated above they do not contribute to our final results.

5 The NDR signal

The observed signal is given by the formula Eq.(16) in our case. In order evaluate it we have to calculate

$$g(z,t) = \langle a^\dagger(z)a^\dagger(z,t)a(z,t)a(z)\rangle, \ a(z) = a(z,t=0),$$

which is transformed according to Eqs.(28,29) into

$$g(z,t) = \langle |\alpha(z,t)|^2|\alpha(z)|^2\rangle = \bar{n} + \langle \epsilon(z,t)\epsilon(z)\rangle, \tag{34}$$

Thus we see that the surplus noise is expressed by the average $\langle \epsilon\epsilon \rangle$ which itself is given by

$$\langle \epsilon(z_1,t_1)\epsilon(z_2,t_2)\rangle = \int_0^\infty d\epsilon_1 \int_0^\infty d\epsilon_2 \epsilon_1 \epsilon_2 R(\epsilon_1, z_1, t)G(\epsilon_1, z_1, t_1|\epsilon_2, z_2, t_2). \tag{35}$$

Here $R(\epsilon, z, t)$ is the physical solution of Eq.(31) The conditional function $G(\ldots|\ldots)$ is a solution of Eq.(31) too:

$$\left(\frac{\partial}{\partial t_2} + \frac{\partial}{\partial z_2}\right)G(\epsilon_1, z_1, t_1|\epsilon_2, z_2, t_2) = -\Gamma\frac{\partial}{\partial \epsilon_2}(\epsilon_2 G) + D\frac{\partial^2 G}{\partial \epsilon_2^2} + \sum_{i=3}^\infty D_i\frac{\partial^i G}{\partial \epsilon_2^i} \tag{36}$$

This function is not equal to zero for $t_2 > t_1$, provided $z_1 = z_2 - t_2 + t_1$. This follows from the transfer equation properties. We must state the boundary conditions:

$$G(\epsilon_1, z = 0, t_1|\epsilon_2, z = 0, t_2) = G_0(\epsilon_1, t_1|\epsilon_2, t_2). \tag{37}$$

Now it is easy to see that the equation for the quantity $\langle \epsilon\epsilon \rangle$ follows from Eqs.(35,36) in the form :

$$\left(\frac{\partial}{\partial t} + \frac{\partial}{\partial z_2}\right)\langle \epsilon(z_2,t_1)\epsilon(z_2,t_2)\rangle = 2\Gamma\langle \epsilon(z_2,t_1)\epsilon(z_2,t_2)\rangle + 2Dl\delta_l(t_2 - t_1). \tag{38}$$

Here the second term on the right arises due to the fact that $\epsilon(t_2)$ is equal to $\epsilon(t_1)$ when $|t_2 - t_1| < l$.

This is an important feature of the time spatial theory. The solution of Eq.(38) has the form:

$$\langle \epsilon(z,t_1)\epsilon(z,t_2)\rangle = \langle \epsilon(0,t_1)\epsilon(0,t_2)\rangle \exp\left(\int_0^z 2\Gamma(z')\,dz'\right) +$$
$$2l\delta_l(t_2 - t_1)\int_0^z D(z')\exp\left(\int_{z'}^z 2\Gamma(z'')\,dz''\right)dz' \tag{39}$$

By using Eq.(32,33) we can obtain the expression for the expectation value of

$$\langle \epsilon(z,0)\epsilon(z,t)\rangle = \langle \epsilon(0,0)\epsilon(0,t)\rangle \left(\frac{I}{I_0}\frac{1+I_0}{1+I}\right)^2 +$$
$$2l\delta_l(t)\frac{\bar{n}I}{(1+I)^2}\left(\frac{N_a}{N}\left(\frac{1}{I_0} - \frac{1}{I} + \ln\left(\frac{I}{I_0}\right)\right) +\right.$$
$$\left.\frac{N_b}{N}\left(I - I_0 + \ln\left(\frac{I}{I_0}\right)\right)\right). \tag{40}$$

For the dimensionless power $I = I(z)$ and $I_0 = I(0)$ we have the relation

$$I - I_0 + \ln\left(\frac{I}{I_0}\right) = A\,z, \tag{41}$$

which follows from the semiclassical equation

$$\frac{dI}{dz} = \frac{AI}{1+I}. \tag{42}$$

Together with all of this we have to keep in mind that the excitation of the active atoms was random. If we want to consider regular pumping we have to substitute

$$\hat{L}_a \rightarrow \hat{L}_a - \frac{1}{2}\hat{L}_a^2, \quad \hat{L}_b \rightarrow \hat{L}_b - \frac{1}{2}\hat{L}_b^2 \tag{43}$$

according to [3]. As a result supplementary terms arise in Eq.(40):

$$\left(-l\delta_l(t)\frac{\bar{n}I}{(1+I)^2}\left(I - I_0 + \ln\left(\frac{I}{I_0}\right)\right)\left(\frac{N_a}{N}\frac{\gamma_b}{\gamma_a + \gamma_b} + \frac{N_b}{N}\frac{\gamma_a}{\gamma_a + \gamma_b}\right)\right) \tag{44}$$

6 The photocurrent spectrum in the saturation regime

Now we have all the possibilities to calculate the expression for the spectrum. However the general result has a complicated form and therefore we are not presenting it here. The most interesting results arise for the saturation regime when $I, I_0 \gg 1$. In this case we have:

$$i_\omega^{(2)} = i_{\text{shot}}^{(2)} \left\{ 1 + 2q\xi_L \frac{W}{W_0} \frac{C_L^2}{C_L^2 + \omega^2} - \left[q \left(1 - \frac{W}{W_0} \right) \frac{\gamma_b}{\gamma_a + \gamma_b} \right] \right\}. \quad (45)$$

This expression corresponds to regular excitation of the upper atomic level ($N_b = 0, N_a = N$). For random pumping the expression within the square brackets vanishes. $W = W(z), W_0 = W(z = 0)$ is the laser light power.

According to Eqs.(39,40) it is necessary to specify the statistical properties of the incident laser light at the front boundary of the amplifying layer. We will suppose that

$$\langle \epsilon(0,0)\epsilon(0,t) \rangle = \bar{n}_0 \xi_0 \exp\left(-\Gamma_0 t\right). \quad (46)$$

Here the LFO parameters $\bar{n}_0, \xi_0, \Gamma_0$ are related to the real laser parameters by

$$\xi_0 \to \xi_L C_L l, \quad \bar{n}_0 \to W_0 l/\omega, \quad \Gamma_0 \to C_L. \quad (47)$$

Here C_L is the spectral width of the laser cavity (I) (see Fig. 1), ξ_L is a Mandel parameter $\langle n_L^2 \rangle - \langle n_L \rangle^2 = \langle n_L \rangle (1 + \xi_L)$, and n_L is the intra-cavity laser photon number. In Fig. 3 the case of effective amplification ($W \gg W_0$) with $q = 1$ and $\xi = -1/2$ (a sub-Poissonian laser) is presented.

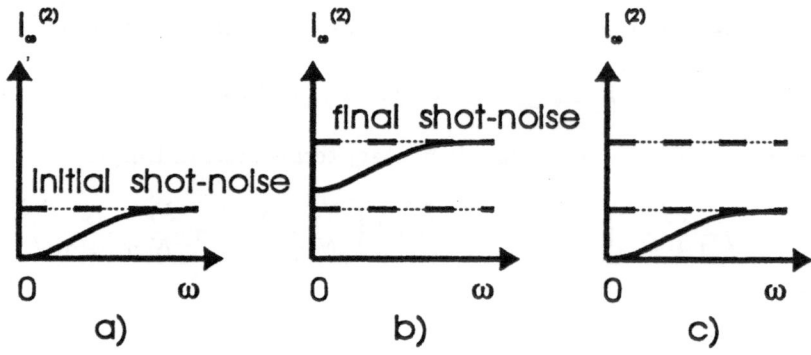

FIGURE 3.

The first plot (Fig. 3a) exhibits the noise of the initial laser light. We see that shot noise is suppressed in the low frequency region. If the amplifying atoms are excited randomly then the initial plot is just shifted together with the level of shot noise (Fig. 3b). Hence the statistics turns out to be almost Poissonian. This conclusion is independent of the initial light statistics.

In case of regular pumping of the amplifying atoms the noise plot remains on the same location (Fig. 3c) but as before the level of shot noise has increased appreciably. This means that on the on hand the statistics turn out to be sub-Poissonian over a wide frequency range (however not more than $\Delta\omega$ according to our initial conditions). On the other hand we see that the quantum properties are preserved under saturated amplification in the medium layer.

7 REFERENCES

[1] Kolobov M. I., Sokolov I. V., Optika i spektr., (1989), 67,123

[2] Golubev Yu. M.,Gorbachev V. N., Sov. Phys. JETP, (1989), 68,267

[3] Golubev Ju. M., Sokolov I. V., Sov. Phys. JETP, (1984), 60,234

[4] Zeiger S. G., Klimontovich Yu. L.,"Wave and Fluctuation Processes in Lasers",M. ,Mir, (1974)

[5] Gevorkjan S. T., Kruchov G. Yu., Zh. Eksp. Teor. Fiz. (1989), 95, 475

[6] Scully M. O., Lamb W. E., Phys. Rev., (1967), 159, 208

[7] Golubev Yu. M., Sov. Phys. JETP, (1974), 38, 228

[8] Glauber R.,"Quantum Optics and Quantum Radiophysics", M., Mir, (1966)

[9] Smirnov D. F., Sokolov I. V., Troshin A. S., Vestnik LGU, (1977), 10, 36

[10] Andreeva T. G., Golubev Yu. M., Sov. Phys. JETP, (1989), 69, 470

Quantum statistics of four-wave mixing of nonclassical light with pump depletion

J. Peřina [1] and J. Křepelka [2]

[1] Department of Optics, Palacký University, Svobody 26, 771 46 Olomouc, Czech Republic
[2] Joint Laboratory of Optics, Palacký University and Physical Institute of Czech Academy of Sciences, 17. listopadu 50, 772 07 Olomouc, Czech Republic

Abstract. Photon statistics and nonclassical behaviour of forward and backward four-wave mixing are examined including the coupling of modes in the approximation of small fluctuations around a stationary point, which is analytically determined and conditions for its stability are obtained. The depletion of pump modes is involved and effects of nonlinear dynamics, nonclassical behaviour of input radiation, external noise and losses are included.

Keywords. Four-wave mixing, quantum statistics, nonclassical light

1 Introduction

The four-wave mixing process was the first to be used for generation of squeezed light using the backward scattering on Na atoms [1] as well as using the forward scattering in optical fibres [2] (for a review, see [3] (Sec.11.5)). It represents also the nonlinear process suitable for phase conjugation of wave fronts of light beams and for various applications in adaptive optics.

In this paper we provide a fully quantum description of the forward as well as backward four-wave interaction [4] using a quantum formulation of propagation [5] and a method of small quantum fluctuations around a stationary point [6]. In this way we can analytically find stationary points of the Heisenberg - Langevian equations, calculate linear operator corrections to the stationary points, examine their stability and calculate quantum statistical properties of the process under examination, including photon - number distributions, their factorial moments, quadrature variances and principal squeezing in the framework of the so-called generalized superposition of coherent fields and quantum noise. We were successful to include nonclassical behaviour of input radiation, lossy mechanism and external noise.

2 Quantum dynamical description

To describe the quantum nonlinear dynamics of the degenerate process (all the frequencies are assumed to be the same) we adopt the effective momentum operator in the interaction picture

$$\hat{G} = \hbar g \hat{a}_1 \hat{a}_2 \hat{a}_3^\dagger \hat{a}_4^\dagger + i\hbar p_3^* \hat{a}_3 + i\hbar p_4^* \hat{a}_4 + \text{h.c.}, \tag{1}$$

where $\hbar$ is the Planck constant divided by 2π, g is the coupling constant proportional to the third-order susceptibility, $\hat{a}_j$ ($\hat{a}_j^\dagger$) are the annihilation (creation) operators of signal modes ($j = 1, 2$) and pump modes ($j = 3, 4$) and p_3, p_4 are external pump forces. The first term in (1) represents the nonlinear dynamics whereas the second and the third ones are the external pump terms. The quantum loss mechanism is adopted in the standard form ([7], [3] Chap. 7). Restricting ourselves to the propagation in the z- direction, the dynamical equations of motion can be written in the form

$$i\hbar \frac{\mathrm{d}\hat{a}_j}{\mathrm{d}z} = \left[\hat{G}, \hat{a}_j \right], \tag{2}$$

where [,] means the commutator. If the counterpropagating beams are assumed, the sign at the corresponding derivatives $\mathrm{d}/\mathrm{d}z$ has to be changed.

Thus for the forward mixing we obtain the dynamical equations

$$
\begin{aligned}
\frac{\mathrm{d}\hat{a}_1}{\mathrm{d}z} &= -\frac{\gamma_1}{2}\hat{a}_1 + ig^* \hat{a}_2^\dagger \hat{a}_3 \hat{a}_4 + \hat{L}_1, \\
\frac{\mathrm{d}\hat{a}_2}{\mathrm{d}z} &= -\frac{\gamma_2}{2}\hat{a}_2 + ig^* \hat{a}_1^\dagger \hat{a}_3 \hat{a}_4 + \hat{L}_2, \\
\frac{\mathrm{d}\hat{a}_3}{\mathrm{d}z} &= -\frac{\gamma_3}{2}\hat{a}_3 + ig \hat{a}_1 \hat{a}_2 \hat{a}_4^\dagger + \hat{L}_3 + p_3, \\
\frac{\mathrm{d}\hat{a}_4}{\mathrm{d}z} &= -\frac{\gamma_4}{2}\hat{a}_4 + ig \hat{a}_1 \hat{a}_2 \hat{a}_3^\dagger + \hat{L}_4 + p_4,
\end{aligned}
\tag{3}
$$

where γ_j, $j = 1, 2, 3, 4$ are the damping constants and $\hat{L}_j$ are the Langevin forces fulfilling the standard Markovian properties

$$
\begin{aligned}
\langle \hat{L}_j(z) \rangle &= \langle \hat{L}_j^\dagger(z) \rangle = 0, \\
\langle \hat{L}_j(z)\hat{L}_k(z') \rangle &= \langle \hat{L}_j^\dagger(z)\hat{L}_k^\dagger(z') \rangle = 0, \\
\langle \hat{L}_j^\dagger(z)\hat{L}_k(z') \rangle &= \gamma_j \langle n_{dj} \rangle \delta_{jk}\delta(z - z'), \\
\langle \hat{L}_j(z)\hat{L}_k^\dagger(z') \rangle &= \gamma_j \left(\langle n_{dj} \rangle + 1 \right) \delta_{jk}\delta(z - z');
\end{aligned}
\tag{4}
$$

here the brackets mean the average over the reservoir, $\langle n_{dj} \rangle$ are the mean numbers of reservoir damping oscillators coupled to the j-th mode, δ is the Dirac function and δ_{jk} is the Kronecker symbol. The lossy reservoir is assumed to have broad spectrum and to be chaotic (having the maximum entropy). If the case of counterpropagating beams is considered, the sign minus is to be added on the left-hand side of the second and fourth equations in (3).

Writing the system of equations (3) in the classical polar form, introducing $a_j = \rho_j \exp(i\varphi_j)$, $\rho_j = |a_j|$, $\varphi_j = \arg a_j$, $j = 1, 2, 3, 4$, we have

$$\frac{d\rho_1}{dz} = -\frac{\gamma_1}{2}\rho_1 + |g|\rho_2\rho_3\rho_4 \sin\vartheta,$$

$$\frac{d\rho_2}{dz} = -\frac{\gamma_2}{2}\rho_2 + |g|\rho_1\rho_3\rho_4 \sin\vartheta,$$

$$\frac{d\rho_3}{dz} = -\frac{\gamma_3}{2}\rho_3 - |g|\rho_1\rho_2\rho_4 \sin\vartheta + |p_3|\cos(\phi_3 - \varphi_3),$$

$$\frac{d\rho_4}{dz} = -\frac{\gamma_4}{2}\rho_4 - |g|\rho_1\rho_2\rho_3 \sin\vartheta + |p_4|\cos(\phi_4 - \varphi_4),$$

(5a)

$$\frac{d\varphi_1}{dz} = |g|\frac{\rho_2\rho_3\rho_4}{\rho_1}\cos\vartheta,$$

$$\frac{d\varphi_2}{dz} = |g|\frac{\rho_1\rho_3\rho_4}{\rho_2}\cos\vartheta,$$

$$\frac{d\varphi_3}{dz} = |g|\frac{\rho_1\rho_2\rho_4}{\rho_3}\cos\vartheta + \frac{|p_3|}{\rho_3}\sin(\phi_3 - \varphi_3),$$

$$\frac{d\varphi_4}{dz} = |g|\frac{\rho_1\rho_2\rho_3}{\rho_4}\cos\vartheta + \frac{|p_4|}{\rho_4}\sin(\phi_4 - \varphi_4),$$

(5b)

where $\vartheta = \varphi_1 + \varphi_2 - \varphi_3 - \varphi_4 + \psi$, ψ being the phase of the coupling constant g and $\phi_{3,4}$ are phases of the pumping powers $p_{3,4}$.

The stationary real amplitudes ρ_{j0}, $j = 1, 2, 3, 4$ of this system can simply be found from the conditions $da_j/dz = 0$, $j = 1, 2, 3, 4$. Taking into account that $\langle \hat{L}_j \rangle = 0$, choosing $\psi = \pi/2$, $\varphi_1 + \varphi_2 - \varphi_3 - \varphi_4 = n\,2\pi$, $n = 0, 1, \dots$ and simplifying calculations to the case $\gamma_1 = \gamma_2 = \gamma_3 = \gamma_4 = \gamma$ and $p_3 = p_4 = p$ they are in the form

$$\rho_{10} = \rho_{20} = \left[\frac{|p|}{|g|\rho_{30}} - \rho_{30}^2\right]^{\frac{1}{2}}, \qquad \rho_{30} = \rho_{40} = \left(\frac{\gamma}{2|g|}\right)^{\frac{1}{2}}$$

(6)

for both the forward and backward scattering, under the condition

$$|p| \geq \frac{\gamma}{2}\rho_{30}.$$

(7)

Denoting the stationary solutions as a_{j0}, we can consider small quantum fluctuations around this stationary point writing $\hat{a}_j(z) = a_{j0} + \delta\hat{a}_j(z)$ and conserving only linear quantum corrections $\delta\hat{a}_j(z)$, this leads to the linear operator equation

$$\frac{d\delta\hat{a}}{dz} = \hat{M}\delta\hat{a} + \hat{L},$$

(8)

where $\delta\hat{a}$ represents the column vector composed of $\delta\hat{a}_j$, $\delta\hat{a}_j^\dagger$, $\hat{L}$ is the column vector composed of $\hat{L}_j$, $\hat{L}_j^\dagger$ and $\hat{M}$ is the corresponding 8×8 - matrix containing the coupling and damping constants and the stationary solution [4]. The equation (8) can be separated for the real amplitudes $\delta\rho_j$ and phases $\delta\varphi_j$ with the corresponding matrices $\hat{M}$ as follows

$$\hat{M}_\rho = \begin{bmatrix} -\frac{\gamma}{2} & |g|\rho_{30}^2 & |g|\rho_{10}\rho_{30} & |g|\rho_{10}\rho_{30} \\ |g|\rho_{30}^2 & -\frac{\gamma}{2} & |g|\rho_{10}\rho_{30} & |g|\rho_{10}\rho_{30} \\ -|g|\rho_{10}\rho_{30} & -|g|\rho_{10}\rho_{30} & -\frac{\gamma}{2} & -|g|\rho_{10}^2 \\ -|g|\rho_{10}\rho_{30} & -|g|\rho_{10}\rho_{30} & -|g|\rho_{10}^2 & -\frac{\gamma}{2} \end{bmatrix}, \quad (9a)$$

$$\hat{M}_\varphi = \begin{bmatrix} -\frac{\gamma}{2} & -|g|\rho_{30}^2 & |g|\rho_{30}^2 & |g|\rho_{30}^2 \\ -|g|\rho_{30}^2 & -\frac{\gamma}{2} & |g|\rho_{30}^2 & |g|\rho_{30}^2 \\ -|g|\rho_{10}^2 & -|g|\rho_{10}^2 & -\frac{\gamma}{2} & |g|\rho_{10}^2 \\ -|g|\rho_{10}^2 & -|g|\rho_{10}^2 & |g|\rho_{10}^2 & -\frac{\gamma}{2} \end{bmatrix} \quad (9b)$$

for forward scattering (for backward scattering the signs are to be changed at the second and fourth lines of these matrices).

Then the operator corrections are obtained in the form

$$\delta\hat{a}_j(z) = \sum_{k=1}^{4} \left[U_{jk}(z)\delta\hat{a}_k(0) + V_{jk}(z)\delta\hat{a}_k^\dagger(0) \right] + \hat{F}_j(z), \quad j = 1, 2, 3, 4, \quad (10)$$

where the dynamical functions $U_{jk}(z)$, $V_{jk}(z)$ and the reservoir operator $\hat{F}_j(z)$ can be obtained explicitly from the solution procedure [8,4].

Concerning the stability of the operator corrections we have to solve the characteristic polynomials $\text{Det}(\hat{M}_{\rho,\varphi} - \lambda\hat{1}) = 0$ leading to the following roots
(i) for forward mixing

$$\lambda_1^{(\rho)} = -\gamma, \quad \lambda_2^{(\rho)} = -\gamma + \frac{|p|}{\rho_{30}}, \quad \lambda_{3,4}^{(\rho)} = -\frac{|p|}{2\rho_{30}}\left[1 \mp \left(1 - \frac{4\gamma^2\rho_{10}^2}{|p|^2}\right)^{\frac{1}{2}}\right],$$

$$\lambda_1^{(\varphi)} = 0, \quad \lambda_2^{(\varphi)} = -\frac{|p|}{\rho_{30}},$$

$$\lambda_{3,4}^{(\varphi)} = -\left(\gamma - \frac{|p|}{2\rho_{30}}\right)\left[1 \mp \left(1 - \frac{\gamma^2}{8(\gamma - |p|/\rho_{30})^2}\right)^{\frac{1}{2}}\right],$$

$$(11a)$$

(ii) for backward mixing

$$\lambda_{1,2,3,4}^{(\rho)} = \pm i\left\{\frac{1}{2}\frac{|p|}{\rho_{30}}\left(\frac{|p|}{\rho_{30}} - \gamma\right)\left[1 \mp \left(1 + \frac{4\gamma^3\rho_{10}^2}{|p|^2(|p|/\rho_{30} - \gamma)}\right)^{\frac{1}{2}}\right]\right\}^{\frac{1}{2}},$$

$$(11b)$$

$$\lambda_{1,2}^{(\varphi)} = 0, \quad \lambda_{3,4}^{(\varphi)} = \pm i\left[\frac{|p|}{\rho_{30}}\left(\frac{|p|}{\rho_{30}} - \gamma\right)\right]^{\frac{1}{2}}.$$

Additionally to the condition (7) we need also the following condition

$$|p| < \gamma \rho_{30} \qquad (12a)$$

to be fulfilled for forward mixing to ensure the stability. The situation is more complicated for backward mixing; however, the change of the signs at some equations of motion leads generally to the change of the quality of the solution (a monotonic solution is transformed to the oscillating one and vice versa).

3 Quantum statistical properties

The quantum statistical properties of light corresponding to the form (10) of the operator solution are described by the normal quantum characteristic function in the Gaussian form as follows [3]

$$
C_{\mathcal{N}}(\{\beta_j\}, z) = \exp \left\{ \sum_{j=1}^{4} \left[-B_j(z)|\beta_j|^2 + \left(\frac{1}{2} C_j(z) \beta_j^{*2} + c.c. \right) \right. \right.
$$

$$
\left. \left. + (\beta_j \xi_j^*(z) - c.c.) \right] + \sum_{i<j} (D_{ij}(z)\beta_i^* \beta_j^* + \bar{D}_{ij}(z)\beta_i \beta_j^* + c.c.) \right\}, \qquad (13)
$$

where β_j, $j = 1, 2, 3, 4$ are complex parameters of the characteristic function, $\xi_j(z)$ are the z-dependent complex amplitudes in the corresponding modes,

$$
\xi_j(z) = \sum_{k=1}^{4} [U_{jk}(z)\xi_k + V_{jk}(z)\xi_k^*] + a_{j0}, \qquad (14)
$$

in relation to (10), ξ_k being the input complex amplitudes corresponding to $\delta \hat{a}_k(0)$, and the quantum noise functions are determined by

$$
B_i(z) = \langle \Delta \hat{a}_i^\dagger(z) \Delta \hat{a}_i(z) \rangle = \langle \hat{F}_i^\dagger(z) \hat{F}_i(z) \rangle - \sum_j |U_{ij}(z)|^2
$$

$$
+ \sum_j \left\{ [|U_{ij}(z)|^2 + |V_{ij}(z)|^2] B_j + [U_{ij}^*(z)V_{ij}(z)C_j^* + c.c.] \right\},
$$

$$
C_i(z) = \langle (\Delta \hat{a}_i(z))^2 \rangle = \langle \hat{F}_i^2(z) \rangle + \sum_j [U_{ij}^2(z)C_j + V_{ij}^2(z)C_j^*
$$

$$
+ 2U_{ij}(z)V_{ij}(z)B_j] - \sum_j U_{ij}(z)V_{ij}(z),
$$

$$
D_{ij}(z) = \langle \Delta \hat{a}_i(z) \Delta \hat{a}_j(z) \rangle = \langle \hat{F}_i(z) \hat{F}_j(z) \rangle
$$

$$+ \sum_k \left\{ U_{ik}(z)U_{jk}(z)C_k + V_{ik}(z)V_{jk}(z)C_k^* + [U_{ik}(z)V_{jk}(z) + V_{ik}(z)U_{jk}(z)]B_k \right\}$$

$$- \sum_k V_{ik}(z)U_{jk}(z), \quad i \neq j,$$

$$\bar{D}_{ij}(z) = -\langle \Delta \hat{a}_i^\dagger(z)\Delta \hat{a}_j(z)\rangle = -\langle \hat{F}_i^\dagger(z)\hat{F}_j^\dagger(z)\rangle$$

$$- \sum_k \left\{ [U_{ik}^*(z)U_{jk}(z) + V_{ik}^*(z)V_{jk}(z)]B_k + V_{ik}^*(z)U_{jk}(z)C_k + U_{ik}^*(z)V_{jk}(z)C_k^* \right\}$$

$$+ \sum_k U_{ik}^*(z)U_{jk}(z), \quad i \neq j, \tag{15}$$

where the input value $B_j = B_j(0) + 1$ is related to antinormal ordering in order to describe initial nonclassical light and $C_j = C_j(0)$ (the input coupling of modes is not included so that $D_{ij}(0) = \bar{D}_{ij}(0) = 0$).

In order to include nonclassical (squeezed and/or sub-Poisson) light with additional external noise, we can choose

$$B_j = \cosh^2 r_j + \bar{n}_{j0}, \quad C_j = \frac{1}{2}\exp(i\theta_j)\sinh 2r_j, \quad j = 1, 2, 3, 4, \tag{16}$$

where r_j are the squeeze parameters, θ_j are the squeeze phases and $\bar{n}_{j0}$ are the external noise components. If $r_j = 0$, we have the input state in the form of the superposition of the coherent state $|\{\xi_j\}\rangle$ and noise $\bar{n}_{j0}$; if $r_j = \bar{n}_{j0} = 0$, the input coherent state is described ($B_j = 1$, $C_j = 0$).

For the photon-number distribution $p(n_j, z)$ and its factorial moments $\langle W_j^k(z)\rangle$ in single modes we can write [3]

$$p(n_j, z) = \frac{1}{(E_j F_j)^{\frac{1}{2}}} \left(1 - \frac{1}{F_j}\right)^{n_j} \exp\left(-\frac{A_{1j}}{E_j} - \frac{A_{2j}}{F_j}\right)$$

$$\times \sum_{k=0}^{n_j} \frac{1}{\Gamma(k + \frac{1}{2})\Gamma(n_j - k + \frac{1}{2})} \left(\frac{1 - \frac{1}{E_j}}{1 - \frac{1}{F_j}}\right)^k$$

$$\times L_k^{-\frac{1}{2}}\left(-\frac{A_{1j}}{E_j(E_j - 1)}\right) L_{n_j-k}^{-\frac{1}{2}}\left(-\frac{A_{2j}}{F_j(F_j - 1)}\right), \tag{17a}$$

$$\langle W_j^k(z)\rangle = k!(F_j - 1)^k \sum_{l=0}^{k} \frac{1}{\Gamma(l + \frac{1}{2})\Gamma(k - l + \frac{1}{2})} \left(\frac{E_j - 1}{F_j - 1}\right)^l$$

$$\times L_l^{-\frac{1}{2}}\left(-\frac{A_{1j}}{E_j - 1}\right) L_{k-l}^{-\frac{1}{2}}\left(-\frac{A_{2j}}{F_j - 1}\right), \tag{17b}$$

where Γ is the gamma function, $L_j^{-\frac{1}{2}}$ are the Laguerre polynomials,

$$E_j = B_j(z) + 1 - |C_j(z)|, \quad F_j = B_j(z) + 1 + |C_j(z)| \tag{18a}$$

represent the quantum noise components and

$$A_{1,2j} = \frac{1}{2}\left[|\xi_j(z)|^2 \mp \frac{1}{2|C_j(z)|}\left(\xi_j^2(z)C_j^*(z) + c.c.\right)\right] \tag{18b}$$

are playing the role of the signal components. The treatment of compound modes, e.g. signal modes (1,2), is more complicated and can be found in [9].

The single-mode quadrature variances are

$$\left\langle \left(\Delta \begin{smallmatrix} \hat{q}_j \\ \hat{p}_j \end{smallmatrix}\right)^2 \right\rangle = 1 + 2[B_j(z) \pm \mathrm{Re}\,C_j(z)] \tag{19a}$$

where $\hat{q}_j = \hat{a}_j + \hat{a}_j^\dagger$, $\hat{p}_j = (\hat{a}_j - \hat{a}_j^\dagger)/i$, and the principal squeeze variance λ_j equals [10]

$$\lambda_j = 1 + 2[B_j(z) - |C_j(z)|]. \tag{19b}$$

This parameter provides the maximum squeezing of vacuum fluctuations provided that $\lambda_j < 1$. For the compound modes (j, k) we have

$$\left\langle \left(\Delta \begin{smallmatrix} \hat{q} \\ \hat{p} \end{smallmatrix}\right)^2 \right\rangle = 2\Big\{1 + B_j(z) + B_k(z) - 2\mathrm{Re}\bar{D}_{jk}(z)$$

$$\pm \mathrm{Re}\big[C_j(z) + C_k(z) + 2D_{jk}(z)\big]\Big\}, \quad j \neq k, \tag{20a}$$

$$\lambda = 2\Big\{1 + B_j(z) + B_k(z) - 2\mathrm{Re}\bar{D}_{jk}(z) - |C_j(z) + C_k(z) + 2D_{jk}(z)|\Big\}, \quad j \neq k, \tag{20b}$$

defining squeezing of vacuum fluctuations in the compound mode (j, k) if $\lambda < 2$. Here $\langle(\Delta\hat{q})^2\rangle = \langle(\Delta\hat{q}_j)^2\rangle + \langle(\Delta\hat{q}_k)^2\rangle + 2\langle\Delta\hat{q}_j\Delta\hat{q}_k\rangle$, $j \neq k$ and similarly for $\langle(\Delta\hat{p})^2\rangle$. In general, considering single modes, the condition $\lambda_j < 1$, $(B_j < |C_j|)$ for squeezing of vacuum fluctuations is not sufficient for sub-Poissonian behaviour of photons [3] (pp. 248-249).

4 Discussion of results

Using the above analytical results, we can perform computations for photon-number distribution, its factorial moments, quadrature and principal squeeze variances for single and compound modes in order to discuss quantum features of light, effects of nonlinear dynamics as well as of input nonclassical light behaviour, losses and external noise. For the computations we choose

$g = i/2$, $\xi_1 = \xi_2 = 0.03$, $\xi_3 = \xi_4 = 0.08$, $p_3 = p_4 = 0.75$, $\gamma = 1$ and
$\varphi_1 = \varphi_2 = \varphi_3 = \varphi_4 = \phi_3 = \phi_4 = 0$. In this case $\rho_{10} = \rho_{20} = 1/\sqrt{2}$,
$\rho_{30} = \rho_{40} = 1$.

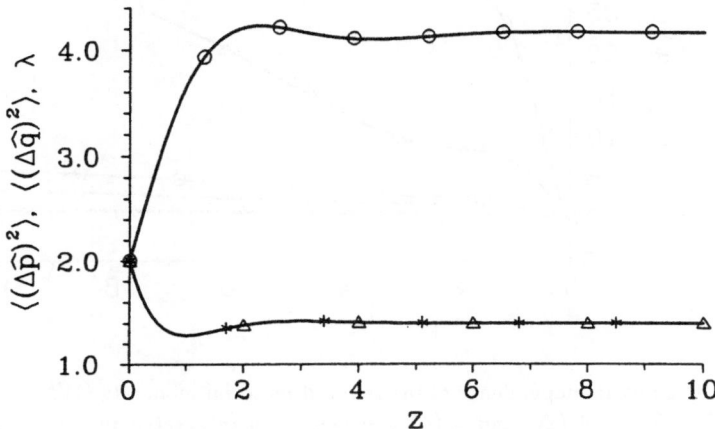

Fig. 1. Quadrature variances $\langle(\Delta\hat{p})^2\rangle(*)$, $\langle(\Delta\hat{q})^2\rangle$ (o) and principal squeeze variance λ (Δ) for the compound signal mode (1,2) in dependence on the travelled
distance z; $g = i/2$, $\xi_1 = \xi_2 = 0.03$, $\xi_3 = \xi_4 = 0.08$, $p_3 = p_4 = 0.75$, $\gamma = 1$,
$r_j = \bar{n}_{j0} = \langle n_{dj}\rangle = 0$, $j = 1, 2, 3, 4$.

In Figure 1 we see the quadrature variances $\langle(\Delta\hat{q})^2\rangle$, $\langle(\Delta\hat{p})^2\rangle$ and principal
squeeze variance λ for the compound signal mode (1,2), the input coherent
field ($r_j = 0$, $j = 1, 2, 3, 4$), without external noise ($\bar{n}_{j0} = 0$, $j = 1, 2, 3, 4$) and
for cold reservoir ($\langle n_{dj}\rangle = 0$, $j = 1, 2, 3, 4$). Strong squeezing ($\langle(\Delta\hat{p})^2\rangle = \lambda <$
2) occurs created by the nonlinear dynamics and the coupling of signal modes
at the beginning of the interaction. Asymptotically, a considerable level of
squeezing is reached, too. The input second-order coherence is conserved in
the interaction for relatively long distances in the compound mode (1,2) as
a consequence of the coupling of the signal modes, which is demonstrated in
Figure 2 (on the contrary, in the single signal modes, coherence is quickly
lost).

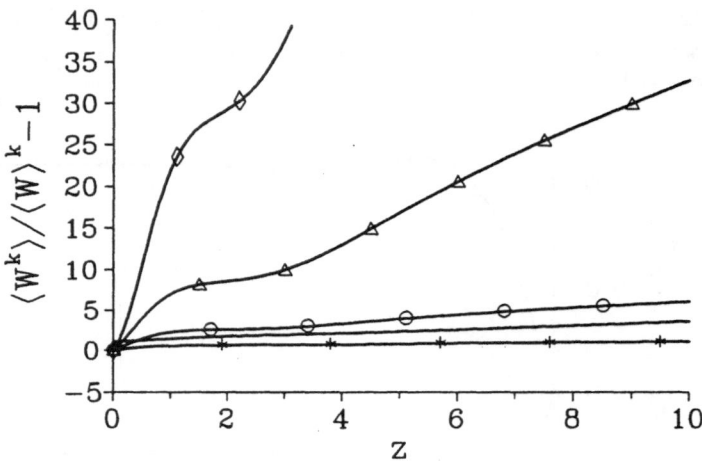

Fig. 2. The spatial dependence of the reduced factorial moments $\langle W^k \rangle / \langle W \rangle^k - 1$, $k = 2$ (∗), 3 (○), 4 (△) and 5 (◊) and the mean integrated intensity $\langle W \rangle$ (full curve without denotation) for the compound signal mode (1,2); the values of the parameters are as in Figure 1.

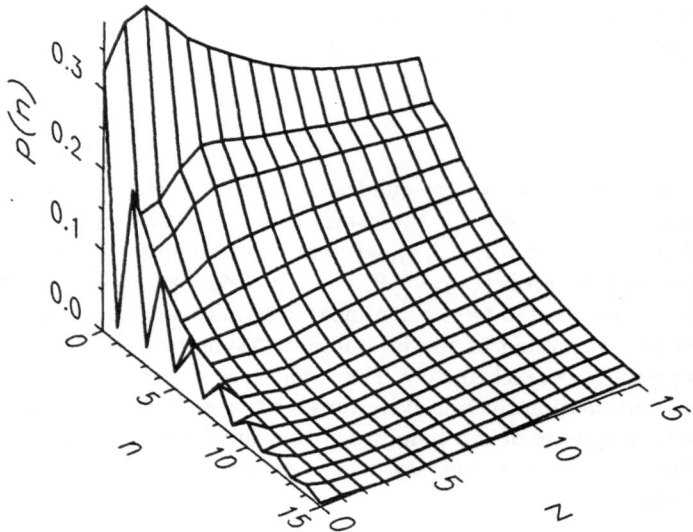

Fig. 3 The photon-number distribution $p(n)$ of input squeezed light ($r_j = 1$, $j = 1, 2, 3, 4$) for compound signal mode (1,2); the values of the parameters are as in Figure 1.

In Figure 3 we show the photon-number distribution for initially squeezed light ($r_j = 1$, $j = 1, 2, 3, 4$) for the compound mode (1,2) again with a typical oscillating behaviour at the beginning. The oscillations, having the fully quantum origin, are successively smoothed out by the nonlinear interaction. Of course, these oscillations are also degraded by the external noise and by the loss mechanism.

We can conclude that the nonlinear dynamics of four-wave mixing provides a great variety of regimes for generation of nonclassical radiation states, particularly if the coupling of modes is involved. The strong initial nonclassical behaviour can be so saturated that the nonlinear dynamics cannot pronounce it more. Even if it is possible to have nonclassical states involving the external noise and loss mechanism, such states are strongly sensitive to these degradation effects.

Acknowledgment. This work was partially supported by an internal grant of the Palacký University in Olomouc.

References

[1] R. E. Slusher, L. W. Hollberg, B. Yurke, J. C. Mertz, J. F. Valley, Phys. Rev. Lett **55** (1985) 2409.

[2] R. M. Shelby, M. D. Levenson, S. H. Perlmutter, R. G. DeVoe, D. F. Walls, Phys. Rev. Lett **57** (1986) 691.

[3] J. Peřina, *Quantum Statistics of Linear and Nonlinear Optical Phenomena*, 2nd ed., Kluwer, Dordrecht-Boston 1991.

[4] J.Peřina, J. Křepelka, J. Mod. Optics **40** (1993) in print.

[5] Y. Ben-Aryeh, A. Lukš, V. Peřinová, Phys. Lett. A **165** (1992) 19.

[6] P. D. Drummond, K. J. McNeil, D. F. Walls, Optics Commun. **28** (1979) 255.

[7] W. H. Louisell, *Quantum Statistical Properties of Radiation*, J. Wiley, New York 1973.

[8] J. Peřina, J. Křepelka, J. Mod. Optics **38** (1991) 2137; **39** (1992) 1029.

[9] M. Kárská, J. Peřina, J. Mod. Optics **37** (1990) 195.

[10] A. Lukš, V. Peřinová, J. Peřina, Optics Commun. **67** (1988) 149.

<u>PART V</u>: Other Fundamentals

PART V: Other Fundamentals

Quantum non-demolition measurements in optics and quantum optical repeaters

Jean-Philippe Poizat, Jean-François Roch,
and Philippe Grangier

Institut d'Optique B.P. 147 - F91403 ORSAY Cedex - France

Abstract

The efficiency of an optical quantum non-demolition (QND) measurement can be characterized using three criteria, which describe respectively the quality of the quantum measurement, the non-destruction of the signal, and the conditional variance of the output signal beam, given the output meter beam (quantum-state-preparation criterion). Quantitative limits can be defined with respect to these criteria, delimiting "classical" and "quantum" domains of operation. We describe the implementation of two experiments which fulfill these criteria, using either three-level atoms inside a doubly-resonant optical cavity, or semiconductors emitters and receivers.

1   Introduction

A quantum measurement usually perturbs the quantity which is measured, adding "back-action noise" to the system under study. However, "quantum non-demolition" (QND) measurements can be performed, leaving the observed quantity unperturbed, while adding the back-action noise into another complementary observable. Quantum non- demolition measurements were first proposed to monitor the motion of mechanical oscillators [1]-[3], but it was then realized that they were easier to implement in the optical domain [4]-[6], [34], [35]. In this case, the measurements are made on a mode of the electromagnetic field considered as a quantum harmonic oscillator, and the coupling responsible for the measurement is created using either χ^2 or χ^3 optical non-linearities [7]-[9]. In this paper, we will focus on the QND measurements of "travelling wave" light beams, rather than fields stored in high Q cavities [10]. Moreover, we will assume that the quantum fluctuations of the beams are very small compared to the mean intensities, so that a linear treatment of these quantum fluctuations is possible [11]. A "travelling wave" optical QND scheme usually involves the coupling of two field modes : one "signal" mode, which is to be measured, and one "meter" mode, which is directly detected and yields some informations about the signal mode. A perfect QND measurement device should leave the measured quadrature component of the signal mode unchanged, and turn the detected meter quadrature component into some kind of "macroscopic copy" of the signal. However, such a device does not exist at present time, and it appears necessary to be able to characterize the "non-ideality" of practical QND schemes. We will briefly recall the criteria [12]-[15] which have been proposed for this characterization, and we will define quantitative limits with respect to these criteria, delimiting "classical" and "quantum" domains of operation. Finally, two experiments, meeting all criteria for QND measurements, will be described in the last sections of the paper.

2 Linear model for the quantum fluctuations

The operators for the various quadrature components of the fields will be written in column matrices in the frequency domain, denoted $F(\omega)$. All over the paper, the subscripts s and m will denote respectively the signal and meter modes·:

$$F(\omega) = \begin{pmatrix} X_s(\omega) \\ Y_s(\omega) \\ X_m(\omega) \\ Y_m(\omega) \end{pmatrix} \tag{1}$$

In a linearized model (relatively to the quantum fluctuations), the relation between the input fluctuations $\delta F^{in}(\omega)$ and the output fluctuations $\delta F^{out}(\omega)$ can be written [11]-[15]:

$$\delta F^{out}(\omega) = \lambda(\omega)\delta F^{in}(\omega) + \delta F^{ad}(\omega) \tag{2}$$

where $\lambda(\omega)$ is a four by four "gain" matrix, and $\delta F^{ad}(\omega)$ is an extra noise added by other degrees of freedom involved in the coupling.

The gain $\lambda(\omega)$ and the covariance matrix of $\delta F^{ad}(\omega)$ are not independent, because the transformation (2) preserves the commutations relations between the input and output modes. In some cases, $\delta F^{ad}(\omega)$ is zero : it is the case for instance for pure "parametric" processes, involving transparent χ^2 or χ^3 media. On the other hand, $\delta F^{ad}(\omega)$ is non- zero if either losses or excess noise in the non linear media are involved, due for instance to resonant atomic non-linearities, or to Brillouin or Raman processes in solids.

In all cases, we will assume that the input beams and the QND device are independent, so that $\delta F^{in}(\omega)$ and $\delta F^{ad}(\omega)$ are not correlated. The fluctuations of the beams will be characterized using symetrically ordered covariances, which fit the Wigner distribution. The output covariance W^{out} is then given as a function of the input covariance W^{in} by [11] :

$$W^{out} = \lambda(\omega) W^{in} \lambda(\omega)^t + W^{ad} \tag{3}$$

The expressions of $\lambda(\omega)$ and W^{ad} are obtained from a quantum treatment for each specific device. This description must ensure the relations between $\lambda(\omega)$ and W^{ad} needed to preserve the commutation relations between the input and output fields [11].

3   Characterization of a QND device as a Quantum State Preparation device

In QND experiments, the performance of the device has usually been assessed from a measurement of some components of the output covariance W^{out}, which contains the information about the noises and the quantum correlations of the two output beams. This procedure allows one to characterize the "quantum state-preparation" (QSP) properties of the device. A convenient quantity for that purpose is the conditional variance [13] W_{QSP} of the output signal (taken here as X_s^{out}), given the output meter (taken here as Y_m^{out}). This quantity is equivalent to a squeezing spectrum, and it can be defined from a suitable combination of the various matrix elements of the output covariance W^{out} :

$$W_{QSP} = W_{X_sX_s}^{out} - |\, W_{X_sY_m}^{out}\,|^2 \,/\, W_{Y_mY_m}^{out} \tag{4}$$

An important advantage of W_{QSP} is that it can be measured directly in an experiment, as a noise level on a spectrum analyser. For that purpose, the signal and meter photocurrents must be recombined, introducing an electronic gain on the meter photocurrent. This gain (or attenuation) must be optimized, in order to obtain the maximum reduction below the shot-noise level of the signal beam. The optimum value of the attenuation depends of the amount of correlation which has been established between the two beams, and it is given by :

$$g(\omega) = W_{X_sY_m}^{out} / W_{Y_mY_m}^{out} \tag{5}$$

The criterion $W_{QSP} < 1$ means that the device is able to establish sub-shot-noise correlations between the two output beams. It is instructive to give the value of W_{QSP} for a very simple model, corresponding to the following "parametric" coupling between the quadratures components X and Y of the signal and meter modes :

$$\begin{aligned} X_s^{out} &= X_s^{in} & Y_s^{out} &= Y_s^{in} + GX_m^{in} \\ X_m^{out} &= X_m^{in} & Y_m^{out} &= Y_m^{in} + GX_s^{in} \end{aligned} \tag{6}$$

This model can be implemented for instance using crossed Kerr effect between the two modes, obtained using two-photon non-linearities [13], [16]-[18]. In this case, X and Y are respectively the amplitude and phase fluctuations, defined relatively to the mean values of the fields. For a shot-noise limited meter beam, one then obtains:

$$W_{QSP} = 1/(1 + G^2) \tag{7}$$

A perfect QSP device is obtained for an infinite value of G. Another useful model is a simple beamsplitter ("linear coupler"), in which the input meter beam can be in a squeezed state [8]. Denoting by r, t the amplitude reflection and transmission coefficients of the beamsplitter (fig.1), the input-output transformation is simply :

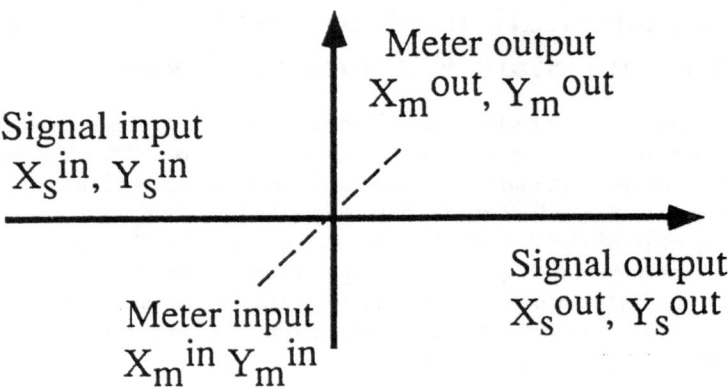

Figure 1: Simplest optical tap : X_s, Y_s and X_m, Y_m are the two quadratures of the signal and meter beams. The input meter beam must be squeezed for QND operation. If the meter beam is simply in the vacuum state, this coupler achieves the lower "classical" limits $W_{QSP} = 1$ and $N_s^{eq} . N_m^{eq} = 1$ (see text).

$$X_s^{out} = t X_s^{in} + r X_m^{in} \qquad X_s^{out} = t Y_s^{in} + r Y_m^{in}$$
$$X_m^{out} = - r X_s^{in} + t X_m^{in} \qquad Y_m^{out} = - r Y_s^{in} + t Y_m^{in} \qquad (8)$$

In that case, the information on the signal amplitude X_s^{in} is obtained on the meter output amplitude X_m^{out}. If the corresponding quadrature of the input meter beam has a noise power squeezing S^{in} , the QSP properties of this scheme is then given by :

$$W_{QSP} = \frac{S^{in}}{(r^2 + t^2 S^{in})} \qquad (9)$$

If the input meter beam is in the vacuum state $S^{in} = 1$, one obtains immediately $W_{QSP} = 1$, for any value of r and t . On the other hand, the inequality $W_{QSP} < 1$ is obtained as soon as the meter beam is squeezed, and perfect QSP is obtained for $S^{in} = 0$. This evidences the close relationship between the conditional variance and a degree of squeezing. We will take therefore the value $W_{QSP} = 1$ as a limit between "classical" and "quantum" regime for an optical coupler, as far as quantum state preparation is concerned. However, this criterion does not characterize completely the QND properties of the device, because one must also prove that the output signal and meter beams really yield an information about the input signal beam[13],[15]. This point will be discussed in the following section.

4 Characterization of a QND device using equivalent input noises

The basic idea that we will use [15] is that two types of quantities can actually be measured experimentally. The first one is the output covariance matrix W^{out}, as stated previously. The second quantity is the gain matrix $\lambda(\omega)$, which represents the transfer function of the system, and can be determined using weak classical probe beams, as shown by Eq.(2). Knowing $\lambda(\omega)$ and W^{out}, one may write linear "estimators" for the measured quadrature component of the signal. For instance, in the parametric case, defined by Eq.(6), one has:

$$G^{-1} Y_m^{out} = X_s^{in} + G^{-1} Y_m^{in} \tag{10}$$

$$G^{-1} W_{Y_m Y_m}^{out} (G^{-1})^* = W_{X_s X_s}^{in} + N_m^{eq} \tag{11}$$

$$N_m^{eq} = |G|^{-2} W_{Y_m Y_m}^{out} - W_{X_s X_s}^{in} \tag{12}$$

The quantity N_m^{eq} appears as the noise added to input signal variance, and is therefore an "equivalent input noise". Here, N_m^{eq} is due to the shot-noise of the meter beam, which appears as a noise even in the parametric case. For real devices, a second noise source is due to the excess noise added by the non-linear system, and a third noise source appears to be the quadrature of the signal conjugated from the observed one. This "unwanted" quadrature can be coupled in the output signal or meter by the non-ideal device. These three degradations of the measurement efficiency can then be included in N_m^{eq}, equivalent input noise on the meter channel [15]-[19]. Similar definition can be also used for relating the output signal to the input signal, corresponding to N_s^{eq}, equivalent input noise on the signal channel.

Another possible characterization of the same properties is obtained by looking at the faithfulness of the transfer of an input signal-to-quantum-noise ratio (SNR or simply R in the equations), toward the signal and meter output channels [7][20][34][35]. The input SNR is defined as the ratio of the power of a classical modulation at a given frequency by the quantum noise power at the same frequency for a given observable X,

$$R = \frac{\langle X \rangle^2}{\langle \Delta X^2 \rangle} , \tag{13}$$

where all quantities are defined in the frequency domain [15]. The signal non-degradation property is therefore evaluated by (the subscript s refers to the signal, and m to the meter)

$$T_s = \frac{R_s^{out}}{R_s^{in}} = \frac{W_s^{in}}{W_s^{in} + N_s^{eq}} , \tag{14}$$

where W_s^{in} is the variance of the input light beam normalized to the shotnoise ($W_s^{in} = 1$ for a coherent state), and N_s^{eq} is the equivalent input noise for the signal [15]. The measurement capability is given by

$$T_m = \frac{R_m^{out}}{R_s^{in}} = \frac{W_s^{in}}{W_s^{in} + N_m^{eq}} , \tag{15}$$

where N_m^{eq} is the equivalent input noise for the meter channel.

Let us finally give the values of N_m^{eq} and N_s^{eq} for our model devices. In the parametric crossed-Kerr effect model, one obtains :

$$N_s^{eq} = 0 \qquad N_m^{eq} = |G|^{-2} \tag{16}$$

In the linear coupler with squeezed meter beam, one has :

$$N_m^{eq} = (t/r)^2 S^{in} \qquad N_s^{eq} = (r/t)^2 S^{in} \tag{17}$$

$$N_s^{eq} N_m^{eq} = (S^{in})^2 \tag{18}$$

If the input meter beam is in the vacuum state $S^{in} = 1$, the product of the equivalent input noises is equal to 1, for any value of r and t .

5 Classical limits for QND measurements

In order to discriminate various modes of operation of a "quantum coupler", it is interesting to know where the "quantum limits" are. As shown above, a useful reference is given by a simple beamsplitter with a meter beam in the vacuum state, for which one has $W_{QSP} = 1$ and $N_s^{eq}.N_m^{eq} = 1$ .We have shown more generally that the domain of operation of a "classical coupler", described as a phase-insensitive device. can be defined by the two following inequalities [19]-[20] :

$$W_{QSP} \geq 1 \tag{19}$$

$$N_s^{eq}.N_m^{eq} \geq 1 \tag{20}$$

where 1 is by convention the standard shot-noise level. The inequality (19) means that "conditional squeezing" is forbidden in the classical domain, just like ordinary squeezing. The inequality (20) means that the *product* of the signal and measurement equivalent noises is restricted : in the classical domain, small signal degradation implies poor measurement. We note that the definition of "classical" used in inequality (20) is not the usual one in quantum optics (based on the properties of the P function), but is rather the one connected to phase-sensitive vs phase-insensitive amplifiers in the theory of linear amplifiers [20].

Another form of the same inequality was introduced in Ref.[21], for the particular case of a beamsplitter, using the signal to noise ratios (SNR) on the signal and meter channels :

$$(SNR)_s^{out} + (SNR)_m^{out} \leq (SNR)_s^{in} \tag{21}$$

or equivalently

$$T_s + T_m \leq 1 \tag{22}$$

This alternative form is equivalent to the previous one, under some conditions which are fulfilled in general. A device violating this "classical" limit will be called a "quantum optical tap". On the other hand, obtaining $W_{QSP} < 1$ is "quantum state preparation". A true QND device will be defined as a quantum optical tap yielding also quantum state preparation. Finally, let us emphasize that the two conditions (16) and (17) are independant, and cannot be linked by any "triangular" inequality. This can be shown by realizing [19] that a "twin-beam" generator will yield W_{QSP} arbitrarily small, but $N_s^{eq}.N_m^{eq} > 1$. On the other hand, it is easy to check [19] that a phase sensitive amplifier followed by a beamsplitter will yield $N_s^{eq}.N_m^{eq}$ arbitrarily small, but $W_{QSP} > 1$. Therefore, both inequalities must be considered to ensure "quantum operation" of a QND device.

6   QND experiments with three-level systems

The basic idea of this experiment is to use the optical Kerr effect to couple the intensity fluctuations of the signal beam to the phase fluctuations of the meter beam ("crossed-phase modulation effect") [9],[22] . The information is then read out using some kind of phase-sensitive (interferometric) detection scheme of the meter beam.

The first attempt to use the optical Kerr effect for QND measurements has been made using non-resonant $\chi^{(3)}$ in optical fibers [4],[23],[24]. Other experiments were done using resonant $\chi^{(3)}$ in solid-state systems [13],[25]. Despite the simplicity of the scheme, it has not been possible to obtain noiseless $\chi^{(3)}$ media: either guided acoustic wave Brillouin scattering (GAWBS) in optical fibers, or spontaneous emission noise in resonant systems, contribute to W^{ad}.

In nearly resonant three-level systems, it is possible to obtain a pure crossed-Kerr coupling, without self-phase modulation, by using two-photon non-linearities [6],[16]-[18]. For a proper choice of the detunings between the two laser beams and the two transitions, the two-photon dispersion can be made dominant, and the ideal transformation given above can be recovered in principle. We will now describe an experiment, using a cascade three-level system in a sodium atomic beam [6],[26][34]. This atomic system has the advantage that the resonant enhancement of the non-linear effect is much larger than in solid- state non-linear systems, and the phase damping of the transitions much smaller.

The experimental set-up used for this experiment (fig.2) is an improvement of those used in Refs [6], [37]. The non-linear coupling of the signal and meter fields is obtained in a beam of three-level atoms in a ladder configuration ($3s_{1/2} - 3p_{3/2} - 3d_{5/2}$ levels of sodium atoms). The atomic beam provides a density of 5.10^{11} atoms/cm^3, over an interaction length of 1 cm, with a Doppler width of about 200 MHz (FWHM). Two cw electronically stabilized dye lasers are tuned around 589 nm (meter beam) for the lower transition and 819.5 nm (signal beam) for the upper one. The non-linearity is enhanced by placing the non-linear medium in a doubly resonant optical cavity (5.2 cm long, 5 cm mirror radius) with a good finesse (100 for each wavelength) and low losses. One of the mirrors is a high reflector at 589 nm ($t^2 = 7.10^{-4}$) and is the input-output mirror for the 819.5 nm beam

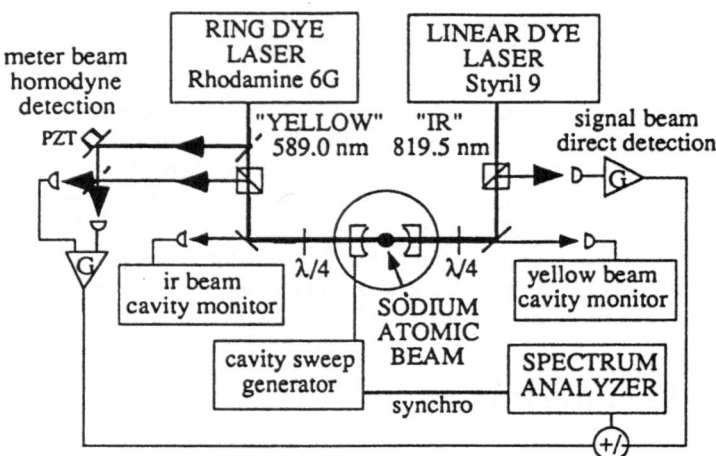

Figure 2: Experimental set-up. Two lasers at 589.0 nm ("signal" beam) and 819.5 nm ("meter"beam) are tuned close to the resonances of the 3s1/2-3p3/2-3d5/2 cascade in a sodium atomic beam. The quarter-wave plates are used to polarize circularly the two beams in the interaction region, and also to separate the input and output beams.

$(t^2 = 7.10^{-2})$ while the other one is high reflector at 819.5 nm $(t^2 = 5.10^{-4})$ and is the input-output mirror for the 589 nm beam $(t^2 = 7.10^{-2})$, so that the cavity is single ended at each wavelength. The total losses are characterized by the on-resonance cavity reflexion coefficient which is 0.95 for the 819.5 nm beam and 0.90 for the 589 nm laser beam. One of the mirror is piezoelectrically mounted, allowing the length of the cavity to be adjusted. In order to have a phase-sensitive low-intensity detection, a homodyne detection with local oscillator power of $800\mu W$ and fringe visibility of 0.92 is used on the meter beam, while the signal output is directly detected. The detectors are $p - i - n$ silicon photodiodes with quantum efficiency of 0.78 at 589 nm and 0.93 at 819.5 nm followed by low-noise preamplifiers.

We use the *ghost transition* configuration of ref [28] (see also ref [38]). The meter beam is tuned around the lower atomic transition and has a very small power ($5\mu W$) compared to the signal beam (about $300\mu W$) which is tuned around the upper transition. In this configuration, the absorption of the signal beam is very small, and two-photon bistability effects [27] are avoided due to the weakness of the signal beam. Therefore, this configuration guaranties a theoretically nearly perfect non-demolition interaction, with quite large coupling between the amplitude of the signal and the phase of the meter beam [28]. The sign of the detunings are chosen in order to be in the most favorable conditions taking into account the hyperfine structure of the sodium atom. Typical Rabi frequencies normalized to the transverse linewidth of the intermediate level are 8 for the meter beam, and 80 for the signal beam, while the detunings are respectively −300 and +100 in the same units (resp. 1.5 GHz and +0.5 GHz in frequency units).

The cavity and the lasers are adjusted so that both fields are simultaneously

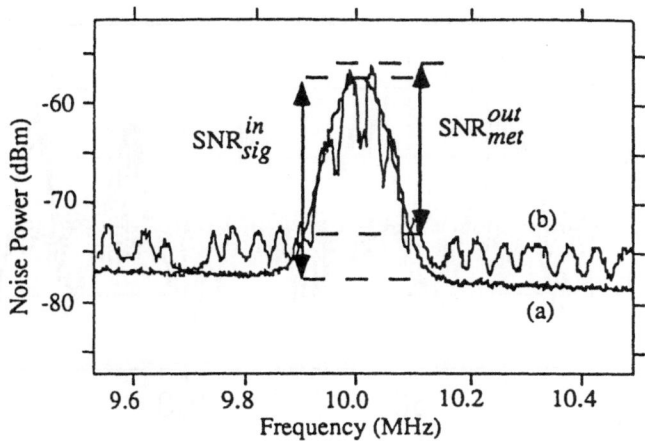

Figure 3: Transfer of the Signal-to-Noise Ratios from the signal input (smooth curve) to the meter output (wigglering curve, due to the scan of the homodyne detection phase). The meter output SNR is obtained as the superior enveloppe of this curve.

resonant in the cavity. A classical amplitude modulation at 10 MHz is applied by an electro optical modulator on the signal beam and the measurement is read out on the phase of the 589 nm laser beam. The levels of the modulations transferred onto the output signal and meter beams are then compared with the signal input modulation level using a spectrum analyser (fig.3). The level of the signal modulation reflected off the offresonant cavity is taken as the reference input modulation level.

Typical results are presented in fig.3. The level difference between the peak and the background noise corresponding to the signal beam (recorded off-resonance) gives the input SNR_{sig}^{in} (trace (a)). The read out of the meter beam is achieved via the homodyne detection, whose phase is continually scanned, giving the fringes in trace (b) of fig.3. The transferred SNR_{met}^{out} is then measured as the level difference between the peak and background noise of the phase quadrature, which corresponds to the superior envelope of the fringes. From fig.3, the raw transfer from the signal onto the meter is of $-3dB$ which gives after corrections due to amplifiers noise a measurement transfer coefficient $T_{met} = 0.45$. On the other hand, no significant degradation of the signal could be observed. Therefore the signal degradation can be estimated according to overall losses due to the imperfection of optical elements along the propagation of the beam. This gives a raw SNR transfer for the signal $T_{sig} = 0.9$. As a consequence, we obtain a sum of the signal and measurement SNR transfers $T_{sig} + T_{met} = 1.35$, which can be seen as a 1.3dB improvement of the overall information transfer of any classical optical tap (a beam splitter for example). A perfect quantum optical tap would have $T_{sig} = T_{met} = 1$ which would give a perfect 3dB improvement, and could also be regarded as a quantum duplicator.

The experimental results concerning the quantum state preparation ability of

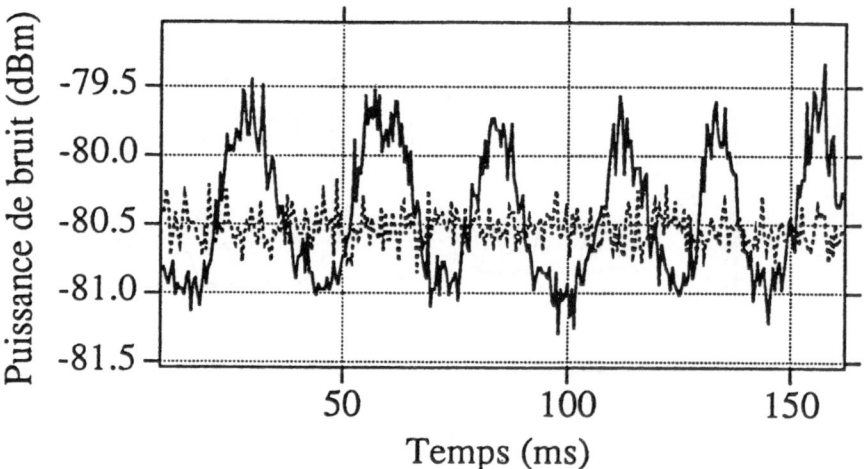

Figure 4: Combined noise levels with 12 dB attenuation of the meter photocurrent. The fringes are due to the phase scan of the homodyne detection, and yield summed and differenced photocurrents (resp. upper and lower envelope of the curve), which go respectively 0.9 dB above and 0.6 dB below the shot-noise-level of the signal alone (flat middle curve).

this device can be characterized by the conditional variance W_{QSP} introduced in section 3. This quantity is measured by the noise reduction of the signal when corrected by the adequately attenuated photocurrent coming from the meter beam [37],[19]. As it can be seen in fig.4, the noise drops by $0.6dB$, leading to a conditional variance $W_{QSP} = 0.85$.

7 Quantum Optical Repeaters using semiconductor emitters and receivers

It has been known for several years that the conversion from light to electrical current and from electrical current to light can be realized below the optical standard quantum limit, using high efficiency semiconductor electro-optical devices. Indeed, the ability of photodiodes to detect sub-shot-noise light by turning it into sub-shot-noise current is a fundamental property of quantum optics [39], and has been experimentally demonstrated by the observation of squeezing [40].

The quantum properties of the inverse transformation are due to the work of Yamamoto and coworkers on semiconductor lasers [21] (see also Ref. [42][43]). More recently, the appearance of high efficiency Light Emitting Diodes (LED) made possible to use these simple devices for generating non-classical light [44][45][46]. A crucial idea in these experiments is that the noise of an electrical current at RF frequencies is usually dominated by thermal (Johnson-Nyquist) noise [48], whose power can be made much smaller than the shot-noise level corresponding to the

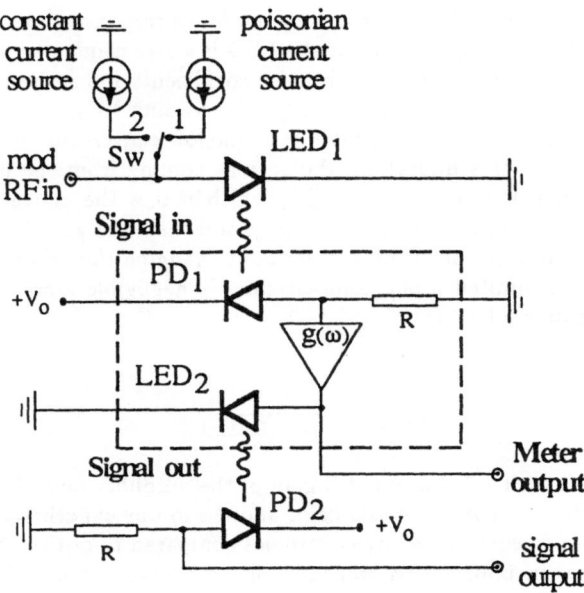

Figure 5: Schematic experimental set-up. The measurement device is enclosed in the dashed box. $LED_{1,2}$ and $PD_{1,2}$ are cooled at 100K. The poissonian current source consists in three parallel connected photodiodes, illuminated by white incandescent lamps. When switch Sw is in position 1, LED_1 is driven by a shotnoise limited current; when in position 2, LED_1 is driven by a constant (filtered) current, and shines therefore squeezed light on PD_1. LED_2 can be also driven by the poissonian current source, in order to get a shotnoise reference on PD_2. Note that, for clarity, all the electrical filtering is not represented in the figure.

same average current intensity [43][44][47]. In this regime, the electrical current behaves classically, and can therefore be measured, duplicated, or amplified without the quantum constraints attached to a light field. If one is concerned by the non-destructive measurement of the intensity of a light beam, one may imagine to convert the light via a photodiode into an electrical current, to measure this current, eventually to amplify it, and to convert it back to a light field using a semiconductor light emitter [49]. This would be a perfect quantum-nondemolition (QND) device [3] if both the conversion rates were unity.

A check of these ideas can be carried out using the experimental set-up schematically depicted in Fig.5. The measurement apparatus itself consists in a large area p-i-n silicon photodiode PD_1 of quantum efficiency $\eta = 0.90 \pm 0.05$, followed by a low noise electronic amplifier, and a light emitting diode LED_2 [34]. The current to current conversion rate of the LED - PD system is $\epsilon = 0.16$ at room temperature and $\epsilon = 0.28$ at 100K . The detectors are placed as close as possible to the LEDs to maximize light collection, and introduced in a liquid nitrogen cryostat. The bandwidth of operation is from 150kHz to 350kHz, the lower limit being due to electronic cut-off, and the upper one to the finite response time of the LEDs. From these numbers, it appears that if the LED is series connected with the photodiode, without any amplifier, the transfer coefficients for a shot-noise-limited input will be respectively $T_s = \eta\epsilon$ and $T_m = \eta$, leading to a rather small effect ($T_s + T_m = 1.15 \pm 0.06$ at 100K) mainly due to the low conversion rate of the LED. In order to overcome this limitation, the current coming from the photodiode is amplified prior to being sent to the LED, which makes the conversion back to light less sensitive to vacuum fluctuations [52], and therefore greatly improves the non-degradation property of the system without changing the others.

For a shot-noise-limited input, and assuming a negligible amplifier noise, the transfer coefficients read

$$T_m = \eta \tag{23}$$

$$T_s = \eta \frac{1}{1 + \frac{1}{g_{\text{eff}}}\frac{(1-\epsilon)}{\epsilon}} . \tag{24}$$

where $g_{\text{eff}} = g^2/G$ with g being the AC gain of the amplifier and G the DC gain. The distinction between AC and DC gain, allowed in our experimental set-up, is an extra degree of freedom given by electronics compared to optics. Note that for practical purposes in information transmission, one has to choose $G = g$, which leads to $g_{\text{eff}} = g$. In the high gain limit ($g_{\text{eff}} \to +\infty$), the sum of these two coefficient goes to $T_s + T_m = 2\eta$, and do not depend on LED_2's conversion rate ϵ anymore. On the other hand, the conditional variance is limited by the conversion rate of LED_2, and is given by

$$W_{QSP} = 1 - \epsilon. \tag{25}$$

In our experiment, the input light beam is the light coming from LED_1, and can therefore be either squeezed (switch Sw in position 2), or shot-noise limited (Sw in position 1). The evaluation of the transfer coefficient is performed by modulating the input signal light (at 250 kHz) by addition of a small RF modulation to LED_1 poissonian driving current. The input signal-to-noise-ratio R_s^{in} is deduced from the direct measurement of R_m^{out} on a spectrum analyser, taking into account the quantum efficiency of PD_1, and the SNR of the signal output R_s^{out} is visualised on the output of PD_2. For high gain ($g_{\text{eff}} = 100$), no distinction between R_m^{out} and R_s^{out} can be made within our experimental precision (about 3%), and therefore $T_s = T_m = \eta = 0.90 \pm 0.05$, i.e. $T_s + T_m = 1.8 \pm 0.1$. We show on figure 6 the experimental evolution of T_s/η with the gain g_{eff}, and compare it with the theoretical expression given by Eq. 24, demonstrating thereby a good theory-experiment agreement.

For the observation of the conditional variance, we have to correct the signal output by substracting the meter output, using a $0 - 180°$ RF power combiner, in order to get the signal fluctuations down to below the shotnoise. This requires a careful adjustment of the relative RF phase, since the fluctuations to be substracted are both about 20dB above the shotnoise. We obtain thus a noise reduction down to 1.2 ± 0.1dB below the shot-noise of the signal beam. This yields $W_{QSP} = 0.76 \pm 0.02$, while the theoretical value is $W_{QSP} = 0.72$.

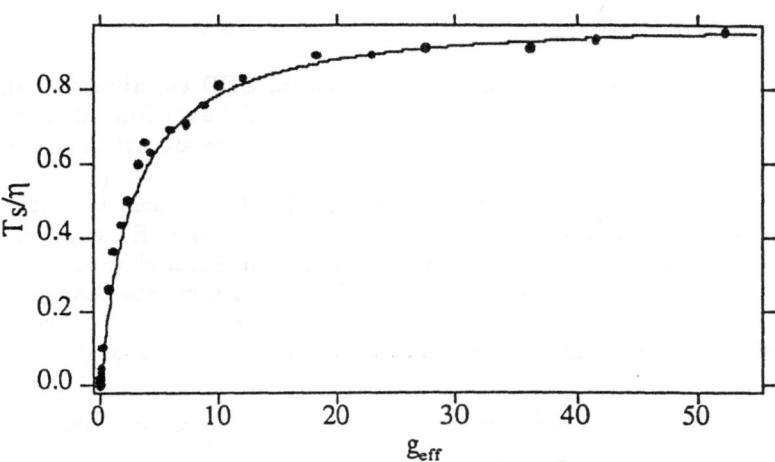

Figure 6: Evolution of the signal transfer coefficient T_s versus the intermediate electronic gain g_{eff}. The experimental points are compared with the theoretical prediction given in the text. The DC currents in the experiment are 25mA for the LEDs and 7mA for the PDs.

Owing to the possibility to have an amplitude squeezed signal input light, when LED_1 is driven by a constant current, we are also able to perform QND measurement of a squeezed light. In the high gain regime, the signal output beam is far above shot noise, but a squeezed output beam can be recovered by decreasing the gain g_{eff} to unity, where our system still operates in the quantum domain, with $T_s + T_m = 1.15$. The input variance W_s^{in} can be inferred from the measured degree of squeezing on the meter channel ($W_m^{out} = 0.80$), as it is usually done for a non-ideal photodetector. The directly measured signal output squeezing is then $W_s^{out} = 0.93$, while the experimental value of the conditional variance is $W_{QSP} = 0.79 \pm 0.02$ (-1.0 ± 0.1dB). All these results clearly show that our scheme does operate in the QND regime, while having a squeezed signal beam, both at the input and the output.

It is worth pointing out that having a signal gain close to one is an extra requirement with respect to the QND criteria, which appears when one is concerned by preserving the squeezing of the input signal. On the other hand, the first QND criteria would rather lead to strong amplification of the intermediate electrical current, because this improves the value of the SNR transfer, by making the system less sensitive to downstream losses. We propose to use the term "quantum repeater" for a device which fulfills the QND criteria with large signal gain, and therefore is not a QND scheme in the original sense. Finally, the value of the conditional variance does not depend on the gain, but only on the quantum efficiency of the reemitting LED and detection system. The perspective of emergence of very high efficiency single-mode LED [53], or alternatively the technological realization of low threshold laser diodes, makes this type of system very promising for a realistic implementation in ultra-low noise optical telecommunication networks.

8 Conclusion

On figure 7, the results obtained in some optical QND experiments realized to present date are plotted as a function of the values of the conditional variance and of $T_s + T_m$. Some other experiments [24][25][54], which are not in the quantum domain, are not represented here.

The best QND result in the usual sense (signal gain close to unity) is the three-level experiment described above (point f), while the LED experiment just described (point g) appears to be the best "quantum repeater" experiment (large signal gain). However, a fair comparison should also take into account also the frequency bandwidth over which the QND effects are achieved. On that respect, pulsed experiments (e.g. point e) have been so far the most successful.

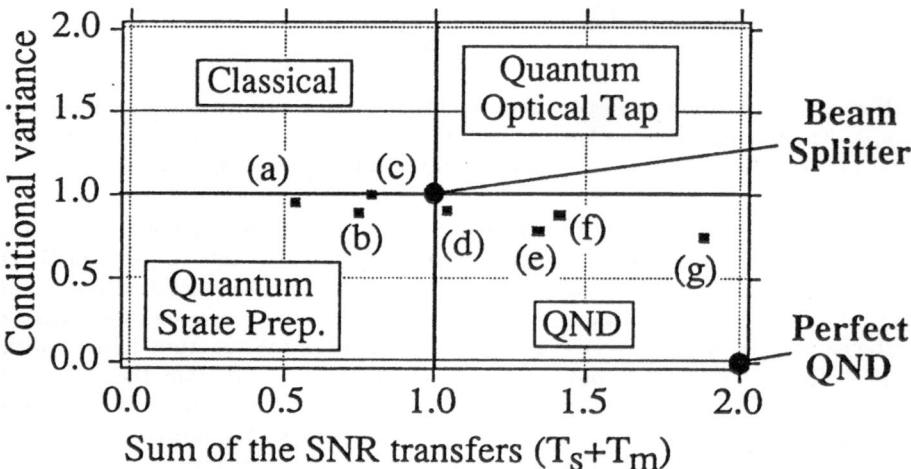

Figure 7: Experimental results plotted in the plane $x = T_s + T_m, y = W_{QSP}$. The point ($x=1$, $y=1$) is simply a beamsplitter (lossless classical coupler, see Fig.1). Points (a), (b), (c), (d), (e) correspond respectively to the experiments described in Ref. (4), (5), (6), (20) and (35). Points (f) and (g) are the results described in the present paper, which fulfill all criteria for QND measurements.

9 Acknowledgements

This work was completed in the framework of the EC Contract ESPRIT III BRA 6934, with partial support from the Centre National d'Etude des Télécommunications (France Telecom). We acknowledge efficient technical assistance of G. Roger, F. Contet, J. C. Rodier and A. Villing. Institut d'Optique Théorique et Appliquée is URA 14 of the Centre National de la Recherche Scientifique.

References

[1] C.M. Caves, K.S. Thorne, R.W.P. Drever, V.D.Sandberg and M.Zimmermann, Rev.Mod.Phys. **52**, 341 (1980).

[2] "Quantum Optics, Experimental Gravitation and Measurement Theory", eds P. Meystre and M.O. Scully (Plenum, New York, 1983).

[3] V.B. Braginsky, Soviet Physics Uspekhi **31**, 836 (1988); V.B. Braginsky, Y.I. Vorontsov and K.S. Thorne, Science **209**, 547 (1980).

[4] M.D. Levenson, R.M. Shelby, M. Reid and D.F. Walls, Phys.Rev.Lett. **57**, 2473 (1986).

[5] A.LaPorta, R.E.Slusher and B.Yurke, Phys.Rev.Lett. **62**, 28 (1989).

[6] P Grangier, J.F. Roch and G. Roger, Phys. Rev. Lett. **66**,1418 (1991).

[7] B. Yurke, J.Opt.Soc.Am B2, 732 (1985).

[8] J.H. Shapiro, Optics Letters **5**, 351 (1980).

[9] G.J. Milburn and D.F. Walls, Physical Review A28, 2065 (1983).

[10] M. Brune, S. Haroche, V. Lefèvre, J.M. Raimond and N. Zagury, Phys. Rev. Lett. **65**, 976 (1990).

[11] J.M. Courty, P.Grangier, L. Hilico, S. Reynaud, Optics Comm. **83** , 251 (1991).

[12] N. Imoto and S. Saito, Physical Review A **39**, 675 (1989).

[13] M.J. Holland, M. Collett, D.F Walls, M.D. Levenson, Phys. Rev A **42**, 2995 (1990).

[14] C.A. Blockley and D.F. Walls, Optics Comm. **79**, 241 (1990).

[15] P. Grangier, J.M. Courty and S. Reynaud, Optics Comm. **89**, 99 (1992).

[16] P.Grangier, J.F. Roch and S. Reynaud, Optics Comm. **72**, 387 (1989).

[17] P.Grangier and J.F. Roch, Quantum Optics **1**, 17 (1989).

[18] P.Grangier and J.F. Roch , Optics Comm. **83**, 269 (1991).

[19] J. F. Roch, Thèse de l'Université Paris XI, (unpublished), (1992).

[20] J.F. Roch, G. Roger, P. Grangier, J.M. Courty and S. Reynaud, Appl. Phys. B **55**, 291 (1992).

[21] Y. Yamamoto, The Transaction of the IEICE, E **73**, 1598 (1990).

[22] N. Imoto, H.A. Haus and Y. Yamamoto, Phys. Rev., A **32**, 2287 (1985).

304

[23] N. Imoto, S. Watkins and Y. Sasaki, Optics Comm. **61**, 159 (1987).

[24] H.A. Bachor, M.D. Levenson, D.F. Walls, S.H. Perlmutter and R.M. Shelby, Phys. Rev. A **38**, 180 (1988).

[25] M.D. Levenson, M.J. Holland, D.F. Walls, P.J. Manson, P.T.H. Fisk, and H.A. Bachor, Phys. Rev. A **44**, 2023 (1991).

[26] J.P. Poizat, M.J. Collett and D.F. Walls, Phys. Rev A **45**, 5171 (1992).

[27] P. Grangier, J. F. Roch, G. Roger, L. A. Lugiato, E. M. Pessina, G. Scandroglio, and P. Galatola, Phys. Rev. A. **46**, 2735 (1992)

[28] K.M. Gheri, P. Grangier, J. Ph. Poizat, and D. F. Walls, Phys. Rev. A. **46**, 4276 (1992).

[29] M.G. Raizen, L.A. Orozco, M. Xiao, T.L. Boyd, and H.J. Kimble, Phys. Rev. Lett., **59**, 198 (1987).

[30] S. Reynaud, C. Fabre, E. Giacobino, and A. Heidmann, Phys. Rev. A **40**, 1440 (1989).

[31] F. Castelli, L.A. Lugiato and M. Vadacchino, Nuovo Cimento, **10D**, 183 (1988).

[32] P. Galatola, L.A. Lugiato, M. Vadacchino, and N.B. Abraham, Opt. Comm.**69**, 414 (1989).

[33] L.M. Narducci, W.W. Edison, P. Furcinitti, and D.C. Eteson, Phys. Rev. A **16**, 1665 (1977).

[34] J.Ph. Poizat and P. Grangier, Phys. Rev. Lett. **70**, 271 (1993)

[35] A. Levenson, I. Abram, P. Fayolle, Th. Rivera, J.C. Garreau and P. Grangier, Phys. Rev. Lett. **70**, 267 (1993)

[36] N. Imoto, H. A. Haus, and Y. Yamamoto, Phys. Rev. A **32**, 2287 (1985).

[37] J. F. Roch, G. Roger, P. Grangier, J. M. Courty, and S. Reynaud, Appl. Phys. B **55**, 291 (1992).

[38] H. A. Bachor and P. T. H. Fisk, Applied Phys. B **49**, 291 (1989).

[39] R.J. Glauber, in Ecole d'Eté des Houches 1964, C. De Witt, A. Blandin and C. Cohen-Tannoudji Eds., (Gordon and Breach, 1965).

[40] Special issues on squeezed light, J. Opt. Soc. Am. B **4**, 1450 (1987), J. Mod. Optics **34** 709 (1987), Applied Physics B **55** 189-303 (1992).

[41] S. Machida and Y. Yamamoto, Optics Comm. **57**, 290 (1986); S. Machida, Y. Yamamoto, and Y. Itaya, Phys. Rev. Lett. **58**, 1000 (1987).

[42] M. C. Teich and B. E. A. Saleh, J. Opt. Soc. Am. B **2**, 275 (1985).

[43] P. R. Tapster, J. G. Rarity, and J. S. Satchell, Europhys. Lett. **4**, 293 (1987).

[44] P. J. Edwards, Int. J. Optoelectronics **6**, 23 (1991); P. J. Edwards and G. H. Pollard, Phys. Rev. Lett. **69**, 1757 (1992).

[45] H.-A. Bachor, P. Rottengatter, and C. M. Savage, Appl. Phys. B **55**, 258 (1992).

[46] E. Goobar, A. Karlsson, G. Björk, and P-J. Rigole, Phys. Rev. Lett. **70**, 437 (1993).

[47] R. H. Koch, D. J. Van Harlingen, and J. Clarke, Phys. Rev. B **26**, 74 (1982).

[48] J. B. Johnson, Phys. Rev. **32**, 97 (1928); H. Nyquist, Phys. Rev. **32**, 110 (1928).

[49] F. Capasso and M. C. Teich, Phys. Rev. Lett. **57**, 1417 (1986).

[50] W. H. Richardson, S. Machida, and Y. Yamamoto, Phys. Rev. Lett. **66**, 2867 (1991).

[51] Y. Yamamoto, Trans. Inst. Electron. Inf. Commun. Eng., Sect. E **73**, 1598 (1990).

[52] H. P. Yuen, Phys. Rev. Lett. **56**, 2176 (1986).

[53] E. Yablonovitch, T. J. Gmitter, R. D. Meade, A. M. Rappe, K. D. Brommer, and J. D. Joanopoulos, Phys. Rev. Lett. **67**, 3380 (1991).

[54] S.R. Friberg, S. Machida and Y. Yamamoto, Phys. Rev. Lett. **69**, 3165 (1992).

Perfect Correlations of Three-Particle Entangled States

K. Wódkiewicz,[1] Liwei Wang,[2] and J. H. Eberly[2]

[1] Institute of Theoretical Physics, Warsaw University,
Warsaw 00-681, Poland, and Center for Advanced
Studies, University of New Mexico,
Albuquerque, NM 87131, USA
[2] Department of Physics and Astronomy, University of
Rochester, Rochester, NY 14627, USA

Abstract. The perfect correlations of three-particle entangled states are discussed in the framework of stochastic models based on local realism. A Bayes analysis of these correlations is performed using the two-point and three-point click-counting distributions. An exactly soluble quantum model in which entanglement of an atomic state with cavity photon leads to a perfectly correlated three-particle state is discussed.

Keywords. Three-particle correlations. Raman atom, local reality.

1. Introduction

Recently it has been shown by Greenberger, Horne and Zeilinger [1] (GHZ) that special entangled states involving three or four particles lead to a much stronger refutation of local realism. In such many-particle correlations only a *single* set of observations is required in order to demolish the local reality assumption. One particularly simple entangled GHZ state of three spin-1/2 particles, named *a*, *b*, and *c* as discussed by Mermin [2], has the following form:

$$|\psi\rangle = \frac{1}{\sqrt{2}}(|+_a, +_b, +_c\rangle - |-_a, -_b, -_c\rangle), \qquad (1.1)$$

where $|+\rangle$ or $|-\rangle$ specifies spin up or down along the appropriate z axis. This entangled state provides an *always* versus *never* test of local realism. A realistic experimental arrangement permitting a three-particle test would be desirable, but straightforward generalizations of two-particle atomic interferometry or photon pairs emitted in a cascade suffer fairly obvious drawbacks. For example a three-photon $J = 0 \rightarrow J = 0$ cascade cannot satisfy dipole selection rules. Only a few specific schemes for the generation of GHZ states have been proposed so far [1,3-4].

It is the purpose of this lecture to discuss GHZ correlations in the framework of intensity-intensity correlations involving single photons. Recently we have presented an exactly soluble Raman model of cavity quantum electrodynamics (CQED) in which entanglement of the atomic state with cavity photon states leads to the state given by Eq. (1.1) [5]. In the framework of this CQED Raman atom a conceptually straightforward test can be designed to measure the three-particle GHZ correlations. We propose an intensity-intensity interference experiment in which GHZ correlations can be measured using methods similar to the one used in the studies of nonclassical effects in processes of parametric down-conversion [6].

We begin our presentation with a statistical analysis of the perfect correlations of three-particle entangled states. We discuss these correlations in the framework of Bayes analysis using the two-point and three-point click-counting distributions that we have recently introduced and applied to the discussion of stochastic quantum jumps in optical transitions [7].

2. Bayes Analysis of Perfect GHZ Correlations

The joint probability for detection of the three particles a, b, and c by three respective polarizers 1, 2 and 3 in an x-y plane perpendicular to the particles' line of flight is:

$$p(\phi_1; \phi_2; \phi_3) = \langle \hat{P}(\phi_1) \otimes \hat{P}(\phi_2) \otimes \hat{P}(\phi_3) \rangle$$
$$= \frac{1}{8}(1 - \cos(\phi_1 + \phi_2 + \phi_3)), \qquad (2.1)$$

where ϕ_1, ϕ_2 and ϕ_3 represent the orientation angles of the detectors. The case where a definite prediction of spin measurement is possible corresponds to $\phi_1 + \phi_2 + \phi_3 = \pi$ with the joint probability equal to 1/4 and to $\phi_1 + \phi_2 + \phi_3 = 0$ with the joint probability equal to zero. These two cases correspond to perfect correlations, i.e., to such correlations when by measuring two spins one can predict with certainty the outcome of the measurement involving the third spin.

While for the two-particle Einstein-Podolsky-Rosen (EPR) spin-singlet state, perfect correlations can be made compatible with a stochastic mode of hidden-variables based on local realism, the three-particle GHZ perfect correlations offer the *never* versus *always* refutation of local realities. In order to fully exhibit this property we shall perform a statistical analysis of the GHZ perfect correlations based on the properties of click-counting distributions (CCD). We select first the case of $\phi_1 + \phi_2 + \phi_3 = \pi$ which corresponds to (x, y, y), (y, x, y) and (y, y, x) measurements of the spins. Using the identity $\hat{P} = \int d\lambda \lambda \delta(\lambda - \hat{P})$ we rewrite the quantum mechanical joint probability in the following form:

$$p(x; y; y) = \int d\lambda_a \int d\lambda_b \int d\lambda_c \lambda_a \lambda_b \lambda_c$$
$$\times P(\lambda_a; \lambda_b; \lambda_c), \qquad (2.2)$$

where the distribution function is defined as

$$P(\lambda_a; \lambda_b; \lambda_c) = \langle \delta(\lambda_a - \hat{P}(x)) \otimes \delta(\lambda_b - \hat{P}(y))$$
$$\otimes \delta(\lambda_c - \hat{P}(y)). \qquad (2.3)$$

Because the projection operators can have their eigenvalues equal to 1 or 0, i.e., can represent only *yes* or *no* answers, the values λ_a, λ_b and λ_c can take only values equal to 1 and 0. GHZ perfect correlations in Eq. (2.2) are described by a bivalued statistical probability distribution function which we shall call CCD. For this particular orientation all correlations can be calculated by a cyclic permutation of the (x, y, y)-indices

$$P(\lambda_a; \lambda_b; \lambda_c) = P(\lambda_a|\lambda_b\lambda_c)P(\lambda_b|\lambda_c)P(\lambda_c), \qquad (2.4)$$

where the distribution $P(\lambda_b|\lambda_c)$ is the conditional of the event λ_b (*yes* or *no*) to occur under the condition that λ_c (*yes* or *no*) have occurred. The distribution $P(\lambda_a|\lambda_b\lambda_c)$ is the conditional of the event λ_a to occur under the condition that λ_b and λ_c

have occurred. The distribution $P(\lambda_c)$ is a one-fold marginal of Eq. (2.3). From the GHZ spin state (1.1) we obtain that: $P(\lambda_c) = 1/2$ and $P(\lambda_b|\lambda_c) = 1/4$ for all values of λ_b and λ_c. The only nonzero elements of the three-fold conditional probability are

$$P(1|11) = P(0|01) = P(1|00) = P(0|10) = 1. \qquad (2.5)$$

This last expression reveals the statistical nature of GHZ correlations, that the value of the third spin is known with probability one if the other two spins have been measured. These statistical picture lead to a simple interpretation of GHZ correlations in terms of random numbers 1 and 0. The quantum mechanical average in this case is represented by an ensemble average of three sequences of random numbers 1 and 0, that are the only possible outcomes of the experiment. On each single polarizer or pair of polarizers the outcomes are completely random and the *yes* or *no* answers occur with equal probability. The perfect GHZ correlations show up in the fact that these completely random sequences are correlated if three polarizers are involved. According to the formulas (2.5) we can write the following sequences of random number for each polarizer:

$$\begin{aligned} \lambda_a &= (0,1,1,0,\ldots) && \text{on } x \\ \lambda_b &= (0,1,0,1,\ldots) && \text{on } y \\ \lambda_c &= (1,1,0,0,\ldots) && \text{on } y. \end{aligned} \qquad (2.6)$$

Note that any subsequence of random numbers composed only of pairs (λ_a, λ_b), (λ_a, λ_c) and (λ_b, λ_c) is completely random, while sequences involving $(\lambda_a, \lambda_b, \lambda_c)$ are perfect in the sense of Eq. (2.5). Note that the three-fold probability distribution fails to satisfy the Markov property that the probability of an event like the registration of each measurement depends only on the preceding outcome and not also on earlier states of the measured system. In short we have:

$$P(\lambda_a|\lambda_b\lambda_c) \neq P(\lambda_a|\lambda_b)P(\lambda_b|\lambda_c). \qquad (2.7)$$

These conclusions are valid for any arbitrary cyclic permutation of the x, y, y polarizers.

The formula (2.2) has a remarkable similarity to a stochastic theory of hidden-variables based on local realism. For such a local hidden-variable (LHV) theory correlations are represented by

$$p_{\mathrm{LHV}} = \int d\lambda_a \int d\lambda_b \int d\lambda_c t(\phi_1, \lambda_a) t(\phi_2, \lambda_a) t(\phi_3, \lambda_a)$$
$$\times P(\lambda_a; \lambda_b; \lambda_c), \tag{2.8}$$

where $P(\lambda_a; \lambda_b; \lambda_c)$ represents a statistical distribution of local hidden variables and where the t-functions describe local realities corresponding to the transmission functions of the corresponding spins through linear polarizers ϕ_1, ϕ_2 and ϕ_3. These transmission functions can be related to the hidden-variable local realities of spin through the following relation $t(\phi, \lambda_i) = \frac{1}{2}(1 + \sigma(\phi, \lambda_i))$ for $i = a, b, c$, where $\sigma(\phi, \lambda_i) = \pm 1$ is an objective spin reality for an arbitrary hidden-parameter λ_i and an arbitrary orientation ϕ of the polarizer. The objective transmission function is also an objective reality with $(t(\phi, \lambda_i))^2 = t(\phi, \lambda_i) = 0, 1$, i.e., *yes* or *no*.

It is clear that the CCD description of perfect correlations for $\phi_1 + \phi_2 + \phi_3 = \pi$ is fully compatible with a LHV theory. The problem starts when we attempt to extend the local and the objective description to GHZ perfect correlations involving $\phi_1 + \phi_2 + \phi_3 = 0$, i.e., $p(x, x, x)$.

From the random sequences (2.6) we have

$$4t_a^x t_b^y t_c^y - 2(t_a^x t_b^y + t_b^y t_c^y + t_a^x t_c^y) + (t_a^x + t_b^y + t_c^y) = 1, \tag{2.9a}$$

where we have used the compact notation: $t_i^\phi = t(\phi, \lambda_i)$. Similar relations hold for the tow remaining combinations involving the measurements (y, x, y) and (y, y, x). Following Mermin's [2] arguments it is easy to show that local realities have to satisfy accordingly the following algebraic property:

$$4t_a^x t_b^x t_c^x - 2(t_a^x t_b^x + t_a^x t_c^x + t_a^x t_c^x) + (t_a^x + t_b^x + t_c^x) = 1, \tag{2.9b}$$

and according to such theories

$$p_{\mathrm{LHV}}(x; x; x) = \frac{1}{4} \tag{2.10}$$

contradictory to the quantum result which leads to $p(x;x;x) = 0$. This can be explained in the framework of our quantum mechanical CCD function (2.2). For the three polarizers in the x-direction it is easy to modify the expression (2.3) and as a result we obtain that for this geometry the GHZ state predicts that: $P(\lambda_c) = 1/2$ and $P(\lambda_b|\lambda_c) = 1/4$ for all values of λ_b and λ_c. This result is the same as in the case of (x, y, y) polarizers. What is different is the three-fold conditional probability which has nonzero elements for the following transitions:

$$P(1|10) = P(1|01) = P(0|11) = P(0|00) = 1. \qquad (2.11)$$

This expression reveals the quantum mechanical nature of GHZ correlations. As in the case of (x, y, y) geometry on each single polarizer or pair of polarizers the outcomes are completely random and the *yes* or *no* answers occur with equal probability. According to the formulas (2.11), these completely random sequences are correlated for three polarizers because the three outcomes are:

$$\begin{aligned}
\lambda_a &= (1, 1, 0, 0, \ldots) && \text{on } x \\
\lambda_b &= (1, 0, 1, 0, \ldots) && \text{on } x \\
\lambda_c &= (0, 1, 1, 0, \ldots) && \text{on } x.
\end{aligned} \qquad (2.12)$$

The reason why these random sequences for the (x, x, x) geometry violate the algebraic property (2.9b) based on local realism is clear from the definition of the CCD given by Eq. (2.3).

The bivalued CCD is positive everywhere, but depends on the polarization directions (x, x, x) or (x, y, y). The distribution function which depends on the orientations of the analyzers (possibly even remote) is nonlocal. In the framework of EPR correlations it is customary to call an analyzer-dependent distribution function a nonlocal function. The nonlocality of this CCD function makes the algebraic property (2.9b) incompatible with quantum prediction even for perfect correlations, i.e., for correlations when the outcome of the measurements can be predicted with certainty. It is worth to point out that the two-point CCD $P(\lambda_b; \lambda_c)$ is local, i.e., is the same for all the orientations of the polarizers. This means that all correlations involving subsequences composed of pairs of random numbers formed from

(2.6) or (2.12) can be described by a LHV based on local realism. The addition of the third particle makes the three-fold probability distribution non-Markovian and nonlocal. This is the statistical essence of the GHZ correlations which we propose to test using interference effects involving single photons.

3. Quantum Interference of Single Photons

Let us consider an interference effect involving two independent single modes a and c of the electromagnetic field. The positive-frequency part of the electric field at a detector can be expressed by the following formula

$$a(\phi) = \frac{1}{\sqrt{2}}(a_a + a_c e^{-i\phi}), \qquad (3.1)$$

where ϕ is a path delay between the two modes described by boson annihilation operators a_a and a_c. The prefactor has been selected in order to preserve the commutation relation $[a(\phi), a^\dagger(\phi)] = 1$. At the detector the field intensity operator (proportional to the number of photons) is given by the following formula:

$$\hat{I}(\phi) = a^\dagger(\phi)a(\phi) = \frac{1}{2}(n_a + n_c + a_a^\dagger a_c e^{-i\phi} + a_c^\dagger a_a e^{i\phi}), \quad (3.2)$$

i.e., an interference of the a-photons with c-photons can take place.

The two independent boson excitations can be used as a Schwinger's representation [8] of a fictitious angular momentum $\vec{J}$, with $j = \frac{1}{2}(n_a + n_b)$ and:

$$J_+ = a_a^\dagger a_c, \quad J_- = a_c^\dagger a_a, \quad J_z = \frac{1}{2}(n_a - n_c). \qquad (3.3)$$

Using this representation we can rewrite the formula for the field intensity in the following form:

$$\hat{I}(\phi) = \frac{1}{2}(n_a + n_c + \vec{J} \cdot \vec{n}(\phi)), \qquad (3.4)$$

with the angular momentum $\vec{J}$ projected on a unit direction $\vec{n}(\phi) = (\cos\phi, \sin\phi, 0)$ in the x-y plane. If the photon space of the modes is restricted by the condition $n_a + n_c = 1$, i.e., interference of only single photons is allowed, the intensity operator is equivalent to a spin-1/2 projection operator, i.e., we have $\hat{I}^2(\phi) = \hat{I}(\phi)$.

Measurements of the interference pattern involving single photons is equivalent to measurements of the spin-1/2 projection operator. The spin orientation ϕ of the polarizer is replaced in the interference effect by a path delay between the two boson modes a and c forming the fictitious spin-1/2 representation of the electromagnetic field intensity operator [9].

4. Generation of GHZ State in a Raman Cavity

The model we are considering consists of one atom and four radiation modes [7, 8, 10]. The radiation modes are associated with the two transverse polarization states of each of two longitudinal modes of a cavity. The modes are pairwise degenerate in frequency: $\omega_a = \omega_c$ and $\omega_b = \omega_d$. The atom has a $J = 1$ ground state and the photons in the cavity modes can induce transitions between the $M = -1$ and $M = +1$ sublevels of the ground state, via virtual circularly polarized transitions to a far-off-resonant upper $M = 0$ level. If the $M = 0$ ground sublevel is initially unoccupied it will not be active at any later time because only fields polarized along z could cause $M = 0$ to $M = 0$ transitions. Thus even though there are three sublevels in the $J = 1$ state, only two of them are participants in the interaction (Fig. 1).

The effective interaction Hamiltonian of such transitions has the following form ($\hbar = 1$).

$$H_{\text{int}} = \lambda(a_a a_c^\dagger + a_b^\dagger a_d)\sigma^\dagger + h.c., \qquad (4.1)$$

As usual we have denoted by σ and $\sigma^\dagger$ the atomic lowering and raising operators and by a_i and $a_i^\dagger$ ($i = a, b, c, d$) the four boson annihilation and creation operators of the cavity modes. The unperturbed states of the free Hamiltonian shall be denoted by $|n_a, n_b, n_c, n_d; \pm\rangle$, where the n_i denote photon numbers and $\pm$ are the atomic indices. We also have assumed that the cavity

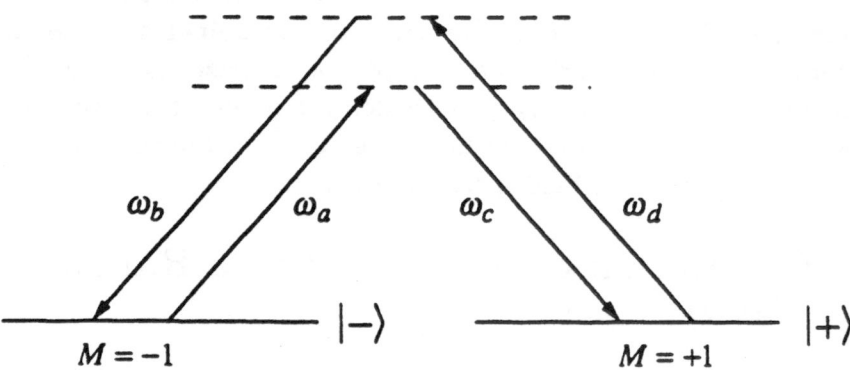

Fig. 1. Transition diagram of our CQED model.

modes support the Raman resonant frequency relations $\omega_d - \omega_b = \omega_a - \omega_c$.

In order to make contact with the spin-1/2 GHZ correlations we shall reformulate the boson interactions in terms of spin-variables given by Eq. (3.3). Using Eq. (3.3) we see that the four independent cavity modes can be used as independent boson components, in Schwinger's representation, to give two angular momentum operators $\vec{J}_1$ and $\vec{J}_2$. Note that $\vec{J}_1$ involves the right-polarized ω_a photons and the left-polarized ω_c photons, while $\vec{J}_2$ involves the left-polarized ω_d photons and the right-polarized ω_b photons.

The third angular momentum involved in our CQED interaction arises naturally from the atomic transition operators. It is well known that this angular momentum is equivalent to a spin-1/2: $S_+ = \sigma^\dagger$, $S_- = \sigma$ and $S_z = \frac{1}{2}\sigma_z$. The CQED Hamiltonian (4.1) expressed in angular momentum variables describes the interaction of a fictitious spin-1/2 (the atom) and a system with two angular momenta $\vec{J}_1$ and $\vec{J}_2$ (combinations of field modes), and has the form:

$$H_{\text{int}} = \lambda(J_{1-} + J_{2-})\sigma^\dagger + h.c. \qquad (4.2)$$

The Hamiltonian of our model is fully soluble, i.e., all energy eigenvectors and eigenvalues can be obtained exactly [11]. But for the purpose of this presentation we shall confine our attention only to the lowest nontrivial *sector* of the Hilbert space of the system. This is because we are interested in the realization of the GHZ state in the framework of this CQED model. From the definitions (3.3) it is clear that $\vec{J_1}$ and $\vec{J_2}$ spaces that correspond to $j_1 = j_2 = 1/2$ involve only photons for which $n_a + n_c = 1$ and $n_b + n_d = 1$.

These CQED states can be denoted by $|m_1, m_2, m_s\rangle$, where the magnetic numbers correspond to spin up or spin down of the angular momenta J_1, J_2 and S. The relation to the field-atom states is straightforward; for example

$$|+, +, +\rangle = |1_a, 0_b, 0_c, 1_d; +\rangle. \tag{4.3}$$

It is not enough to say that three spin-1/2 particles exist. The difficult task is to show how the three particles interact in a physically realizable way and in a way that permits GHZ *always-never* correlations to be observed. The generation and the detection of the GHZ state can be achieved in two experimental stages.

In the first stage of the experiment we prepare an initial cavity state by depositing one photon in the ω_a mode and one photon in the ω_d mode. This can be achieved, for example, by passing an atom which can undergo a two-photon cascade spontaneous emission through the cavity. Then, an active atom prepared initially in the state $|\text{ATOM}\rangle = |-\rangle$ is injected into the cavity. As a result of the injected atom, the initial state in the cavity is no longer stable and will evolve in time. As a result of the interaction given by the Hamiltonian (4.2) the only states $|\text{ATOM}\rangle \otimes |\text{FIELD}\rangle$ that are dynamically accessible in the cavity form the following chain:

$$|1_a, 1_b, 0_c, 0_d; +\rangle \leftrightarrow |1_a, 0_b, 0_c, 1_d; -\rangle$$
$$\leftrightarrow |0_a, 0_b, 1_c, 1_d; +\rangle, \tag{4.4a}$$

which, in our spin-1/2 notation, is the same as

$$|+, -, +\rangle \leftrightarrow |+, +, -\rangle \leftrightarrow |-, +, +\rangle. \tag{4.4b}$$

This chain of states spans a closed sector in the Hilbert space. Due to the interaction, only states from the chain (4.4) will occur in the cavity. These states will oscillate with this *vacuum* Rabi frequency. One can show that after a time $\lambda\sqrt{2}t = \pi/2$ the initial state $|+, +, -\rangle$ in the chain will become:

$$|\text{FIELD}\rangle \otimes |+\rangle = -\frac{i}{\sqrt{2}}(|+, -\rangle + |-, +\rangle) \otimes |+\rangle. \qquad (4.5)$$

Thus, as a result of the atomic pumping, the field in the cavity is in a superposition state of one photon in each of ω_a and ω_b modes and one photon in each of the ω_c and ω_d modes.

We suppose that the evolution time $t = \pi/2\sqrt{2}\lambda$ is also the atomic transit time through the cavity. After the atom leaves the cavity, the cavity field (4.5) cannot evolve further. Note that the state of the emerging atom can be monitored to verify the field state. A second atom is prepared (e.g. by passage through a Ramsey zone) in a dipole coherent state:

$$|\text{ATOM}\rangle = \frac{1}{\sqrt{2}}(|+\rangle - |-\rangle), \qquad (4.6)$$

and then this atom passes through the cavity. As a result of the interaction of the second atom with the prepared state of the field, the chain given by Eq. (4.4) is supplemented by an additional chain of states given by:

$$|1_a, 1_b, 0_c, 0_d; -\rangle \leftrightarrow |0_a, 1_b, 1_c, 0_d; +\rangle$$
$$\leftrightarrow |0_a, 0_b, 1_c, 1_d; -\rangle, \qquad (4.7a)$$

and this is equivalent to

$$|+, -, -\rangle \leftrightarrow |-, -, +\rangle \leftrightarrow |-, +, -\rangle. \qquad (4.7b)$$

The state of the field in the cavity evolves during the passage of the second atom. The atomic state evolves as well, and after the atomic transition time the combined $|\text{ATOM}\rangle \otimes |\text{FIELD}\rangle$ state has become:

$$|\text{FINAL}\rangle = \frac{1}{\sqrt{2}}(|+, +, -\rangle - |-, -, +\rangle). \qquad (4.8)$$

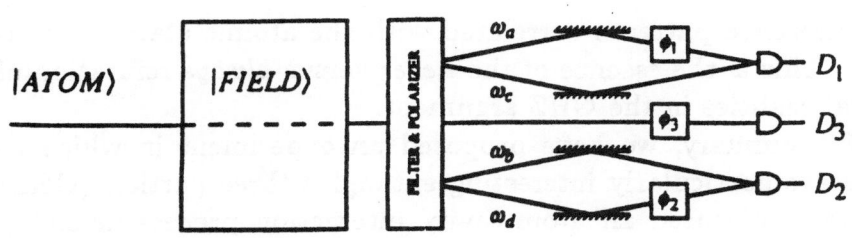

Fig. 2. Schematic description of the GHZ correlations involving detectors D_1, D_2 and D_3.

The |FINAL⟩ state is therefore an entangled state and is in fact just the desired GHZ state (1.1), with a change of the atomic labels from |+⟩ and |−⟩. We see that in this CQED scheme we can achieve a fully dynamical generation of perfect GHZ states.

In this second stage of our experiment we first separate the ω_a and ω_c photons from the ω_b and ω_d photons emerging from the cavity by a proper set of filters and polarizers and place three detectors D_1, D_2 and D_3. The ω_a and ω_c modes are mixed with an angle ϕ_1 and the intensity $\hat{I}(\phi_1)$ (see Eq. (3.2)) is detected by D_1. The ω_b and ω_d modes are mixed with an angle ϕ_2 and the corresponding intensity $\hat{I}(\phi_2)$ is detected at D_2. The atomic state is rotated through angle ϕ_3 and is detected by D_3 (Fig. 2).

Joint clicks at the detectors D_1, D_2 and D_3 reveal the following joint correlation function

$$p(\phi_1; \phi_2; \phi_3) = \langle \hat{I}(\phi_1) \hat{I}(\phi_2) \hat{P}(\phi_3) \rangle, \qquad (4.9)$$

where $\hat{I}(\phi_1)$ and $\hat{I}(\phi_2)$ are the fictitious spin-1/2 operators corresponding to the four modes in the cavity and $\hat{P}(\phi_3)$ is a spin-1/2 projector corresponding to the two-level atom interacting with the cavity field.

As a result of this identification the function $p(\phi_1, \phi_2, \phi_3)$ gives the *joint* probability for a detection involving two interference patterns characterized by ϕ_1 and ϕ_2 and the atomic spin orientation characterized by ϕ_3. Perfect GHZ correlations lead for $\phi_1 + \phi_2 + \phi_3 = \pi$ to $p = 1/4$, while for $\phi_1 + \phi_2 + \phi_3 = 0$ the joint probability is equal to zero. This means that for this particular orientation in the x-y plane no joint detection of photons in the interference patterns correlated with the atomic state is possible. This is the essence of the *never* versus *always* refutation of local realities in the GHZ argument.

In summary, we have proposed an experiment in which to observe particularly interesting entangled three particle (GHZ) states. We used an atom-cavity interaction process involving four modes of the electromagnetic field. By detecting appropriately selected mode pairs, making use of the known cavity-atom dynamical evolution and choosing conveniently the atomic transition time through the cavity, we have obtained exactly the desired GHZ spin-1/2 correlations. We note that our proposed experiment not only suggests a method for realization of these so-far unobserved states, but does so in a distinctly unusual way within the framework of previous tests of the violation of local realism, i.e., by combining five completely distinct physical systems (atom and four field modes) to make the three *spins*.

5. Acknowledgment

This work was supported by the U. S. National Science Foundation grants INT 89-05178 and PHY 91-11562, and the Polish KBN Programs 204279101 and 204269101.

6. References

[1]. D. M. Greenberger, M. Horne, and A. Zeilinger in *Bell's Theorem, Quantum Theory, and Conceptions of the Universe*, edited by M. Kafatos (Kluwer Academic, Dordrecht, the Netherlands, 1989), pp. 107; D. M. Greenberger, M. Horne, A. Shimony, and A. Zeilinger, Am. J. Phys. **58**, 1131 (1990).

[2]. N. D. Mermin, Am. J. Phys. **58**, 731 (1990); Phys. Today **43** (6), 9 (1990).

[3]. M. D. Reid and W. J. Munro, Phys. Rev. Lett. **69**, 997 (1992).

[4]. D. N. Klyshko, Phys. Lett. A **172**, 399 (1993).

[5]. K. Wódkiewicz, Liewei Wang and J. H. Eberly, Phys. Rev. A (1993) in print.

[6]. Z. Y. Ou, and L. Mandel, Phys. Rev. Lett. **61**, 50 (1988); ibid. **61**, 54 (1988).

[7]. K. Wódkiewicz and J. H. Eberly, Ann. Phys. (New York), **216**, 268 (1992).

[8]. J. Schwinger, in *Quantum Theory of Angular Momentum*, edited by L. C. Biederharn and H. Van Dam (Academic, New York, 1965), p. 229.

[9]. C. Su and K. Wódkiewicz, Phys. Rev. A **44**, 6097 (1991).

[10]. Liwei Wang, R. R. Puri and J. H. Eberly, Phys. Rev. A **46**, 7192 (1992).

[11]. Liwei Wang and J. H. Eberly, Phys. Rev. **A47** (in press 1993).

Irreversibility in Quantum Dynamical Processes

Stig Stenholm

The Academy of Finland and
Research Institute for Theoretical Physics
P.O.Box 9, Siltavuorenpenger 20C
SF-00014 University of Helsinki
Finland

1 Introduction

Many physical applications are today concerned with the investigation of time dependent phenomena. Recent developments in laser technology has provided good pulses with a duration reaching from the ns to the fs range. Thus, to investigate the actual time development of a physical system is of the utmost importance.

To integrate the time dependent Schrödinger equation is made possible by present powerful computers but only in special cases. In particular, the inclusion of dissipative processes adds considerably to the requirements of the numerical treatments. However, in many physical applications, this feature is essential or unavoidable in physically interesting situations. Thus it is an important task to investigate the possibilities to approach irreversible time evolution in quantum mechanics.

Quantum optics has recently developed a simulation method, [1] and [2], which simplifies the numerical treatment of systems with spontaneous decay. This approach considers the quantum ensemble to be formed by a collection of case histories which are realizations of the stochastic content in quantum mechanics. Earlier treatments using this philosophy are in [3] and [4]. The dynamic reduction program [5]-[7] regards these histories as constituting really occurring events in Nature, and this view has recently been discussed by other authors as well [8]. The procedure can also be formalized into the mathematical framework of stochastic processes [9].

The mathematical theory of dynamical semigroups has concerned itself with the most general evolution equation which generates a group satisfying reasonable physical requirements; see Davis [10]. An approach more accessible to physicists has been given by Sudarshan et al. [11]-[12]. A review of the field is given by Alicki and Lendi [13].

In connection with our work on laser-induced dynamic processes in molecules [14], I have considered various aspects of the introduction of irreversibility into the quantum time evolution equations. In these lectures I summarize some of the main arguments and my understanding of the present situation. My treatment is heuristic rather than rigorous, and no exact validity conditions for the conclusions are attempted.

Section 2 derives a master equation from an approach assuming the occurrence of brief encounters with intervening external systems. This can be a model for atomic collisions [15]-[16] or photon counting [2] and [17]. This forces the master equation into a specific form, which has certain physical implications.

Section 3 tries to derive the most general linear generator for a dynamical semigroup. In the first part, I present some heuristic arguments giving the desired form of the equation and connect this to the conventional procedure of using a bath assumption for the introduction of irreversibility. Finally I summarize the formal mathematical approach without going into any detail. It seems that it is useful for the physicist to know this, even if the detailed proofs are neither transparent nor useful.

Finally, in Sec.4 I give two examples. The first concerns spontaneous emission which fits into the formalism eminently well; in fact most applications I know have been concerned with this case. The second application is the familiar case of linear friction. This physically well understood case turns out to differ from the general approach in significant ways. As I do not know if this problem is an essential or an accidental feature of the theory, I find it important to present the case and invite a discussion about this question.

2 Disruptive quantum evolution

2.1 Origin of the master equation

We assume that the system develops under the influence of a Schrödinger equation with the Hamiltonian H. Then, in a time interval Δt, the density matrix changes by

$$\Delta \rho_1 = -i[H, \rho]\Delta t \ . \tag{1}$$

In these notes, I assume time to be measured in inverse energy units, so that Planck's constant does not appear in the time evolution equations.

As an alternative to the time evolution (1), we assume that the system is subjected to a series of fast disruptive interventions, which in a short time can effect a significant change in the state. Such events may be e.g. the encounter with colliding particles or registering counters. The events are supposed to be fast with respect to the time scale of the evolution (1), and each intervention is taken to cause a change of the density matrix at the instant t by

$$\rho(t + \delta t) = S \ \rho(t) \ S^\dagger \ , \tag{2}$$

where δt is a small interval of time, see Refs [15] and [16]. The S-matrix characterizing the intervention is expressed in terms of the T-operator according to

$$S = 1 + iT \ . \tag{3}$$

The intervening processes have a distribution of parameters characterized by the label j, and each type occurs with the rate r_j. In a time interval Δt, the average change of the density matrix caused by the interventions is given by

$$\begin{aligned}
\Delta \rho_2 &= \rho(t + \Delta t) - \rho(t) \\
&= \sum_j r_j \Big(i(T_j \rho - \rho T_j^\dagger) + T_j \rho T_j^\dagger \Big) \Delta t \\
&= \Big[i(\overline{T}\rho - \rho \overline{T^\dagger}) + \overline{T \rho T^\dagger} \Big] \Delta t \ .
\end{aligned} \tag{4}$$

Assuming that the two causes of time evolution, (1) and (4), act independently, we obtain a coarse grained master equation in the form

$$\frac{d}{dt}\rho = -i[H, \rho] + i(\overline{T}\rho - \rho \overline{T^\dagger}) + \overline{T \rho T^\dagger} \ . \tag{5}$$

The intervening entities may well carry away information about the system; they are, in fact, supposed to model probes and detectors.

Thus we cannot assume that the S-matrix in (2) is unitary, but it has to be isometric, i.e. preserve the norm of the states implying $S^\dagger S = 1$. With the T-matrix in (3) this is equivalent with

$$i(T - T^\dagger) = -T^\dagger T \tag{6}$$

In the master equation (5) this serves to preserve the normalization

$$\frac{d}{dt} Tr\ \rho = Tr\left[\rho\left(i(\overline{T} - \overline{T^\dagger}) + \overline{T^\dagger T}\right)\right] = 0\ , \tag{7}$$

which is necessary for the probabilistic interpretation of the density matrix.

Separating

$$T = T_{\text{real}} + iT_{\text{imag}}\ , \tag{8}$$

we find with (7) the final form of the master equation

$$\frac{d}{dt}\rho = -i[(H - T_{\text{real}}), \rho] - \frac{1}{2}\overline{T^\dagger T}\rho - \frac{1}{2}\rho\overline{T^\dagger T} + \overline{T\rho T^\dagger}\ . \tag{9}$$

The real part of T is seen to renormalize the energies of the system and the imaginary part can be expressed with the aid of the relation (6). Next we assume that the operator T can be expressed in terms of operators τ_ν with one-dimensional range (see Ref [18]) such that

$$T = \sum_\nu \lambda_\nu \tau_\nu\ , \tag{10}$$

$$T^\dagger T = \sum_\nu \lambda_\nu^2 \tau_\nu^\dagger \tau_\nu\ .$$

The latter is a spectral resolution of the positive hermitean operator $T^\dagger T$, and $\tau_\nu^\dagger \tau_\nu$ is a spectral projector. If we now suppose that the average in (9) acts to destroy the coherence between the various members in the sum over ν, we can set

$$\overline{\tau_\nu \rho \tau_\mu^\dagger} \propto \delta_{\nu\mu}\ . \tag{11}$$

With this assumption, the master equation can be written in the form

$$\frac{d}{dt}\rho = -i[H, \rho] - \frac{1}{2}\sum_\nu \lambda_\nu^2\left(\tau_\nu^\dagger \tau_\nu \rho + \rho \tau_\nu^\dagger \tau_\nu - 2\tau_\nu \rho \tau_\nu^\dagger\right)\ . \tag{12}$$

This is exactly the form of the master equation which can be used in a stochastic simulation [1], which, in turn, can be given an interpretation in terms of actually occurring processes [8]. The possibility rests on two conditions. Firstly, the random phase assumption (11) must hold. Secondly, it seems that the assumption about objective independent events assumes that the processes generated by the operators τ_ν end up in orthogonal subspaces.

2.2   Main properties of the master equation

The purity of the density matrix is measured by the quantity $Tr(\rho^2)$. The more mixed a state is, the smaller is this quantity. We calculate its time derivative using (9)

$$\frac{d}{dt}Tr\rho^2 = 2\left[Tr\left(\rho T\rho T^\dagger - T^\dagger T\rho^2\right)\right] .$$ (13)

Starting with a pure state $|\psi>$ and introducing the projector

$$\Pi = 1 - |\psi><\psi| ,$$ (14)

we find from (13)

$$\begin{aligned} \frac{d}{dt}Tr\rho^2 &= -2<\psi|T^\dagger\Pi T|\psi> \\ &= -2\parallel \Pi T|\psi>\parallel^2 \le 0 . \end{aligned}$$ (15)

Thus a pure state cannot be sustained, and the density matrix tends to become more and more mixed with time.

Let us next assume that our master equation describes a counting type of occurrence. This is the case with a series of encounters with collision partners or a sequence of photon counting events. Then there exists a set of projectors P_N such that

$$\sum_N P_N = 1 ,$$ (16)

and

$$P_N H P_M = H(N)\delta_{NM} .$$ (17)

The variable N can then be taken to obey a superselection rule, which allows us to interpret it as a classical quantity, which we can inspect at will [19].

Such a sequence of classical events can proceed in one direction only. This singles out a direction of time; thus the intervening interactions have to obey the relation

$$P_N T P_M = t(M)\delta_{N,M+1}$$
$$P_N T^\dagger P_M = t^\dagger(M)\delta_{N,M-1} .$$

(18)

In this case the density matrix remains diagonal in the superselection variable and with the notation

$$P_N \rho P_N = \rho_N ,$$

(19)

we obtain the evolution equation

$$\frac{d}{dt}\rho_N = -i[H(N),\ \rho_N] - \frac{1}{2}t^\dagger(N)t(N)\rho_N - \frac{1}{2}\rho_N t^\dagger(N)t(N)$$
$$+t(N-1)\rho_{N-1}t^\dagger(N-1) .$$

(20)

For each value of N, this describes the Schrödinger time evolution in the corresponding Hilbert space, and the operators $t(N)$ transfer populations to ever larger values of N.

The probability to be in the space characterized by N is given by

$$n_N = Tr\ \rho_N .$$

(21)

To see the emergence of a rate process from the Eq.(20) we make the ansatz

$$\rho_N = n_N r^0 ,$$

(22)

where r^0 is some equilibrium density matrix with unit trace. Inserting (22) into Eq.(20) and taking the trace, we obtain the rate equation

$$\frac{d}{dt}n_N = -W_N n_N + W_{N-1} n_{N-1} ,$$

(23)

where

$$W_N = Tr\left[r^0 t^\dagger(N)t(N)\right] \geq 0$$

(24)

are the transition probabilities between the subspaces with different N. The rate equation (23) clearly describes a stochastic process. In certain situations, we can thus interpret the quantum time evolution in terms of a stochastic sequence of events, just as quantum complementarity asserts. In his Como lecture [20], Bohr stated that the continuous Schrödinger evolution and the sudden jump interpretation are complementary.

The rate equations (23) can be regarded as the first member in a sequence of rate equations similar to the hierarchies used in the statistical mechanics description of transport processes.

2.3 Some special cases

We begin by considering a physical quantity A which is conserved under the Hamiltonian evolution generated by H. When we apply the master equation (9), we find that the expectation value changes at the rate

$$\frac{d}{dt} < A >= -\frac{1}{2}Tr\left[\rho\left(AT^\dagger T + T^\dagger T A - 2T^\dagger A T\right)\right] . \tag{25}$$

The quantity A retains its expectation value if its linear transformation

$$L_T(A) = [A, T^\dagger]T \tag{26}$$

is purely antihermitean

$$[L_T(A)]^\dagger = -L_T(A) . \tag{27}$$

Similarly it follows from (25) that the variable A is damped if

$$2Re(L_T(A)) > 0 . \tag{28}$$

For a damped radiation mode in a cavity, whose quanta are created by the operator $b^\dagger$, the energy $H = \omega b^\dagger b$ is conserved. For the dissipation we take $T = b$ [21] and find

$$L_b(H) = \omega b^\dagger b \geq 0 . \tag{29}$$

Thus the cavity mode experiences damping of its energy.

In the dynamic state reduction program [5]-[7], the dissipative mechanism is supposed to couple to the position variable, $T = \lambda q$. Omitting the Hamiltonian part, we can then write the master equation in the form

$$\frac{d}{dt}\rho = -\frac{\lambda^2}{2}[q,[q,\rho]] \ . \tag{30}$$

This gives in the position representation the solution

$$< q|\rho|q' > \propto \exp\left[-\frac{\lambda^2}{2}(q-q')^2 t\right] \ . \tag{31}$$

This form destroys the quantum coherence between spatially separated positions and tends to localize the quantum system.

However, the Hamiltonian for most systems defined in ordinary configuration space are of the form

$$H = \frac{p^2}{2M} + V(q) \ , \tag{32}$$

which gives from (26)

$$L_q(H) = [H,q]q = -\frac{1}{2M}\left(1 + i(pq + qp)\right) \ . \tag{33}$$

This has a negative real part, and from Eq.(25) we conclude that the energy is steadily increased by this interaction. This has been used as an argument against the dynamic reduction program [22].

3 Dynamical semigroups

3.1 The most general Markovian time evolution

The time development of a quantum mechanical state ρ is called Markovian if its time derivative at time t depends only on the state $\rho(t)$ at that time. In addition, we require the time evolution equation to be linear, in order to preserve the mixture of states. Only in that case can a state of the form

$$\rho(t) = a\rho_1(t) + b\rho_2(t) \tag{34}$$

retain this form during the course of time. The time evolution equation is thus

$$\frac{d}{dt}\rho = L\rho \ , \tag{35}$$

where L is a linear transformation on the operators ρ. In order to obtain stationarity, we assume that L contains no explicit time dependence.

The equation (35) generates a semigroup of transformations $\{\Lambda_t\}$, wich can be written formally as

$$\rho(t) = \Lambda_t \rho(0) = e^{Lt} \rho(0) \tag{36}$$

with $\Lambda_t \Lambda_s = \Lambda_{t+s}$. Because the state $\rho^\dagger$ is the same as the state ρ, the physical interpretation of the formalism requires that L is hermitean within the scalar product $Tr(A^\dagger B)$, and consequently we require

$$Tr\ \rho_1 L \rho_1 = Tr(L^\dagger \rho_1)\rho_1 = Tr\ \rho_1 (L\rho_1) \tag{37}$$
$$\frac{d}{dt}\rho^\dagger = L^\dagger \rho^\dagger = L\rho^\dagger \ .$$

Let us now consider the class of all permissible linear transformations $\mathcal{L}$ on the states ρ defined on a Hilbert space $\mathcal{H}$. This class can be generated by the following set of postulates:

(a) The transformation $c\rho$ belongs to $\mathcal{L}$ if $c \in \mathbb{C}$.
(b) The transformation $B\rho$ belongs to $\mathcal{L}$ if B operates on $\mathcal{H}$.
(c) The transformation ρC belongs to $\mathcal{L}$ if C operates on $\mathcal{H}$.
(d) If L_1 and L_2 belong to $\mathcal{L}$, then $L_1 + L_2$ belongs to $\mathcal{L}$.
(e) If L_1 and L_2 belong to $\mathcal{L}$, then $L_1 \circ L_2$ belongs to $\mathcal{L}$.
(f) The identity transformation $1\rho = \rho$ belongs to $\mathcal{L}$.

Using these rules we can construct all possible linear transformations to be of the form

$$L\rho = c\rho + B\rho + \rho C + \sum_{i,j} \lambda_{ij} K_1^i \rho K_2^j \ . \tag{38}$$

All transformations generated by the postulates (a)-(f) are of this form. **Note:** We do not include among our linear transformations the complex conjugation which, in fact, is antilinear.

If we apply hermitean conjugation to Eq.(38) we obtain

$$(L\rho)^\dagger = c^* \rho + \rho B^\dagger + C^\dagger \rho + \sum_{i,j} \lambda_{ij}^* K_2^{j\dagger} \rho K_1^{i\dagger} \ , \tag{39}$$

and requiring this to be the same as the original equation we find

$$C^\dagger = B \; ; \; B^\dagger = C \; ; \; \lambda_{ij}^* = \lambda_{ji} \; ; \; K_1^{i\dagger} = K_2^i \; ; \; K_2^{j\dagger} = K_1^j \text{ and } c = c^* \; .(40)$$

With these results, we can formulate [11]-[13]:

Theorem 1. The most general probability-conserving linear Markovian time evolution is described by an equation of the form

$$\frac{d}{dt}\rho = -i[H,\rho] - \frac{1}{2}\sum_{i,j}\lambda_{ij}K_j^\dagger K_i \rho - \frac{1}{2}\sum_{i,j}\lambda_{ij}\rho K_j^\dagger K_i$$

$$+ \sum_{i,j}\lambda_{ij}K_i\rho K_j^\dagger \; , \tag{41}$$

where $[\lambda_{ij}]$ is a hermitean matrix.

Proof: From Eqs.(38) and (39) we find the relation

$$\frac{d}{dt}Tr \; \rho = Tr\left[\rho\left(c + B + B^\dagger + \sum_{i,j}\lambda_{ij}K_j^\dagger K_i\right)\right] \; . \tag{42}$$

When probability is conserved, this derivative vanishes for all ρ which with

$$B = R - iH \tag{43}$$

gives

$$c + 2R = -\sum_{i,j}\lambda_{ij}K_j^\dagger K_i \; . \tag{44}$$

Consequently Eq.(41) follows. The Hamiltonian time evolution is generated by the part $H = -Im \; B$; we can add any scalar shift to this without affecting the equation.

In order to be able to interpret ρ as a density matrix at all times, its eigenvalues must remain positive. We resolve the initial ρ into its spectral components by writing

$$\rho(0) = \sum_\eta |\eta > \eta < \eta| \; , \tag{45}$$

where $\eta \geq 0$. We introduce the spectral representation

$$\rho(t) = \sum_\eta |\eta(t) > \eta(t) < \eta(t)| \tag{46}$$

of the density matrix at a later time. The basis $\{|\eta(t)>\}$ is generated from the original basis $\{|\eta>\}$ by a unitary transformation

$$|\eta(t)> = e^{iS(t)}|\eta> \ . \tag{47}$$

The functions $\eta(t)$ have to remain nonnegative in order to allow a probability interpretation of ρ. Introducing (47) into (41) we find the time evolution equation

$$\sum_\eta |\eta(t) > \dot{\eta}(t) < \eta(t)| = -i\left[(H + i\dot{S}) \ , \ \sum_\eta |\eta(t) > \eta(t) < \eta(t)|\right]$$

$$-\frac{1}{2}\sum_{i,j}\lambda_{ij}\left(K_j^\dagger K_i \sum_\eta |\eta(t) > \eta(t) < \eta(t)| \right. \tag{48}$$

$$+ \sum_\eta |\eta(t) > \eta(t) < \eta(t)|K_j^\dagger K_i - 2K_i \sum_\eta |\eta(t) > \eta(t) < \eta(t)|K_j^\dagger\bigg) \ .$$

The states $|\eta(t) >$ are orthonormal for all times. Thus by taking the expectation value of Eq.(48) in the state $|\mu(t) >$ we find

$$\dot{\mu}(t) = -\sum_{i,j}\lambda_{ij} < \mu(t)|K_j^\dagger K_i|\mu(t) > \mu(t) +$$

$$+ \sum_{i,j}\lambda_{ij}\sum_\eta \eta(t) < \mu(t)|K_i|\eta(t) >< \eta(t)|K_j^\dagger|\mu(t) > \ . \tag{49}$$

The eigenvalue $\mu(t)$ starts off nonnegative, and it remains so if, every time it reaches zero, it can only grow. Let us write the Eq.(49) for such a point in time when $\mu(t) = 0$ and get

$$\dot{\mu}(t) = \sum_\eta \eta(t) \sum_{i,j}\lambda_{ij} < \mu(t)|K_i|\eta(t) >< \eta(t)|K_j^\dagger|\mu(t) > \geq 0 \ . \tag{50}$$

The coefficients in (50) are not known, but in order for $\mu(t)$ to stay positive for any combination of coefficients $\eta(t)$, we require the matrix $[\lambda_{ij}]$ to be positive definite. This is the final condition we need to impose on the time evolution in Eq.(41). It is, however, customary to diagonalize the positive matrix $[\lambda_{ij}]$, and introduce its eigenvalues β_i^2 to obtain

$$\frac{d}{dt}\rho = -i[H,\rho] - \frac{1}{2}\sum_i \beta_i^2 K_i^\dagger K_i\rho - \rho\frac{1}{2}\sum_i \beta_i^2 K_i^\dagger K_i \tag{51}$$

$$+ \sum_i \beta_i^2 K_i\rho K_i^\dagger \ .$$

This is the final master equation; its form is just like the one we discussed in Sec.21. There we also noted that it is in the correct form for a stochastic simulation.

3.2 Reduced dynamics

Assume that we are considering the time evolution of a system described in the direct product of two Hilbert spaces $\mathcal{H}_1 \otimes \mathcal{H}_2$. We describe the general time evolution in this product space by the isometric transformation $\mathcal{U}_t$ acting on both spaces. The initial state is taken to be factorable $\rho_1 \otimes \rho_2$. The time evolved density matrix is then found to be

$$\rho(t) = \mathcal{U}_t \rho_1 \otimes \rho_2 \mathcal{U}_t^\dagger . \tag{52}$$

This is easily seen to preserve the trace and the positivity of the eigenvalues of ρ. We can take this transformation as the prototype for the reduced dynamics in the Hilbert space $\mathcal{H}_1$ when this is coupled to an external system $\mathcal{H}_2$, which is often taken to act as a reservoir.

We now use (52) to define the dynamic time evolution reduced to the space $\mathcal{H}_1$ by introducing the linear transformation [12]

$$L_t(\rho_1) = Tr_2\left(\mathcal{U}_t \rho_1 \otimes \rho_2 \mathcal{U}_t^\dagger\right) . \tag{53}$$

This is immediately seen to be trace preserving and hermitean,

$$\left(L_t(\rho_1)\right)^\dagger = L_t(\rho_1) . \tag{54}$$

We introduce a complete set of orthonormal projectors $\{\Pi_\alpha\}$ in $\mathcal{H}_2$ which satisfy

$$\sum_\alpha \Pi_\alpha = 1 . \tag{55}$$

With these we expand

$$\mathcal{U}_t = \sum_\alpha U_\alpha^1(t)\Pi_\alpha , \tag{56a}$$

$$U_\alpha^1(t) = Tr_2 \mathcal{U}_t \Pi_\alpha . \tag{56b}$$

Inserting this expansion into (53), we find the result

$$L_t(\rho_1) = \sum_\alpha \eta_\alpha^2 \; U_\alpha^1(t)\rho_1 U_\alpha^1(t)^\dagger \; , \tag{57}$$

where

$$\eta_\alpha^2 = Tr_2\rho_2\Pi_\alpha \tag{58}$$

is the positive expansion coefficient of ρ_2 in the basis Π_α.

It is easy to see that the transformation (57) preserves the positivity of the density matrix, because if $\rho = A^\dagger A$ we have for any state $|\psi>$

$$< \psi|L_t(\rho_1)|\psi> = \sum_\alpha \eta_\alpha^2 \; \| AU_\alpha^1(t)|\psi>\|^2 \geq 0 \; . \tag{59}$$

It is also easy to see that the transformations

$$V_\alpha = \eta_\alpha U_\alpha^1(t) \tag{60}$$

satisfy the relation

$$\begin{aligned}
\sum_\alpha V_\alpha^\dagger V_\alpha &= \sum_\alpha \eta_\alpha^2 Tr_2\Big(\mathcal{U}_t^\dagger\Pi_\alpha\Big) Tr_2\Big(\mathcal{U}_t\Pi_\alpha\Big) \tag{61}\\
&= Tr_2\Big[\sum_\beta \Pi_\beta \; \eta_\beta^2 \; \mathcal{U}_t^\dagger \sum_\alpha \Pi_\alpha Tr_2\Big(\mathcal{U}_t\Pi_\alpha\Big)\Big]\\
&= Tr_2\Big[\sum_\beta \Pi_\beta \; \eta_\beta^2 \; \mathcal{U}_t^\dagger\mathcal{U}_t\Big] = Tr_2\rho_2 = 1 \; .
\end{aligned}$$

We now introduce a complete basis of operators $\{K_i\}$ in the Hilbert space $\mathcal{H}_1$. We can choose

$$K_0 = 1 \; , \tag{62}$$

and then the other operators can be chosen traceless. We expand the operators (56b) in these and obtain [11]

$$U_\alpha^1(t) = \sum_i u_\alpha^i(t)K_i \; . \tag{63}$$

Using this expansion in (57) and separating the terms with index zero we can write an expression for the coarse grained time derivative of the density matrix

$$\frac{d\rho}{dt} \simeq \frac{\rho(t+\Delta t)-\rho(t)}{\Delta t} = \frac{c_{00}(\Delta t)-1}{\Delta t}\rho + \sum_i {}' \frac{c_{i0}(\Delta t)}{\Delta t}K_i\rho$$

$$+\rho \sum_j {}' \frac{c_{0j}(\Delta t)}{\Delta t}K_j^\dagger + \sum_{i,j} {}' \frac{c_{ij}(\Delta t)}{\Delta t}K_i\rho K_j^\dagger , \qquad (64)$$

where

$$c_{ij}(t) = \sum_\alpha \eta_\alpha^2 \, u_\alpha^i(t)u_\alpha^j(t)^\dagger = \left(c_{ji}(t)\right)^* . \qquad (65)$$

The primed sums in (64) go over all nonzero indices. It is easy to see that letting Δt go to zero in (64) we obtain a time evolution equation of the form (41). However, the origin of the time evolution operator $\mathcal{U}_t$ in some dynamic theory makes it difficult to justify the existence of the limit $\Delta t \to 0$.

If we, on the other hand, make the assumption common in reservoir derivations of master equations, that the system $\mathcal{H}_2$ is large and has a dense spectrum, we can choose Δt long enough to exceed the correlation time of the system $\mathcal{H}_2$, and then we find the asymptotically valid result

$$c_{ij}(\Delta t) = \sum_\alpha \eta_\alpha^2 \, u_\alpha^i(\Delta t)u_\alpha^j(\Delta t)^\dagger \to \lambda_{ij}\Delta t , \qquad (66)$$

which holds for times Δt large with respect to the reservoir time scales. Adding the condition that the trace be conserved, we can directly from Eq.(64) obtain an evolution equation of the form

$$\frac{d\rho}{dt} \simeq \frac{\rho(t+\Delta t)-\rho(t)}{\Delta t} = -i[H,\rho] -$$

$$-\frac{1}{2}\sum_{i,j} {}' \lambda_{ij}\left(K_j^\dagger K_i\rho + \rho K_j^\dagger K_i - 2K_i\rho K_j^\dagger\right) . \qquad (67)$$

This is exactly the master equation derived in Sec. 3.1.

In the stochastic simulation approach, [9], the limit $\Delta t \to 0$ can be taken, because the functions $u_\alpha^i(\Delta t)$ are replaced by stochastic variables which in the limit behave like $\sqrt{\Delta t}$. This assumption does, however, not derive from any quantum dynamical model but consists in the imbedding of quantum theory into a stochastic process.

3.3 Complete positivity

In the mathematical literature the generators of dynamical semigroups are derived from a condition called **complete positivity** (denoted by CP later). This generalizes the concept of a positive linear transformation which preservs the positivity of eigenvalues for the elements on which it acts. The mathematical theory is worked out on the elements of a C^* algebra, and uses a rigorous mathematical formulation of the dynamical setting [10]. To try to follow this here would become too dreary, and the procedure does not seem to be very illuminative for the physical understanding anyway. Thus I will only formulate the setting and quote the main results of relevance for the present work.

A linear transformation $L(A)$ is acting on some algebraic elements A, which we can think of as operators on a Hilbert space. The transformation L is extended to an operator L_n acting on elements A_n defined using the original algebra as

$$A_n = \begin{bmatrix} A_{11}\ A_{12} \cdots\cdots\ A_{1n} \\ A_{21}\ A_{22} \cdots\cdots\cdots\ A_{2n} \\ \cdots\cdots\cdots\cdots\cdots\cdots \\ A_{n1}\ A_{n2} \cdots\cdots\cdots\ A_{nn} \end{bmatrix}, \tag{68}$$

where each entry A_{ij} is a member of the original set of elements. The transformation L_n is defined on these objects as

$$L_n(A_n) = \begin{bmatrix} L(A_{11})\ L(A_{12}) \cdots\cdots\ L(A_{1n}) \\ L(A_{21}) \cdots\cdots\cdots\cdots\ L(A_{2n}) \\ \cdots\cdots\cdots\cdots\cdots\cdots\cdots \\ L(A_{n1}) \cdots\cdots\cdots\cdots\ L(A_{nn}) \end{bmatrix}. \tag{69}$$

The transformation L is said to be CP if L_n is positive for all n. This procedure is claimed to be abstracted from considerations like those in the previous section, but from a physical point of view the procedure is arbitrary and can only be justified by its mathematical expedience.

We are going to present the main results in the form of Theorems starting with:

Theorem 2: The operator $L(A) = V^\dagger A V$ is CP.
Proof: We introduce a vector **x** with elements in the Hilbert space where the operators A act by setting

$$\mathbf{x} = \{x_1, x_2, \dots x_n\} \tag{70}$$

and the scalar product

$$\ll \mathbf{y}|\mathbf{x} \gg = \sum_{i=1}^{n} < y_i|x_i > \ . \tag{71}$$

An operator A_n is positive if its elements can be written in the form

$$A_{ij} = B_i^\dagger B_j \ , \tag{72}$$

and the general positive operator can be expressed as a sum of terms of this type. For each term we obtain the result

$$\ll L_n(A_n)\mathbf{x}|\mathbf{x} \gg = \sum_{i,j} < L(A_{ij})x_j|x_i > = \sum_{i,j} < L(B_i^\dagger B_j)x_j|x_i >$$
$$= \sum_{i,j} < V^\dagger B_i^\dagger B_j V x_j|x_i > = \| \sum_i B_i V x_i \|^2 \geq 0 \ , \tag{73}$$

This proves that a transformation of this type is CP.

The converse result:

Theorem 3: Any CP transformation is of the form

$$L(A) = \sum_\alpha V_\alpha^\dagger A V_\alpha \ . \tag{74}$$

is much harder to prove; see Ref. [10]. Lindblad [23] proves the following:

Theorem 4: If $D(A)$ is the infinitesimal generator of a semigroup, there exists an operator K and a CP transformation $J(A)$ such that

$$J(A) = D(A) - KA - AK^\dagger \ . \tag{75}$$

Separating the imaginary part of K, and applying the requirement of the conservation of the trace, we end up with the usual equation of motion

$$\dot{A} = i[H, A] + V^\dagger A V - \frac{1}{2} V^\dagger V A - \frac{1}{2} A V^\dagger V \ . \tag{76}$$

Because the operators A correspond to the observables of quantum theory, this equation is given in the Heisenberg picture. The Schrödinger time evolution for the state ρ is exactly of the form obtained earlier. This concludes our summary of the derivation based on complete positivity. I hope that the more simple minded approaches in the two previous sections may serve to show the physical significance of the various features involved in the formal approach.

4 Examples

4.1 Spontaneous emission - the trivial case

For many time dependent problems we can restrict the theoretical treatment to a discussion of two levels only $\{|i >; i = 1, 2\}$. The two levels are coupled by space and time dependent laser fields, and they experience different potentials $\mathcal{U}_i$. The effective Hamiltonian can then be written in the form

$$H = \frac{p^2}{2M} + (\mathcal{U}_2(q) \quad + \quad \hbar\omega)|2 >< 2| + \mathcal{U}_1(q)|1 >< 1| + \qquad (77)$$

$$+ \quad V(q,t)\Big(|1 >< 2| + |2 >< 1|\Big) \, ,$$

where $\hbar\omega$ is the energy separation between the levels. Hamiltonians of this type are used in the theories of laser cooling, laser-induced molecular processes and many others.

Spontaneous emission between the levels is effected by the operator

$$A = |1 >< 2| \, . \qquad (78)$$

Including these spontaneous decay processes, we obtain the evolution equation

$$\frac{d}{dt}\rho = -i[H,\rho] - \frac{\lambda^2}{2}\Big(A^\dagger A\rho + \rho A^\dagger A - 2A\rho A^\dagger\Big) \, . \qquad (79)$$

This is of the form discussed in the previous sections, and it can directly be simulated by a stochastic process. Most applications of the simulation approach have been concerned with spontaneous emission. The operator $A^\dagger A$ is the projector on the upper level, which guarantees that only this is affected by spontaneous decay.

The master equation (79) makes it possible to introduce spontaneous emission into all processes describable by a Hamiltonian of the form (77). In particular, we have investigated its effects on pulsed excitation of wave packets and level crossings. Both the direct solution of the master equation and the stochastic simulation process has been applied. In the former case we generate the ensemble averages directly, and in the latter case we build up the ensemble by constructing its different members from individual realizations of the quantum process. Both methods have been found to agree, as is to be expected.

4.2 Friction - the odd case

In many applications, the degrees of freedom described by the Hamiltonian (77) make up only a small fraction of all molecular degrees of freedom. Even in moderately large molecules, the remaining dynamical variables have a dense enough spectrum to make their treatment as a reservoir sensible. They may then be described as noise sources acting on the dynamical variables of interest, which is known to makes these decày in an irreversible way.

The motion of a system embedded in a reservoir is expected to obey the phenomenological equations

$$\dot{q} = -\frac{p}{M}$$
$$\dot{p} = -\gamma p - \frac{\partial \mathcal{U}}{\partial q} + \mathcal{F} , \tag{80}$$

where $\mathcal{F}$ is a Langevin force deriving from the fluctuations of the environment causing the friction described by γ. This force contributes a Brownian motion which is characterized by the diffusive spreading of the momentum according to

$$< p^2 >= 2Dt . \tag{81}$$

This is seen to increase the kinetic energy of the system steadily.

In quantum theory, we expect Eqs.(80) to be the correct Heisenberg equations of motion, and they have been derived in this way by e.g. Senitzky [24]. The Langevin force $\mathcal{F}$ then becomes a quantum noise source.

In quantum optics, a damped electromagnetic eigenmode in a cavity is known to be described by the master equation

$$\frac{d}{dt}\rho = -[\omega b^\dagger b, \rho] - \frac{\gamma}{2}\left(b^\dagger b\rho + \rho b^\dagger b - 2b\rho b^\dagger\right) , \tag{82}$$

where $b^\dagger$ creates the photons of the mode [21]. For simplicity I have written down only the zero temperature version of the equation.

If we decompose the annihilation operator into q and p components according to

$$b = \sqrt{\frac{M\omega}{2}}\left(q + i\frac{p}{\omega M}\right) , \tag{83}$$

we find that both the expectation value of q and that of p are damped at the rate γ. For an electromagnetic mode this is correct, as the only difference between the two components is determined by the arbitrarily chosen reference phase for the electric field.

For a mechanical system, on the other hand, a real difference exists between q and p. They damp differently according to the system (80). There can be no physical process which forces the system to the origin of the q coordinate. The correct master equation for this case is easily seen to be of the form

$$\frac{d}{dt}\rho = -i[H,\rho] - i\frac{\gamma}{2}[q,[p,\rho]_+] - D[q,[q,\rho]] \ . \tag{84}$$

An equation of this form was derived by Agarwal [25] and discussed in detail by Caldeira and Leggett [26].

The equation (84) is seen to give the average values of q and p damping terms in agreement with our expectations (80). In addition, it contains the effect of the environment fluctuations because we can derive

$$\frac{d}{dt} < p^2 > = -2\gamma < p^2 > +2D \ . \tag{85}$$

The last term gives diffusive heating in accordance with the expectation in Eq.(81). When an Einstein relation holds between D and γ, we can obtain the correct equipartition energy for a free particle system from the steady state solution of Eq.(85).

The problem with the equation (84) is, however, that it is not of the form derived in Secs. 2 and 3. The last term is as expected, in fact, it was used as an example in Sec. 2.3. But the term proportional to γ in (84) is not of such a form. In the present case, it is required by a fluctuation-dissipation theorem when the master equation is derived by a perturbative elimination of an unobserved reservoir system. There exists, actually, a whole class of master equations like (84), which are not of the form discussed so far in these notes.

If we want to use a master equation as the basis for a stochastic simulation, we need to rewrite it in the form (12). The last term in Eq.(84) is already such and hence we concentrate on the term proportional to γ; all other terms are not written out. Using the relation of q and p to the annihilation operator from (83), we can rewrite the term according to

$$\frac{d}{dt}\rho = -i\frac{\gamma}{2}[q,[p,\rho]_+]$$ (86)

$$= -i\frac{\gamma}{4}[(pq+qp),\rho] - i\frac{\gamma}{4}[[q,p],\rho]_+ - i\frac{\gamma}{2}\left(q\rho p - p\rho q\right)$$

$$= -i\frac{\gamma}{4}[(pq+qp),\rho] + \frac{\gamma}{4}[[b,b^\dagger],\rho]_+ - \frac{\gamma}{2}\left(b^\dagger\rho b - b\rho b^\dagger\right)$$

$$= -i\frac{\gamma}{4}[(pq+qp),\rho] - \frac{\gamma}{4}\left(b^\dagger b\rho + \rho b^\dagger b - 2b\rho b^\dagger\right)$$

$$+\frac{\gamma}{4}\left(bb^\dagger\rho + \rho bb^\dagger - 2b^\dagger\rho b\right).$$

The first term can be included in the Hamiltonian, and the following terms appear to be of the form which would allow a stochastic simulation. However, the last term poses a problem, its sign shows that it occurs with a negative probability; compare Eq. (12). It cannot correspond to a stochastic event and no simulation is possible. Thus the general theory developed earlier in these notes is not applicable to the case of friction.

At the time of writing these notes, I do not know how to reconcile the discrepancy between the physically well understood case of friction with the general theory, which according to Sec. 3 is expected to be of rather universal validity. I hope that some clarification of this problem may add further understanding to the urgent problem of including dissipative effects into our dynamic models for quantum time evolution.

Acknowledgements

Many of the topics presented in these notes have been discussed with Kalle-Antti Suominen, Barry Garraway and Wai Lai. I owe them my sincerest thanks for many useful insights and inspiring questions. In addition, I want to thank Steve Barnett for having helped me clarify the theoretical description of friction on several occasions.

References

[1] J.Dalibard, Y.Castin and K. Mølmer, Phys.Rev.Lett. **68**, 580 (1992) and K.Mølmer, Y.Castin and J.Dalibard, JOSA **B**, **10**, 524 (1993)

[2] H.J.Carmichael, **ULB Lectures in Nonlinear Optics, Lecture Notes in Physics**, (Springer Verlag, Heidelberg) to be published; and L.Tian and H.J.Carmichael, Phys.Rev. **A46**, R6801 (1992)

[3] J.Javanainen, Phys.Rev.**A33**, 2121 (1986)

[4] P.Zoller, M.Marte and D.F.Walls, Phys.Rev.**A35**, 198 (1987)

[5] P.Pearle, Phys.Rev.**A39**, 2277 (1989)

[6] G.C.Ghirardi, P.Pearle and A.Rimini, Phys.Rev.**A42**, 78 (1990)

[7] G.C.Ghirardi, in **Quantum Chaos - Quantum Measurements**, ed. P.Cvitanovic, I.Percival and A.Wirzba (Kluwer Academic Press, Dordrecht, 1992), page 305

[8] N.Gisin and I.C.Percival, J.Phys.A:Math.Gen. **25**, 5677 (1992) and Physics Letters, **A167**, 315 (1992)

[9] C.W.Gardiner, A.S.Parkins and P.Zoller, Phys.Rev.**A46**, 4363 (1992) and R.Dum, A.S.Parkins, P.Zoller and C.W.Gardiner, Phys.Rev.**A46**, 4382, (1992)

[10] E.B.Davies, **Quantum Theory of Open Systems**, (Academic Press, London, 1976)

[11] V. Gorini, A.Kossakowski and E.C.G.Sudarshan, J.Math.Phys. **17**, 821 (1976)

[12] E.C.G.Sudarshan, Phys.Rev. **A46**, 37 (1992)

[13] R.Alicki and K.Lendi, **Quantum Dynamical Semigroups and Applications**, Lecture Notes in Physics 286, (Springer-Verlag, Berlin, 1987)

[14] K.-A. Suominen, B.M. Garraway anad S. Stenholm, Phys.Rev. **A45**, 3060 (1992), B.M. Garraway and S. Stenholm, Opt.Commun., **83**, 349 (1991) and Phys.Rev. **A46**, 1413 (1992)

[15] V.A.Alekseev, T.L.Andreeva and I.I.Sobel'man, Soviet Phys. JETP **35**, 325 (1972)

[16] P.R.Berman,Phys.Rev.**A5**, 927 and **A6**, 2157 (1972)

[17] M.D.Srinivas and E.B.Davis, Opt.Acta **28**, 981 (1981)

[18] S.Stenholm, Physica Scripta, to appear

[19] J.M.Jauch, **Foundations of Quantum Mechanics**, (Addison-Wesley, New York 1968), section 6-9

[20] N.Bohr, Nature, **121**, 580 (1928)

[21] W.H.Louisell, **Quantum Statistical Properties of Radiation** (J.Wiley, New York 1973)

[22] L.E.Ballentine, Phys.Rev.**A43**, 9 (1991)

[23] G.Lindblad, Commun.Math.Phys. **48**, 119 (1976)

[24] I.R.Senitzky, Phys.Rev. **119**, 670 (1960) and **124**, 642 (1961)

[25] G.S.Agarwal, Phys.Rev. **A4**, 739 (1971)

[26] A.O. Caldeira and A.J. Leggett, Physica **121A**, 587 (1983)

On Lasing Without Inversion Within the Sodium D_1 Line

Marlan O. Scully, Edward S. Fry, Georg M. Meyer, Dmitri E. Nikonov, and Shi-Yao Zhu

Department of Physics, Texas A&M University, College Station, TX 77843, USA and Texas Laser Laboratory, Houston Advanced Research Center, 4800 Research Forest Drive, The Woodlands, TX 77381, USA

Abstract. We present and analyze a simple scheme demonstrating lasing without inversion (LWI) within the Na D_1 line, i.e., within the hyperfine manifolds of the $3\,^2P_{1/2}$ and $3\,^2S_{1/2}$ states.

Keywords. Atomic coherence, hyperfine structure, lasing without inversion, optical pumping, population trapping, sodium

Recently, a new LWI scheme has been proposed, analyzed and experimentally demonstrated using the Na D_1 line [1]. Here, we optimize this scheme and study further ways of achieving LWI. In order to put the present approach in perspective, let us recall that LWI typically [2,3] envisions one intense laser field (the prep field) in order to prepare the medium so as to cancel absorption on a weak probe, which is to be amplified even though there is no population inversion, see Figs. 1, 2.

The optical pumping LWI scheme depends on first locking in a population in the ground state manifold with, say, σ^+ radiation at frequencies ν_1 and ν_2 and then pumping with σ^- radiation so as to excite some atoms to the $3\,^2P_{1/2}$

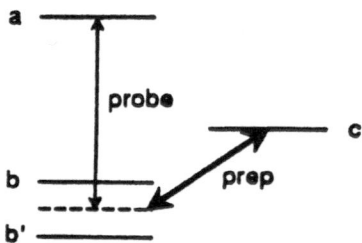

Fig. 1. LWI via quantum coherence. A Raman like prep field generates coherence which cancels absorption for a probe field.

343

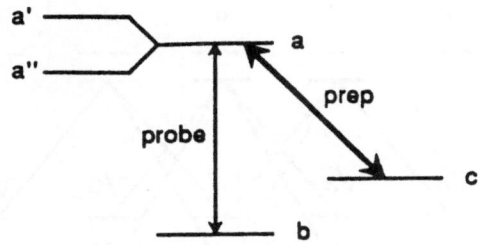

Fig. 2. LWI via quantum interference. Here the prep field can be thought of as generating Rabi side band states a' and a'' and thus providing two paths $b \to a' \to c$ and $b \to a'' \to c$. Their interference can cancel absorption.

manifold as in Figs. 3, 4. The excited population gives rise to a transient LWI gain [1].

We designate the upper $F = 2$ level as a, the upper $F = 1$ level as a', the lower $F = 2$ level as b, and the lower $F = 1$ level as b'. The corresponding sublevels are indicated by a subscript giving their magnetic quantum number m_F. If two σ^+ fields couple the lower levels b and b' to the upper level a', part of the population is coherently trapped in the pairs of lower states with $m_F = -1$ and $m_F = 0$.

After establishing this population trapping, we can excite atoms to the upper levels by applying one σ^- pump field at ν_2 as in the experiment of Ref. [1]. At the same time, however, some coherence between the level pair b_0 and b'_0 is destroyed because population is pumped out of b_0. The destruction of

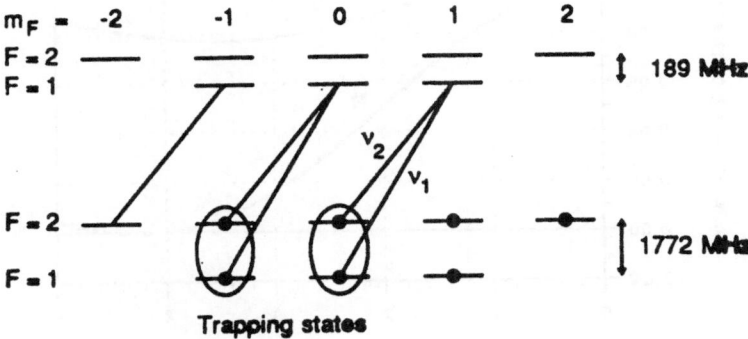

Fig. 3. Due to atomic coherence effects, right circularly polarized light (σ^+) locks population in two pairs of ground states (denoted b_{-1}, b'_{-1} and b_0, b'_0 as in Fig. 4) and pumps part of their population to other ground states.

344

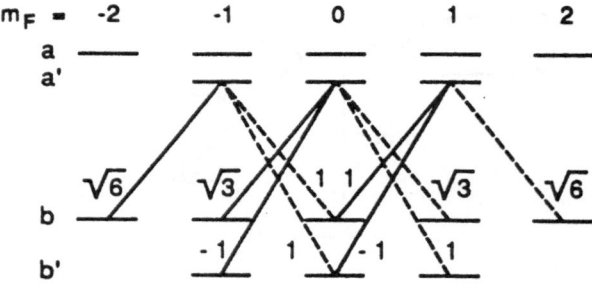

Fig. 4. After locking in ground state population with σ^+ light, as in Fig. 3, we now turn on left circularly polarized light (σ^-), which is tailored to promote atoms from the states b_2, b_1, b_1' and only conditionally from b_0 and b_0'. The relative dipole matrix elements for the σ^+ and σ^- transitions are given.

the low-frequency coherence can be avoided by using two σ^- fields at ν_1 and ν_2, which couple b_0' and b_0 to a_{-1}', respectively.

A driven transition $b \rightarrow a$ is characterized by the phase factor product $p_{ab} = \exp(i\varphi_{ab})\exp(i\phi)$, where φ_{ab} is the phase of the coupling constant g_{ab} and ϕ is the phase of the driving field. If the phase factor product $p_{a_1' b_0} p^*_{a_1' b_0'}$ for the two transitions driven by the σ^+ fields is equal to the one for the transitions driven by the σ^- fields, $p_{a_{-1}' b_0} p^*_{a_{-1}' b_0'}$, the two σ^- fields tend to

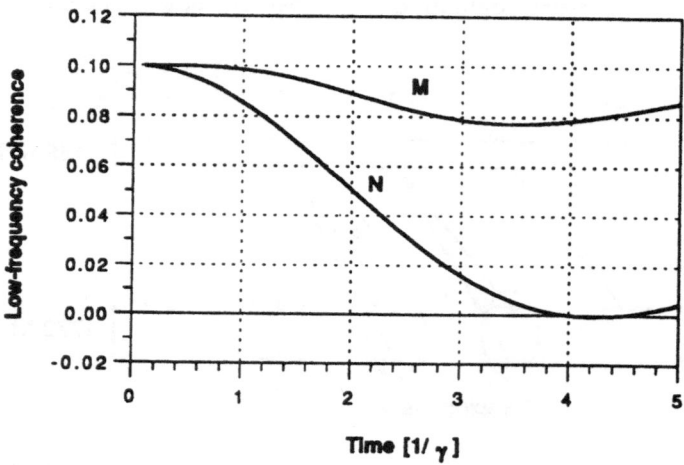

Fig. 5. The coherence between the lower levels b_0 and b_0' as a function of time (in units of the radiative decay time γ) after switching on σ^- light in the cases of matching phases (the curve labelled M) and non-matching phases (labelled N).

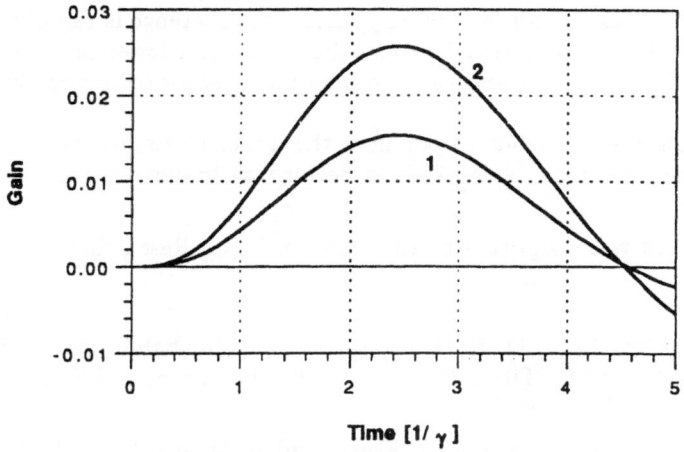

Fig. 6. The weighted sum of the optical polarizations of all σ^+ transitions from b' to a' (the curve labelled 1) and from b to a' (labelled 2) as a function of time (in units of the radiative decay time γ) after switching on σ^- light.

build up the same coherence between b_0 and b_0' as the σ^+ fields. Therefore, in the case of matching phases, the coherence between b_0 and b_0' will be essentially preserved when the σ^- light is switched on. If the two products have opposite signs, the σ^- fields tend to build up the opposite coherence between b_0 and b_0'. In this case of non-matching phases, switching on the σ^- fields will destroy the coherence built up by the σ^+ fields.

Here, the relative signs of the dipole matrix elements for the σ^+ and σ^- transitions play an important role. The coupling coefficients are given in Fig. 4, where the phase conventions from Ref. [4] have been adopted. In the Na D_1 line, we have $\varphi_{a'_{-1}b_0} = \varphi_{a'_{-1}b_0'} = \varphi_{a_1'b_0} = 0$ and $\varphi_{a_1'b_0'} = \pi$. When the "trapping" σ^+ fields at ν_1 and ν_2 have the same phase while the "excitation" σ^- fields at ν_1 and ν_2 have a phase difference of π, the phase factor products for the σ^+ and σ^- fields are equal. Therefore, no atoms are excited by the σ^- fields from the states b_0 and b_0', and the full coherence between b_0 and b_0' will be preserved. (The upper levels are only populated from the lower levels b_2, b_1, and b_1'.) This will lead to maximum cancellation of absorption from b_0 and b_0' and, consequently, to larger gain than in the case of just one σ^- pump field.

A numerical simulation of the time evolution of the density matrix for the atoms confirms the above consideration of the phase-sensitive response of the atoms. All applied fields were taken to be of equal intensity; for simplicity, collisional relaxation was disregarded. It is found that in the case of matching phases (same phases for the σ^- fields and opposite phases for the σ^+ fields in the above situation) the coherence between b_0 and b_0' is essentially preserved, while in the case of non-matching phases, the coherence is rapidly destroyed (Fig. 5). The polarization between the upper and lower levels shows gain on both frequencies ν_1 and ν_2 for some time after switching on the σ^- light (Fig. 6).

By using two σ^- fields to populate the upper levels, we can preserve the maximum coherence. This will be a better way to demonstrate LWI.

This work was supported by the Office of Naval Research.

[1] E. S. Fry, X. Li, D. E. Nikonov, G. G. Padmabandu, M. O. Scully, A. V. Smith, F. K. Tittel, C. Wang, S. R. Wilkinson, and S. Y. Zhu (to be published in Phys. Rev. Lett.).

[2] S. Y. Zhu, M. O. Scully, H. Fearn, and L. M. Narducci, Z. Phys. D **22**, 483 (1992).

[3] A. Imamoğlu, J. E. Field, and S. E. Harris, Phys. Rev. Lett. **66**, 1154 (1990).

[4] R. N. Zare, *Angular Momentum* (Wiley, New York, 1988).

Lecture Notes in Physics

For information about Vols. 1–384
please contact your bookseller or Springer-Verlag

New Series m: Monographs